T0344273

A NON-HAUSDORFF COMPLETION

The Abelian Category of
C-complete Left Modules over
a Topological Ring

A NON-HAUSDORFF COMPLETION

The Abelian Category of
C-complete Left Modules over
a Topological Ring

Saul Lubkin

University of Rochester, USA

 World Scientific

NEW JERSEY · LONDON · SINGAPORE · BEIJING · SHANGHAI · HONG KONG · TAIPEI · CHENNAI

Published by

World Scientific Publishing Co. Pte. Ltd.

5 Toh Tuck Link, Singapore 596224

USA office: 27 Warren Street, Suite 401-402, Hackensack, NJ 07601

UK office: 57 Shelton Street, Covent Garden, London WC2H 9HE

Library of Congress Cataloging-in-Publication Data
Lubkin, Saul, 1939–
    A non-Hausdorff completion : the Abelian category of C-complete left modules over a topological ring / Saul Lubkin, University of Rochester, USA.
    pages cm
    Includes bibliographical references.
    ISBN 978-9814667388 (hardcover : alk. paper)
    1. Topology. 2. Abelian categories. 3. Algebra, Homological. 4. Commutative algebra.
5. Topological rings. I. Title.
    QA611.L83 2015
    512'.44--dc23
                            2015001145

British Library Cataloguing-in-Publication Data
A catalogue record for this book is available from the British Library.

Printed in Singapore

Dedication: To my wife, Maxine, who encouraged me to complete this book.

Acknowledgement: This manuscript was put into LaTeX by my former graduate student, Chad Gratton, to whom I am very grateful.

Preface/Introduction

Suppose that we have a commutative ring A and an ideal I in A. Then we have the well-known *I-adic completion* $M^{\wedge I}$ of any left A-module M,

$$M^{\wedge I} = \varprojlim_{n \geq 1} M/I^n M .$$

The assignment: $M \rightsquigarrow M^{\wedge I}$ is an additive functor, that in general is neither left nor right exact; the usual completion functor fails to have many useful properties, that often make computation difficult.

In this book, we introduce a new functor, $C^I(M)$, the *C-completion of M with respect to the ideal I*. Actually we make this construction in far greater generality—if A is *any* not-necessarily-commutative topological ring with identity such that the topology is given by right ideals and if M is any abstract left A-module, then we define $C(M)$. $C(M)$ can be defined quickly as being the zeroth derived functor of the usual completion functor, $M \rightsquigarrow M^\wedge$. For example, if we choose P_1, P_2 projective left A-modules and an exact sequence

$$P_1 \to P_2 \to M \to 0 ,$$

then $C(M) = \mathrm{Cok}(P_1^\wedge \to P_2^\wedge)$.

In all cases, the functor $M \rightsquigarrow C(M)$ is right exact. However, unlike $M^\wedge$ $C(M)$ is rarely Hausdorff (not even if the topological ring A is a complete discrete valuation ring that is not a field). Hence $C(M)$ can be thought of as being "a Non-Hausdorff Completion of the abstract left A-module M."

Although $C(M)$ and the traditional $M^\wedge$ are in general different, one can recover $M^\wedge$ from $C(M)$. E.g., under mild conditions,

$$C(M)/(\text{divisible elements}) \approx M^\wedge .$$

Thus, $M^\wedge$ can be thought of as being a weaker construction than $C(M)$.

In addition, since the functor C is a right exact functor, it has higher derived functors. These are the higher C-completions, $C_i(M), i \geq 0$. ($C_0(M) = C(M)$.) These are used to construct spectral sequences, that are very useful in computing $C(M)$ and $C_i(M), i \geq 0$.

If A is a topological ring such that the topology is given by right ideals and M is an abstract left A-module, then we define the notion of an *infinite sum structure* on the abstract left A-module M. Basically, if $(a_i)_{i \in I}$ are elements in $A^\wedge$ that converge to zero, and if $(m_i)_{i \in I}$ are any elements of the left A-module M, then an infinite sum structure tells us how to define

$$\sum_{i \in I} a_i m_i \in M \ .$$

A left A-module M, together with an infinite sum structure, is called *a C-complete* left A-module. And we define the notion of an *infinitely linear function* between two C-complete left A-modules. For example, if M is any abstract left A-module, then both $M^\wedge$ and $C(M)$ have natural such structures, and, therefore, are naturally C-complete left A-modules, and the natural map: $C(M) \to M^\wedge$ is infinitely linear. The category of all C-complete left A-modules and infinitely linear functions turns out to be a very interesting abelian category, which we shall denote $\mathscr{C}_A$.

It should be noted that, under reasonably mild conditions—e.g., if the topological ring A is such that the topology is given by denumerably many two-sided ideals $I_i, i \geq 1$, each of which is finitely generated as right ideal, and such that I_i^2 is open, $i \geq 1$, then the category $\mathscr{C}_A$ turns out to be a full exact abelian subcategory of the category $\mathscr{M}_A$ of all abstract left A-modules—that is, every linear map of C-complete left A-modules is then automatically infinitely linear. In all cases, whatever the topological ring A, the "stripping functor": $\mathscr{C}_A \rightsquigarrow \mathscr{M}_A$ that to each C-complete left A-module associates the corresponding abstract left A-module, is always exact and faithful, and preserves direct products. In particular, $\mathscr{C}_A$ is always an exact abelian subcategory of $\mathscr{M}_A$.

We now summarize these constructions, and others, in more detail. For the rest of this Preface, we will refer only to topological rings A such that the topology is given by right ideals.

In Chapter 2, a C-*complete left A-module* is defined to be an abstract left A-module together with an infinite sum structure. For example, if M is an abstract left A-module, then $M^\wedge$ is a C-complete left A-module in an obvious way. In fact, every C-complete left A-module is isomorphic to the cokernel of an infinitely linear map: $F^\wedge \to G^\wedge$, where F and G are free left A-modules. (Note: The map $F^\wedge \to G^\wedge$ need not come from a map in $\mathscr{M}_A : F \to G$.) In Example 3 of Chapter 2, we construct a complete submodule N of $(\mathscr{O}^{(\omega)})^\wedge$, where $A = \mathscr{O}$ is any c.d.v.r. that is not a field, such that $(\mathscr{O}^{(\omega)})^\wedge/N$ is not Hausdorff. However, of course, it is C-complete, for the infinite sum structure inherited from $(\mathscr{O}^{(\omega)})^\wedge$.

In Corollary 2.3.10 of Chapter 2, we show that if A is commutative, then the A-module $Hom_{\mathscr{C}_A}(M, N)$, where $M, N \in \mathscr{C}_A$ also has a natural structure of C-complete A-module (it is given by the infinite sum structure inherited from N^M). In Remark 8 of Section 3 of Chapter 2, if the topological ring A is commutative, if $M, N, L \in \mathscr{C}_A$ and if $f : M \times N \to L$ is a function, then we define what it means for f to be *infinitely bilinear*, and we use this to define $M \bigotimes_A^C N$, the C-*complete tensor product of M and N*. Also, if A is commutative, then we define

$$Hom_{\mathscr{C}_A} : \mathscr{C}_A^0 \times \mathscr{C}_A \rightsquigarrow \mathscr{C}_A \ .$$

The functors $Hom_{\mathscr{C}_A}$ and $\bigotimes_A^C$ are adjoint:

$$Hom_{\mathscr{C}_A}(M \overset{C}{\underset{A}{\bigotimes}} N, L) \approx Hom_{\mathscr{C}_A}(M, Hom_{\mathscr{C}_A}(N, L)) \ .$$

This is an isomorphism of functors from $\mathscr{C}_A^0 \times \mathscr{C}_A^0 \times \mathscr{C}_A$ into the category of sets (even into the category $\mathscr{C}_A$).

Always, $\mathscr{C}_A$ is abelian and is closed under infinite direct products and inverse limits, and the "stripping functor": $\mathscr{C}_A \rightsquigarrow \mathscr{M}_A$ is exact and preserves arbitrary direct products and inverse limits. Also, $\mathscr{C}_A$ has enough projectives.

Infinite direct sums and arbitrary direct limits also always exist in $\mathscr{C}_A$—but they are very different from the usual construction in $\mathscr{M}_A$: these constructions are pathological in $\mathscr{C}_A$.

Some interesting details: Every finitely presented abstract left $A^\wedge$-module has a natural structure as C-complete left A-module. And, if M is any abstract left A-module, then $C(M)$ can be characterized as being the universal C-complete left A-module into which M maps by a homomorphism of abstract left A-modules. And, the functor $C : \mathscr{M}_A \rightsquigarrow \mathscr{C}_A$ preserves arbitrary direct limits.

In Chapter 3, Section 2, we study the divisible part of C-complete left A-modules. For example, if the topology of A is given by denumerably many open right ideals, and if M is a C-complete left A-module, then M always has no non-zero infinitely divisible elements (i.e., there is no non-zero submodule of M that is divisible). And then also for every abstract left A-module M, $C(M) = 0$ *iff* $M^\wedge = 0$ *iff* M is A-divisible. And then also for every $M \in \mathscr{C}_A$ we have the short exact sequence:

$$0 \to (\text{div } M) \to M \to M^\wedge \to 0$$

in $\mathscr{C}_A$, where $(\text{div } M)$ denotes the divisible part of M; and if $N \in \mathscr{M}_A$, then we have the short exact sequence

$$0 \to \text{div } (C(N)) \to C(N) \to \widehat{N} \to 0$$

in $\mathscr{C}_A$. Of course, these hypotheses are very mild, and hold in all serious current applications to algebraic geometry and commutative algebra. And then,

$$N^\wedge = C(N)/(\text{divisible elements}) \ ,$$

so that $N^\wedge$ is "$C(N)$ made Hausdorff", for all abstract left A-modules N.

Note: If the topology of A is the right t-adic for some element $t \in A$, such that, e.g., *either* t is not a left divisor of zero, *or* A is right Noetherian, then

$$Ker(M \to C(M)) = \{\text{infinitely } t - \text{divisible elements of } M\} \ .$$

And

$$M/(\mathrm{div}\ M) \hookrightarrow M^{\wedge},$$

$$M/(\text{infinitely divisible part of } M) \hookrightarrow C(M),$$

$$M^{\wedge} = \frac{C(M)}{\left[\dfrac{\mathrm{div}\ (M)}{(\mathrm{inf.div.}M)}\right]} \ .$$

In Chapter 4, we study the higher C-completions, $C_i(M), i \geq 0$ —these are the left derived functors of the functor C from $\mathcal{M}_A$ into $\mathcal{C}_A$—or, equivalently, of the usual A-adic completion functor, $M \rightsquigarrow M^{\wedge}$, from $\mathcal{M}_A$ into $\mathcal{C}_A$.

If B_* is any non-negatively indexed chain complex of abstract left A-modules, then we have the two spectral sequences in the category $\mathcal{C}_A$ starting with

$$^I E^1_{p,q} = C_q(B_p)$$

and

$$^{II} E^2_{p,q} = C_p(H_q(B_*))\,,$$

both abutting at the same sequence $K_n, n \geq 0$, in $\mathcal{C}_A$ (but with different filtrations). From these, we deduce the *spectral sequence of the C-completion: B_** as above, if also

$$C_i(B_q) = 0, i \geq 1, q \geq 0\,, \tag{$*$}$$

then we have a first quadrant homological spectral sequence:

$$E^2_{p,q} = C_p(H_q(B_*)) \Rightarrow H_n(C(B_*)),\ n \geq 0\,.$$

Note: Condition $(*)$ holds if the topology of A is given by denumerably many right ideals, and if B_i is left flat as A-module, all $i \geq 0$.

The above spectral sequence is very important in many computations involving cohomology of completions and p-adic cohomology of algebraic varieties and schemes. For example, the short exact sequence (I.8) of [PPWC], and of [COC], Chapter 2, is a very special case of this spectral sequence.

A corollary of the spectral sequence: If the topology of A is given by denumerably many right ideals $I_1 \supset I_2 \supset I_3 \supset \cdots$, then we have the short exact sequence

$$0 \to \left(\varprojlim_{j \geq 1}{}^1 \mathrm{Tor}^A_1(A/I_j, M)\right) \to C(M) \to \widehat{M} \to 0\,,$$

for every abstract right A-module M. And, if the I_j are two-sided ideals that are finitely generated as right ideals, and if I_j^2 is open for all $j \geq 1$ (i.e., if for all $j \geq 1$, $I_j^2 \supset I_k$ for some $k \geq j$), then for every C-complete left A-module M, we have that

$$\mathrm{div}(M) \approx \varprojlim_{j \geq 0}{}^1 \mathrm{Tor}^A_1(A/I_j, M)$$

as C-complete left A-modules.

If the topology of A is given by denumerably right ideals, and if M is a flat left A-module, then

$$C(M) = M^{\wedge} \text{ and } C_i(M) = 0, i \geq 1 .$$

As a special case:

If A is a commutative ring, and t is an element that is not a divisor of zero, and if the topology of A is the t-adic topology, then

$$\text{div}(C(M)) = \varprojlim_{i \geq I} \left(\text{precise } t^i\text{-torsion in } \frac{M^{\wedge}}{M} \right)$$

where "{precise t^i-torsion in an A-module N}" means "Ker $(t^i : N \to N)$", and where, if $M \in \mathcal{M}_A$, then "$M^{\wedge}/M$" is shorthand for "$(\text{Cok}(M \to M^{\wedge}))$". And, in this case $\text{Ker}(M \to C(M)) = \{\text{infinitely } t\text{-divisible elements in } M\}$, for all abstract A-modules M.

In Chapter 5, we study direct sums and direct limits in $\mathcal{C}_A$. As we have noted above, direct sum is usually very different from the direct sum of abstract left A-modules and is not exact. Because of its unusual behavior, we use the symbol

$$\int_{i \in I} M_i$$

to denote the direct sum of objects M_i in $\mathcal{C}_A, i \in I$. The notation

$$\bigoplus_{i \in I} M_i$$

will mean the direct sum in $\mathcal{M}_A$—i.e., as abstract left A-modules, ignoring the infinite sum structures. Under mild conditions, we have that

$$\int_{i \in I} M_i = C\left(\bigoplus_{i \in I} M_i \right) \text{ in } \mathcal{C}_A ,$$

whenever $M_i \in \mathcal{C}_A$, all $i \in I$. And $C\left(\varinjlim_{i \in D} M_i \right)$ is the direct limit in $\mathcal{C}_A$ of any direct system $(M_i, \alpha_{ij})_{i,j \in D}$ of objects and maps in $\mathcal{C}_A$—where, as usual $\varinjlim_{i \in D} M_i$ denotes the direct limit in $\mathcal{M}_A$, ignoring the C-complete left A-module structures of the M_i, $i \in i$.

The functor C from $\mathcal{M}_A$ into $\mathcal{C}_A$ always preserves arbitrary sums and direct limits; in particular, we have that

$$C\left(\bigoplus_{i \in I} M_i \right) = \int_{i \in I} C(M_i) ,$$

for all $M \in \mathcal{M}_A$ and all sets I.

The natural map from the direct sum into the direct product in $\mathcal{C}_A$:

$$\int_{i \in I} M_i \longrightarrow \prod_{i \in I} M_i$$

is almost never injective. In fact, under very mild conditions,

$$\mathrm{div}\left(\int_{i\in I} M_i\right) = \mathrm{Ker}\left(\int_{i\in I} M_i \longrightarrow \prod_{i\in I} M_i\right),$$

and this is often non-zero. For example, if $A = \mathscr{O}$ is a c.d.v.r. not a field with uniformizing parameter t, then

$$\mathrm{div}\left(\int_{i\geq 1} \mathscr{O}/t^i\mathscr{O}\right) = \mathrm{Ker}\left(\int_{i\geq 1} \mathscr{O}/t^i\mathscr{O} \longrightarrow \prod_{i\geq 1} \mathscr{O}/(t^i\mathscr{O})\right)$$

and is non-zero. And, therefore, the C-complete $\mathscr{O}$-module

$$\int_{i\geq 1} \mathscr{O}/t^i\mathscr{O} = C\left(\bigoplus_{i\geq 1}(\mathscr{O}/t^i\mathscr{O})\right)$$

is a C-complete $\mathscr{O}$-module that is not complete.

In §5.6, since the direct sum $\int_{i\in I} M_i$ in the category $\mathscr{C}_A$ is usually not exact, but is always right exact, and since $\mathscr{C}_A$ always has enough projectives, we define and study the *higher direct sums*

$$(M_i)_{i\in I} \rightsquigarrow \int_{i\in I}^n M_i, \quad n \geq 0,$$

which are by definition the higher left derived functors of $\int_{i\in I}$. For example, we always have the first quadrant homological spectral sequence in the category $\mathscr{C}_A$,

$$E^2_{p,q} = \int_{i\in I}^p C_q(M_i) \Rightarrow C_n\left(\bigoplus_{i\in I} M_i\right) \tag{$*$}$$

where $M_i \in \mathscr{C}_A$, all $i \in I$.

And, under mild conditions on A, the natural infinitely linear function:

$$C_p\left(\bigoplus_{i\in I} M_i\right) \longrightarrow \int_{i\in I}^p M_i$$

is an isomorphism, all $p \geq 0$, whenever $M_i \in \mathscr{C}_A$, all $i \in I$. And, under the same mild conditions, the spectral sequence $(*)$ simplifies to

$$E^2_{p,q} = C_p\left(\bigoplus_{i\in I} C_q(M_i)\right) \Rightarrow C_n\left(\bigoplus_{i\in I} M_i\right).$$

Sometimes, however, the infinite direct sum $\int_{i\in I} M_i$ in $\mathscr{C}_A$ *is* exact: For example, if A is a right t-adic ring (meaning that there is an element $t \in A$ such that the topology of A has an open neighborhood base at zero consisting of the right ideals $t^i A, i \geq 0$), and such that the t-torsion is bounded below (meaning that there is an integer $n \geq 1$ such that $t^m a = 0$ in A implies that $t^n a = 0$, all $a \in A$, all $m \geq 1$), then, for every set I, the I-fold direct sum

$$\int_{i \in I} : \quad \mathscr{C}_A^I \rightsquigarrow \mathscr{C}_A$$

is exact.

In §5.7, we study some of the consequences of the fact that $\int_{i \in I}$ is usually not exact.

An abelian category $\mathscr{A}$ obeys the *Eilenberg–Moore Axiom* $(P1)$ *iff* denumerable direct products exist and the functor "denumerable direct product": $\mathscr{A}^\omega \rightsquigarrow \mathscr{A}$ is exact. $\mathscr{A}$ obeys the *Eilenberg–Moore Axiom* $(P2)$ *iff* denumerable direct products exist, and if also whenever

$$\cdots \to A_{i+1} \to A_i \to \cdots \to A_1$$

is an inverse system in which all the maps are epimorphisms, then the induced map:

$$\left[\varprojlim_{i \geq 1} A_i\right] \longrightarrow A_1$$

is an epimorphism. $\mathscr{A}$ *obeys the E-M Axiom* $(S1)$ (resp. $(S2)$) *iff* the dual category $\mathscr{A}^0$ obeys the $(P1)$ (resp. $(P2)$). We show the well-known facts that $(P2) \Rightarrow (P1)$, and that $(S2) \Rightarrow (S1)$; and that if $\mathscr{A}$ has enough injectives, and if denumerable direct sums exist, and if denumerable direct limit is exact, then $\mathscr{A}$ obeys $(S2)$, and therefore also $(S1)$.

Now, if A is any ring, and t is an element in the center of A that is not nilpotent, and if we give A the t-adic topology, then the abelian category $\mathscr{C}_A$ does not obey the EM Axiom $(S2)$.

In fact, if A is a commutative ring and $t \in A$, then t is not nilpotent *iff* $\mathscr{C}_A$ does not obey EM $(S2)$.

And, we also show that, given a ring A, and an element t in the center of A that is not nilpotent, and such that the t-torsion is bounded below, then if we give A the t-adic topology, we have that the abelian category $\mathscr{C}_A$ obeys the Eilenberg–Moore Axiom $(S1)$ but not $(S2)$. Since $\mathscr{C}_A$ does not obey $(S2)$, it follows that it also does not have enough injectives. And, in this case, $\mathscr{C}_A$ is an exact full abelian subcategory of the category of left A-modules $\mathscr{M}_A$.

As a special case, if $\mathcal{O}$ is a complete discrete valuationg ring not a field, then the full subcategory $\mathscr{C}_\mathcal{O}$ of the category of $\mathcal{O}$-modules obeys $(S1)$ but not $(S2)$, and also does not have enough injectives (but does have enough projectives). Such examples are hard to come by, and $\mathscr{C}_\mathcal{O}$ is a pretty natural such example.

Contents

Chapter 1

Admissible Topological Rings

1.1 Completions of Topological Groups, Rings, and Modules

A *topological additive group* is an additive group $(A, +)$ together with a topology on the set A, such that the sum function $+ : A \times A \to A$, that maps every $(x, y) \in A \times A$ into $x + y$ is continuous for the product topology, and such that the negative $- : A \to A$, that maps every $x \in A$ into $-x$, is continuous. A *topological ring* A is a ring $(A, +, \cdot)$ together with a topology on the set A, such that $(A, +)$ together with this topology is a topological group and such that the multiplication $\cdot : A \times A \to A$ that maps every $(x, y) \in A \times A$ into $x \cdot y$ is continuous. If A is a topological ring, then a *topological left module* over A is a left module $(M, +, \cdot)$ over A, together with a topology on the set M, such that $(M, +)$ is a topological additive group, and such that the scalar product $\cdot : A \times M \to M$ that maps every $(a, x) \in A \times M$ into $a \cdot x \in M$, is continuous for the product topology on $A \times M$.

If A is a topological additive group, then A has a natural structure as a uniform space – namely, for every open neighborhood V of zero in A, let $U_V = \{(x, y) \in A \times A$ such that $y - x \in V\}$. Then $\{U_V : V$ is an open neighborhood of zero in $A\}$ is the base for a uniformity on the set A, so that A becomes a uniform space. (Of course, the associated topology of the uniformity just constructed and the underlying topology of the topological group A always coincide.) Therefore we have the *completion* $\widehat{A}$ of the topological group A, which also has a natural structure as a topological group. If A is a *topological ring*, and M is a topological left module over A, then the completion $\widehat{M}$ of the uniform space of M has a natural structure as a topological left module over the completion $\widehat{A}$ of A. (Of course, if A is an additive topological group then the uniformity of $\widehat{A}$, considered as a completion of the uniform space of A,

1

coincides with the uniformity on $\widehat{A}$ induced by virture of $\widehat{A}$ being a topological group. Similarly for if $\widehat{A}$ is a topological ring; and similarly for topological left modules.)

Of course, a topological ring or a topological module has an underlying topological additive group (by just ignoring the multiplication, or scalar multiplication, respectively). If A is a topological group, then the topology on A is completely determined by the additive group structure and the neighborhood system of zero (in fact, even the uniformity of A is completely determined by these). An additive topological group is said to be, such that *the topology is given by open subgroups* iff there is a complete system of neighborhoods of zero each of which is a subgroup. When this is the case, any complete system of neighborhoods of zero, each of which is a subgroup, is called *a complete system of open subgroups*.

(Remark: A neighborhood of zero in a topological group that is a subgroup is always necessarily open. This justifies the last bit of terminology introduced.)

A topological ring is said to be, such that the topology is given by left ideals (respectively: right ideals; two-sided ideals) iff there exists a complete system of neighborhoods of zero, each of which is a left ideal (respectively: a right ideal; a two-sided ideal). Such a collection is called *a complete system of open left* (respectively *right*; *two-sided*) *ideals*. A topological left module M over a topological ring A is said to be, such that *the topology is given by left submodules* iff there exists a complete system of neighborhoods of zero, each of which is a left submodule. Such a collection is called *a complete system of open left submodules*.

Note: In the above definition of topological group (resp: ring, ring, ring; left module) such that the topology is given by subgroups (resp: left ideals; right ideals; two-sided ideals; submodules), notice that no assumption is made on whether or not there exists a *denumerable* neighborhood base at zero.

Topological rings such that the topology is given by left ideals (or right ideals, or two-sided ideals) can of course be regarded as a special case of topological additive groups such that the topology is given by subgroups. Similarly for topological modules such that the topology is given by left submodules.

Let A be a topological additive abelian group such that the topology is given by subgroups, and let $\mathscr{F}$ be any complete system of open subgroups in A. Then the completion $\widehat{A}$ is canonically isomorphic, as a topological group, to $\varprojlim_{F \in \mathscr{F}} A/F$, where the inverse limit is taken over $\mathscr{F}$ regarded as a directed set with the ordering, reverse inclusion; where A/F is given the discrete topology for all $F \in \mathscr{F}$; and where the topology on the inverse limit group is the one induced from the product topology in $\prod_{F \in \mathscr{F}} A/F$.

Remark: A *map* of topological groups is a homomorphism of groups that is continuous. With this notion of "map", topological groups form a category. Then the topology on the above inverse limit can be described as the unique topology such that the indicated inverse limit is the inverse limit in the category of topological groups.

If A is a topological ring such that the topology is given by two-sided ideals, and if $\mathscr{F}$ is a complete system of open two-sided ideals, then A/F is a ring with the discrete topology for all $F \in \mathscr{F}$ and the completion $\widehat{A}$ as a topological ring is canonically isomorphic to $\varprojlim_{F \in \mathscr{F}} A/F$ in the category of topological rings (with maps the continuous homomorphism of rings). Similarly, if M is a left topological module over a topological ring A, and if the topology of M is given by left submodules, then if $\mathscr{F}$ is any complete system of open left submodules in M, then $\widehat{M} \approx \varprojlim_{N \in \mathscr{F}} M/N$ canonically in the category of left topological A-modules, and continuous homomorphisms of left A-modules, where M/N is given the discrete topology for all $N \in \mathscr{F}$. If A is a topological additive group (or a topological ring; or a topological left module over a topological ring), then A is *complete* iff A is complete as uniform space. Equivalently, iff the natural map: $A \to \widehat{A}$ is a bijection. By the above it follows, in the special case that A is an additive abelian group such that the topology of A is given by open subgroups, that if $\mathscr{F}$ is any complete system of open subgroups for A, then, in this case, A is complete iff the natural mapping: $A \to \varprojlim_{F \in \mathscr{F}} A/F$ is a bijection (in which case it is necessarily an isomorphism of topological groups). Of course, if A is a topological ring such that the topology is given by two-sided ideals, and if $\mathscr{F}$ is a complete system of open ideals, then that bijection is of course an isomorphism of topological rings. Similarly for topological left modules such that the topology is given by left submodules.

Let A be an additive topological group, and let $\mathscr{F}$ be any complete system of neighborhoods of zero (not necessarily subgroups). Then consider all indexed families $(x_F)_{F \in \mathscr{F}}$ of elements of A indexed by the set $\mathscr{F}$. Every such family can of course be regarded as a net in the topological space A, indexed by the directed set $\mathscr{F}$ (where $\mathscr{F}$ is ordered by reverse inclusion). We will call such an indexed family a *quick Cauchy net* — or, more precisely, a *quick Cauchy net with respect to the neighborhood base* $\mathscr{F}$ — iff, whenever $F, G \in \mathscr{F}$ with $F \subset G$, we have that $x_G - x_F \in G$. Then it is easy to see that the topological group A is complete iff every quick Cauchy net has a unique limit. (In fact, this condition is seen immediately to be equivalent to the condition that every Cauchy net converges uniquely.)

Another characterization of completeness: A and $\mathscr{F}$ as above, the topological group A is complete iff it is Hausdorff, and every quick Cauchy net has at least one limit.

<u>Remarks:</u>

1. If A is any topological group, and $\mathscr{F}$ is any complete system of open neighborhoods of zero, such that $\cup_{F \in \mathscr{F}} F = A$, then a model for the completion $\widehat{A}$ of A is the quotient set of the set of all quick Cauchy nets by the equivalence relation: $(x_F) \sim (y_F)$ iff $(x_F - y_F)_{F \in \mathscr{F}}$ converges to zero (equivalently, iff for every $F \in \mathscr{F}$, we have that $x_F - y_F \in F - F$, where $F - F = \{h - k : h, k \in F\}$).

2. If the topology of A is given by open subgroups, and if $\mathscr{F}$ is a complete system of open subgroups in the topological group A such that the union of $\mathscr{F}$ is A, then the quick Cauchy nets with respect to $\mathscr{F}$ form a group (a subgroup of $A^{\mathscr{F}}$), and the completion $\widehat{A}$ is then the quotient group of the group of quick Cauchy nets by the subgroup consisting of those quick Cauchy nets that converge to zero — i.e., by $\{(x_F)_{F \in \mathscr{F}} : x_F \in F, \text{ all } F \in \mathscr{F}\}$.

3. More generally, if X is any uniform space, and if $\mathscr{F}$ is any filter base for the uniformity of X, then one can define a "quick Cauchy net" in X to be any indexed family $(x_F)_{F \in \mathscr{F}}$ of elements of X indexed by $\mathscr{F}$, such that, whenever $F, G \in \mathscr{F}$ and $F \subset G$, we have that $(x_F, x_G) \in G$. Then the uniform space X is complete iff every quick Cauchy net has a unique limit. Equivalently, iff X is Hausdorff, and every quick Cauchy net has at least one limit.

   Whether or not X is complete, if $\cup_{F \in \mathscr{F}} F = A \times A$, then a model for the completion of X is, the quotient space of the set of all quick Cauchy nets by the equivalence relation: $(x_F) \sim (y_F)$ iff for all $F \in \mathscr{F}$, we have that $(y_F, x_F) \in F^{-1} \circ F \circ F$, (where, of course, F^{-1} means $\{(y, x) \in X : (x, y) \in F\}$).

Let A be a topological additive group. Suppose that A has a denumerable neighborhood base at zero. Then as a special case of the preceding considerations, A is complete iff every Cauchy sequence has a unique limit iff every quick Cauchy sequence has a unique limit. If A is a topological additive group having a denumerable neighborhood base at zero, and if $U_1 \supset U_2 \supset U_3 \supset \ldots$ is a denumerable base at zero, then A is complete iff every infinite series $\sum_{i \geq 1} a_i$ such that $\sum_{i \leq k \leq j} a_k \in U_i$, all i, j with $j \geq i \geq 1$, has a unique limit. Equivalently, iff A is Hausdorff, and every such series converges. We will sometimes call such an infinite series a *quick Cauchy series*, or more precisely a *quick Cauchy series with respect to the base $U_1 \supset U_2 \supset U_3 \supset \ldots$*. Thus, a topological additive group A having a denumerable base of neighborhoods of zero is complete iff A is Hausdorff, and every quick Cauchy series (with respect to any fixed open neighborhood base at zero $U_1 \supset U_2 \supset U_3 \supset \ldots$) converges.

Notice also, that if A is a topological additive group such that the topology is given by open subgroups, and such that there exists a denumerable neighborhood base at zero, then there exists a denumerable neighborhood base at zero $U_1 \supset U_2 \supset U_3 \supset \ldots$ such that U_i is an open subgroup, for all integers $i \geq 1$. Then the notion of "quick Cauchy series" with respect to this denumerable base simplifies to: "series $\sum_{i \geq 1} a_i$ such that $a_i \in U_i$, for all $i \geq 1$".

Exercise 0.

1. Let A be a ring. Then to give a topology on A such that A becomes a topological ring, and such that the topology is given by right ideals, is equivalent to giving a set $\mathscr{F}$ of right ideals in A, such that:

    (1) $I \in \mathscr{F}$, J a right ideal in A, $I \subset J$ implies $J \in \mathscr{F}$
    (2) $I \in \mathscr{F}$, $c \in A$ implies there exists $J \in \mathscr{F}$ such that $c \cdot J \subset I$.
    (3) $I, J \in \mathscr{F}$ implies $I \cap J \in \mathscr{F}$

2. Let A be a ring, and let τ be a topology on A such that (A, τ) is a topological ring. Then there exists a finest topology τ^{rt} on A, such that τ^{rt} is coarser than τ, and such that (A, τ^{rt}) is a topological ring such that the topology is given by right ideals.

    Explicitly, if for every neighborhood V of zero in A with respect to τ, we let V^{rt} be the right ideal generated by V, then a complete system of neighborhoods of zero in A for the topology τ^{rt} is $\{V^{rt} : V$ is a neighborhood of zero in A for the topology $\tau\}$.

    An alternate description of τ^{rt} is: $\{$right ideals in A that are open for the topology $\tau\}$.

3. Let A be a topological additive group and let $\widehat{A}$ be the completion of A. Then, if H is an open subgroup of A, show that the closure in $\widehat{A}$ of the image of H is an open subgroup H' of $\widehat{A}$; and that H' with the topology induced from $\widehat{A}$ is the completion of the topological group H.

4. Let A be a topological additive group and let $\widehat{A}$ be the completion of A. Then, if H is a closed subgroup of A, show that if H' is the closure of the image of H in $\widehat{A}$, then H' is a closed subgroup of $\widehat{A}$; and that H' with the topology induced from $\widehat{A}$ is the completion of the topological group H.

Lemma 1.1.1. *Let A be a topological additive group such that the topology is given by open subgroups and let B be a quotient group. Then the group B, together with the quotient topology from A, is a topological group. If A is complete and if B is Hausdorff, and if A has a denumerable neighborhood base at zero, then B is complete.*

Proof. The first assertion is obvious. Let $U_1 \supset U_2 \supset U_3 \supset \ldots$ be a denumerable neighborhood base at zero in A consisting of open subgroups. Then the images in B, $U_1' \supset U_2' \supset U_3' \supset \ldots$ are also a denumerable neighborhood base at zero in B. If $\sum_{i \geq 1} b_i$ is any quick Cauchy series in B, so that $b_i \in U_i'$, then choose $a_i \in U_i$ representing b_i, for all $i \geq 1$. Then $\sum_{i \geq 1} a_i$ is a quick Cauchy series in A. Since A is complete, this series in A converges. If a is the limit, then the image of a in B is a limit of $\sum_{i \geq 1} b_i$. $\square$

Corollary 1.1.2. *Let A and B be topological additive groups with A complete and B Hausdorff. Suppose that the topology of A is given by denumerably many open subgroups. Let $f : A \to B$ be a continuous homomorphism. Regard the image $f(A)$ as a topological group by giving it the quotient topology from A (i.e., the topology such that a subset of $f(A)$ is open iff the preimage in A is open). Then the topological group $f(A)$ is complete.*

Proof. The inclusion $f(A) \to B$ is continuous. Since B is Hausdorff, it follows that $f(A)$ is Hausdorff (since, if x is in the closure of zero in $f(A)$, then x is in the closure of zero in B, and therefore $x = 0$). Therefore the corollary follows from Lemma 1.1.1.   □

Corollary 1.1.3. *Let A and B be topological additive groups. Suppose that the topology of A is given by denumerably many open subgroups. Suppose that we have $f : A \to B$ a continuous epimorphism of topological abelian groups such that B has the quotient topology from A. Then the induced mapping: $\widehat{f} : \widehat{A} \to \widehat{B}$ is a continuous epimorphism, and we have that $\widehat{B}$ has the quotient topology from $\widehat{A}$.*

Proof. Let $A = U_1 \supset U_2 \supset U_3 \supset \ldots$ be a neighborhood base at zero in A consisting of open subgroups. Then since B has the quotient topology from A, $B = f(U_1) \supset f(U_2) \supset f(U_3) \supset \ldots$ is a neighborhood base at zero in B, consisting of open subgroups. If x is any element of $\widehat{B}$, then there exist elements $b_i \in f(U_i)$, $i \geq 1$, such that $x = \sum_{i \geq 1} b_i$ in $\widehat{B}$. Let $a_i \in U_i$ be such that $b_i = f(a_i)$. Then $u = \sum_{i \geq 1} a_i$ is an element of $\widehat{A}$ that maps under $\widehat{f}$ into x. Therefore $\widehat{f} : \widehat{A} \to \widehat{B}$ is surjective.

For every integer $i \geq 1$, the restriction of $f : U_i \to f(U_i)$ is likewise surjective. By the first part of the Corollary, it follows that $\widehat{U_i} :\to \widehat{f(U_i)}$ is surjective. But $\widehat{U_i}$ (resp. $\widehat{f(U_i)}$) is the closure of the image of U_i (respectively $f(U_i)$) in $\widehat{A}$ (respectively $\widehat{B}$). Since {closures of images of U_i in $\widehat{A}$: $i \geq 1$} is an open neighborhood base for $\widehat{A}$, and {closures of images of $f(U_i)$ in $\widehat{B}$: $i \geq 1$} is an open neighborhood base at zero in $\widehat{B}$ (and we have just observed that $\widehat{f}$ maps the ith element of the former onto the ith element of the latter, for $i \geq 0$), it follows that $\widehat{B}$ has the quotient topology from $\widehat{A}$.   □

1.2   Completion of an Abstract Left Module Over a Topological Ring

Let A be a ring and let M be a left A-module. If S is a subset of A, then $S \cdot M$, *the additive subgroup of M generated by* { $s \cdot m$: $s \in S, m \in M$ }, denotes { $\sum_{1 \leq i \leq n} s_i m_i$: n is a non-negative integer, $s_1, \ldots, s_n \in S$ and $m_1, \ldots, m_n \in M$}. This is, of course, the smallest subgroup of M containing $s \cdot m$ for all $s \in S$ and all $m \in M$. If we regard A as a left module over itself, then $S \cdot A$ as defined above is the right ideal generated by S. For the moment, let $S^{rt} = S \cdot A$ denote the right ideal generated by S, for any subset S of A. Then, if M is any left A-module, then $S \cdot M = S^{rt} \cdot M$. Suppose now that A is a topological ring, and that M is a module over the ring A (an *abstract* module over the underlying abstract ring A, ignoring the topology of A). Then it is easy to see that {$S \cdot M$: S is a neighborhood of zero in A} is a base for the neighborhood system at zero of a topology on M, such that the left module M becomes a topological left A-module. We call this topology on the left A-module M the A-*adic topology on M.

Notice that, since $S \cdot M = S^{rt} \cdot M$, it follows that, if τ is the topology of A, and if τ^{rt} is the topology on the underlying abstract ring of A described in Exercise 0 of §1, then the topology induced on the abstract A-module M by τ is the same as the topology induced by τ^{rt}. Therefore, in studying the A-adic topology on an abstract left A-module M, we do not reduce the generality by insisting that the topology of the ring A be given by right ideals. (In fact, if we regard A as a left-module over itself, then the A-adic topology of the left A-module A induced by τ is τ^{rt}, as described in Exercise 0 of §1.) Notice that, since $S \cdot M = S^{rt} \cdot M$, it follows that, if τ is the topology of A, and if τ^{rt} is the topology on the underlying abstract ring of A described in Exercise 0 of §1, then the topology induced on the abstract A-module M by τ is the same as the topology induced by τ^{rt}. Therefore, in studying the A-adic topology on an abstract left A-module M, we do not reduce the generality by insisting that the topology of the ring A be given by right ideals.

Therefore, in studying the topology induced on left modules over a topological ring, we can and do henceforth insist that the topology of the topological ring be given by right ideals. Therefore, for the rest of this chapter, all topological rings are assumed to be such that the topology is given by right ideals.

Remarks

1. If M is a left A-module, then the A-adic topology on M is of course given by subgroups. However, for S an open right ideal in A, the subgroup $S \cdot M$ is not in general a left submodule of A. Therefore, when the topology of A is not given by *two-sided ideals*, the topology of M need not in general be given by left submodules. (E.g., if we regard A as a left submodule over itself, then the A-adic topology of the left module A is given by left submodules iff the topology of the ring A is given by two-sided ideals.)

2. Of course, however, if the topology of A is given by two-sided ideals, then the induced topology on every left A-module M is given by left submodules.

If A is any ring such that the topology is given by right ideals, and M is any left A-module, then we have the A-adic topology on M, and therefore the completion $\widehat{M}$ of M as a topological A-module. Since the topology of M is given by open subgroups, this is isomorphic as an abelian group to $\varprojlim\limits_{S \in \mathscr{R}} M/SM$, where $\mathscr{R}$ is the set of open right ideals in A. (If the topology of A is given by two-sided ideals, then this isomorphism is also an isomorphism of left A-modules.)

If M is an abstract left A-module, then we say that M is A-*adically complete* iff M is complete for the A-adic topology. Equivalently, iff the natural homomorphism: $M \to \widehat{M}$ is an isomorphism of abelian groups.

One should be careful about the topologies on $\widehat{M}$. On the one hand, $\widehat{M}$ is a left A-module, and we therefore have the A-adic topology on $\widehat{M}$. On the other hand, $\widehat{M}$ also has another, in general coarser, topology, by virtue of being the completion of M as a topological left A-module. We will call this latter topology the *uniform topology* on $\widehat{M}$. For this latter topology, a complete system of open neighborhoods of zero are the subgroups $\{\text{Ker}(\widehat{M} \to M/SM) \colon S \text{ is an open ideal}$ in $A\}$ — clearly a coarser topology than the A-adic topology on $\widehat{M}$, for which a complete system of open neighborhoods of zero are the subgroups $\{S \cdot \widehat{M} \colon S \text{ is an}$ open right ideal in $A\}$. We will show shortly that under relatively mild finiteness assumptions on the topological ring A (conditions that hold automatically, e.g., if the topology of A is given by denumerably many two-sided ideals and if the square of every open ideal is open), then these two topologies on $\widehat{M}$ coincide. However, if no finiteness conditions are made on A, then this need not be so even for a free module of denumerable rank over A, see the Example at the end of this section. (In this Example, $\widehat{M}$ is not even complete in its A-adic topology. That is, the A-adic completion of M is not A-adically complete.)

Lemma 1.2.1. *Let A be a topological ring such that the topology is given by right ideals, and let M be an abstract left A-module. Let S be any open right ideal in A. Then the following three conditions are equivalent:*

- ① *There exists a topology τ on the abelian subgroup $S \cdot \widehat{M}$ of $\widehat{M}$, such that $(S \cdot \widehat{M}, \tau)$ is a complete topological abelian group, and such that the homomorphisms: $S \cdot M \to (S \cdot \widehat{M}, \tau)$, $(S \cdot \widehat{M}, \tau) \to (\widehat{M}, unif)$, are continuous.*

- ② *$S \cdot \widehat{M}$ is closed in $\widehat{M}$ for the uniform topology of $\widehat{M}$.*

- ③ *$S \cdot \widehat{M}$ is open in $\widehat{M}$ for the uniform topology of $\widehat{M}$.*

When these three equivalent conditions hold, the topology τ described in ① is necessarily uniquely determined. τ is the induced topology from the uniform topology of $\widehat{M}$. And $(S \cdot \widehat{M}, \tau)$ is the completion of the topological group $S \cdot M$. Note: In the statement of the Lemma, the topology on the open subgroup $S \cdot M$ of M is taken, as always, to be the induced topology from M (for the A-adic topology of M). (Thus a complete system of open neighborhoods of zero in $S \cdot M$ is $\{S' \cdot M \colon S' \text{ is an open right ideal in } A \text{ contained in } S\}$.)

Proof. Let $\iota \colon M \to \widehat{M}$ denote the natural map, and let $\overline{\iota(S \cdot M)}$ denote the closure of $\iota(S \cdot M)$ in $\widehat{M}$ for the uniform topology. Regard $\overline{\iota(S \cdot M)}$ as a topological group by giving it the induced topology from the uniform topology of $\widehat{M}$. Then since $S \cdot M$ is an open subgroup of M, by the third part of Exercise 0 of §1, we have that $\overline{\iota(S \cdot M)}$ is the completion of $S \cdot M$ as a topological group. Clearly, $S \cdot \widehat{M} \subset \overline{\iota(S \cdot M)}$. Therefore, if ① holds, then we have the commuative diagram of topological groups and continuous homomorphisms

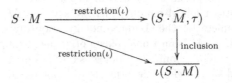

By condition ①, $(S \cdot \widehat{M}, \tau)$ is complete. Therefore, since $\overline{\iota(S \cdot M)}$ is a completion of $S \cdot M$, by the universal mapping property characterization of the completion, there exists a unique continuous homomorphism of topological abelian groups $\alpha : \overline{\iota(S \cdot M)} \to (S \cdot \widehat{M}, \tau)$ such that the diagram

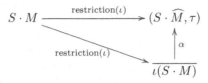

is commutative. Then the composite:

$$(1) \qquad \overline{\iota(S \cdot M)} \xrightarrow{\alpha} (S \cdot \widehat{M}, \tau) \xrightarrow{\text{inclusion}} \overline{\iota(S \cdot M)}$$

is a continuous endomorphism of the completion $\overline{\iota(S \cdot M)}$ of $S \cdot M$ compatible with the identity of $S \cdot M$. By universality of the completion, it follows that this composite is the identity of $\overline{\iota(S \cdot M)}$. Therefore the second of the maps in the composite is an epimorphism; i.e., $S \cdot \widehat{M} = \overline{\iota(S \cdot M)}$. By Exercise 0, part 3, of §1, $\overline{\iota(S \cdot M)}$ is open in $\widehat{M}$ for the uniform topology. Therefore $S \cdot \widehat{M}$ is open in $\widehat{M}$ for the uniform topology. Therefore ① implies ③.

③ implies ② is trivial (since in any topological group, an open subgroup is always closed).

On the other hand, suppose that condition ② holds. By Exercise 0, part 3, of §1, $\overline{\iota(S \cdot M)}$ is the completion of $S \cdot M$. Clearly $\iota(S \cdot M) \subset S \cdot \widehat{M} \subset \overline{\iota(S \cdot M)}$. Therefore $S \cdot \widehat{M} = \overline{\iota(S \cdot M)}$. Then define the topology τ of $S \cdot \widehat{M}$ to be the topology of $\overline{\iota(S \cdot M)}$. Then condition ① is satisfied. Therefore condition ② implies condition ①.

Thus conditions ①, ②, and ③ are indeed equivalent. Suppose all these conditions hold. Then, in the course of showing that ① implies ③, we have observed that the composite of the sequence (1) of continuous homomorphisms is the identity of $\overline{\iota(S \cdot M)}$, and therefore that the inclusion: $(S \cdot \widehat{M}, \tau) \to \overline{\iota(S \cdot M)}$ is bijective. Since both maps in (1) are continuous, it follows that α and this inclusion are homeomorphisms and inverse to each other. In particular, $\alpha = inclusion = identity$, and $(S \cdot \widehat{M}, \tau) = \overline{\iota(S \cdot M)}$ as topological groups. Therefore τ is the topology of $\overline{\iota(S \cdot M)}$, i.e., is the induced topology from $\widehat{M}$. On the other hand, since $\overline{\iota(S \cdot M)}$ is the completion of $S \cdot M$, and $(S \cdot \widehat{M}, \tau) = \overline{\iota(S \cdot M)}$, it follows likewise that $(S \cdot \widehat{M}, \tau)$ is the completion of $S \cdot M$. □

Corollary 1.2.2. *Let the hypotheses be as in Lemma 1.2.1. Suppose, in addition, that the topological ring A has a denumerable neighborhood base at zero. And suppose that the open right ideal S is right finitely generated (i.e., there exists an integer $d \geq 0$ and elements $x_1, \ldots, x_d \in S$ that generate S as a right ideal – i.e., such that $S = x_1 A + \ldots + x_d A$); and in fact, such that there exists a finite set $x_1, \ldots, x_d$ of generators for S as a right ideal, such that, for every* *open right ideal S' in A, the right ideal $x_1 S' + \ldots + x_d S'$ is open in A. Then the three equivalent conditions of Lemma 1.2.1 hold; and therefore so do the conclusions of the last paragraph of Lemma 1.2.1.*

Proof. Let $x_1, \ldots, x_d$ be a finite set of generators of the right ideal S as in the hypotheses. Then by Exercise 0, first conclusion, of §1, for every open right ideal S' in A, for $i = 1, \ldots, d$, there is an open right ideal S_i'' such that $x_i S_i'' \subset S'$. Letting S'' be the intersection of the S_i'', $1 \leq i \leq d$, we obtain an open right ideal S'' such that $x_1 S'' + \cdots + x_d S'' \subset S'$. By hypothesis, $x_1 S'' + \cdots + x_d S''$ is an open right ideal in A. Therefore, the set of all $x_1 S' + \ldots x_d S'$ such that S' is an open right ideal in A, is an open base for the topology of A.

We have that $S \cdot M = x_1 M + \ldots + x_d M$ and $S \cdot \widehat{M} = x_1 \widehat{M} + \ldots + x_d \widehat{M}$. Consider the epimorphism of abelian groups $\theta : (\widehat{M})^d \to S \cdot \widehat{M}$ that maps $(m_1, \ldots, m_d) \in (\widehat{M})^d$ into $x_1 m_1 + \ldots + x_d m_d$. Regard $\widehat{M}$ as a complete topological abelian group with the uniform topology, and regard $(\widehat{M})^d$ as a complete topological abelian group with the product topology. Let τ be the quotient topology on $S \cdot \widehat{M}$ from $(\widehat{M})^d$ by means of the epimorphism θ. Then, by Corollary 1.1.2 of §1 applied to $\theta : (\widehat{M})^d \to \widehat{M}$, it follows that $(S \cdot \widehat{M}, \tau)$ is a complete topological abelian group. Since $\theta : (\widehat{M})^d \to \widehat{M}$ is continuous (since $\widehat{M}$ for the uniform topology is a left topological A-module), and since $S \cdot \widehat{M} = \text{Im} \, \theta$, and since τ is the topology induced from the quotient topology for the epimorphism induced by $\theta : (\widehat{M})^d \to \text{Im} \, \theta$, it follows that the inclusion: $(S \cdot \widehat{M}, \tau) \to \widehat{M}$ is continuous (for the uniform topology on $\widehat{M}$).

An open neighborhood system at zero for the topological group $(S \cdot \widehat{M}, \tau)$ is $\{(x_1 S' + \ldots + x_d S') \cdot \widehat{M} : S'$ is an open right ideal in A such that $x_1 S' + \ldots x_n S' \subset S\}$. By hypothesis, for every open right ideal S' in A, the right ideal $x_1 S' + \ldots + x_d S'$ is open in A. But then $(x_1 S' + \ldots + x_d S') \cdot M$ is open in $S \cdot M$, and maps into $(x_1 S' + \ldots + x_d S') \cdot \widehat{M}$ under ι. Therefore the restriction of ι, from $S \cdot M$ into $(S \cdot \widehat{M}, \tau)$ is continuous.

Therefore all of the hypotheses of condition ① of Lemma 1.2.1 hold. Therefore the Corollary follows from Lemma 1.2.1.   □

<u>Example 1.</u> If A is a topological ring such that the topology of A is given by denumerably many two-sided ideals, and if S is an open right ideal in A that is right finitely generated, and such that, for every open two-sided ideal S' in A, we have that $S \cdot S'$ is open in A, then the hypotheses of Corollary 1.2.2 hold.

Proof. Let $x_1, \ldots, x_d$ be a finite set of generators for S as a right ideal. Let S' be any open two-sided ideal in A. Then it suffices to show that (1): $x_1 S' + \ldots + x_d S'$ is open in A. In fact, since S' is a two-sided ideal, $A \cdot S' = S'$, so that the right ideal (1) is equal to $x_1 A \cdot S' + \ldots + x_d A \cdot S'$ which in turn is equal to $S \cdot S'$; and this latter is open by the hypotheses of the Example.   □

Theorem 1.2.3. *Let A be a topological ring, such that the topology of A is given by denumerably many two-sided ideals $S_1 \supset S_2 \supset S_3 \supset \ldots$, such that*

1. *S_i^2 is open for all $i \geq 1$ (equivalently, $i \geq 1$ implies there exists $j \geq 1$ such that $S_i^2 \supset S_j$),*

2. *S_i is finitely generated as a right ideal, for all $i \geq 1$.*

Then for every abstract left A-module M, the A-adic topology and the uniform topology on $\widehat{M}$ coincide. In particular, $\widehat{M}$ is complete in its A-adic topology. In addition, for every integer $i \geq 1$, we have that $S_i \cdot \widehat{M}$ is the closure of the image of $S_i \cdot M$ in $\widehat{M}$. Also $S_i \cdot \widehat{M}$ with the induced topology from $\widehat{M}$ is the A-adic completion of the left A-module $S_i \cdot M$ for all integers $i \geq 1$.

Note: If the hypotheses on the topological ring A are weakened to: "...the topology of A is given by denumerably many right ideals S_i, $i \geq 1$, such that (1) For every $i \geq 1$, there exists a finite set $x_{i1}, \ldots, x_{id_i}$ of elements of S_i that generate S_i as a right ideal, and such that, for every open right ideal S' in A, we have that the right ideal $x_{i1}S' + \ldots + x_{id_i}S'$ is open in A. (Equivalently: "...and such that, for every integer $j \geq 1$, the right ideal $x_{i1}S_j + \ldots + x_{id_i}S_j$ is open in A)...", then all the conclusions of the above theorem hold, for all abstract left A-modules M.

Remark. In the special case that the topology of A is given by two-sided ideals, then the hypotheses of the above Note are equivalent to those of the Theorem.

Proof. First, by Example 1 above, the hypotheses of the Theorem imply the hypotheses of the Note. Then, by Corollary 1.2.2, whenever the hypotheses of the Note hold, then for $i \geq 1$, we have, for every abstract left A-module M, that the open right ideal S_i obeys the hypotheses of Corollary 1.2.2 (with S_i replacing S). Therefore, by the conclusions of Corollary 1.2.2, we have in particular that $S_i \cdot \widehat{M}$ is closed in $\widehat{M}$ for the uniform topology of $\widehat{M}$, and therefore that $S_i \cdot \widehat{M}$ is the closure of the image of $S_i \cdot M$ in $\widehat{M}$ for the uniform topology of $\widehat{M}$. But, since $\widehat{M}$ is the completion of M, a complete system of neighborhoods of zero in $\widehat{M}$ for the uniform topology is $\{\overline{S_i \cdot M} : i \geq 1\}$ (where "$\overline{S_i \cdot M}$" is the closure of the image of $S_i \cdot M$ in $\widehat{M}$ for the uniform topology — this is because $\{S_i \cdot M : i \geq 1\}$ is a complete system of neighborhoods of zero in M). We have just seen that $\overline{S_i \cdot M} = S_i \cdot \widehat{M}$, for $i \geq 1$. Therefore, $\{S_i \cdot \widehat{M} : i \geq 1\}$ is a complete system of neighborhoods of zero for the uniform topology of $\widehat{M}$. Since this latter is by definition a complete system of neighborhoods for the A-adic topology of $\widehat{M}$, it follows that the A-adic and uniform topologies in $\widehat{M}$ coincide.

Therefore in particular $\widehat{M}$ is complete for the A-adic topology. Also, in the course of the proof, we have observed that the closure of the image of $S_i \cdot M$ in $\widehat{M}$ is $S_i \cdot \widehat{M}$, and the conclusions of Lemma 1.2.1 likewise imply that $S_i \cdot \widehat{M}$ with the induced topology from $\widehat{M}$ is the A-adic completion of $S_i \cdot M$. $\square$

Remark 1. Under the hypotheses of Lemma 1.2.1, suppose that the open right ideal S is right finitely generated, and let $t_1, \ldots, t_d$ be any finite set of right generators for the right ideal S. Then the three equivalent conditions of Lemma 1.2.1 are also equivalent to the condition:

④ Let J be a directed set and let $(x_{ij})_{(1 \le i \le d, j \in J)}$ be any doubly indexed sequence of elements of M, such that the net $(t_1 x_{1j} + \ldots + t_d x_{dj})_{j \in J}$ is Cauchy in M. Then there exists a doubly indexed sequence $(y_{ij})_{1 \le i \le d, j \in J}$ of elements in M, such that each of the nets: $(y_{ij})_{j \in J}$ is Cauchy in M for the uniform topology, $1 \le i \le d$, and such that $t_1 (\lim_{j \in J} y_{ij}) + \ldots + t_d (\lim_{j \in J} y_{dj}) = \lim_{j \in J} (t_1 x_{1j} + \ldots + t_d x_{dj})$ in $\widehat{M}$, the limits being taken for the uniform topology of $\widehat{M}$.

Note: In condition ④, one can insist, if one wishes, that the directed set J be any fixed neighborhood base at zero for the topological ring A ordered by reverse inclusion. (In particular, if A admits a denumerable neighborhood base one can insist that J be the positive integers with the usual ordering.)

Proof. Clearly, in the circumstances of either condition ④ or of the note to condition ④, the set of all $\lim_{j \in J} (t_1 x_{1j} + \ldots + t_d x_{dj})$ for such $(x_{ij})_{1 \le i \le d, j \in J}$ is precisely the closure of the image of $S \cdot M$ in $\widehat{M}$. While the set of all $t_1 (\lim_{j \in J} y_{1j}) + \ldots + t_d (\lim_{j \in J} y_{dj})$ for all such $(y_{ij})_{1 \le i \le d, j \in J}$ is exactly $S \cdot \widehat{M}$. Therefore, condition ④, or condition ④ with the restrictions of the Note, is equivalent to the condition that: _The closure for the uniform topology of $\widehat{M}$ of the image of $S \cdot M$ in $\widehat{M}$ is $S \cdot \widehat{M}$._
Since $S \cdot \widehat{M}$ contains the image of $S \cdot M$ and is always contained in the closure of the image of $S \cdot M$, this latter is equivalent to: $S \cdot \widehat{M}$ _is closed in $\widehat{M}$ for the uniform topology of $\widehat{M}$._
But this latter condition is condition ② of Lemma 1.2.1.   □

Remark 2. Under the hypotheses of Lemma 1.2.1, consider the special case in which the abstract A-module M is a free module, say one with basis indexed by a fixed infinite set J; i.e., $M \approx A^{(J)}$. Then in this case the three conditions of Lemma 1.2.1 are equivalent to:

(F_J) Let $(x_j)_{j \in J}$ be any indexed family of elements of S that converges to zero in A. (I.e., such that every open right ideal in A contains x_j for all but finitely many $j \in J$.) Then there exists an integer $d \ge 1$, elements $t_1, \ldots, t_d \in S$, and a doubly indexed family $(y_{ij})_{1 \le i \le d, j \in J}$ of elements of A such that the family $(y_{ij})_{j \in J}$ converges to zero in A for each $i, 1 \le i \le d$, and such that $x_j = t_1 y_{1j} + \ldots + t_d y_{dj}$ for all $j \in J$.

Notes

1. If the right ideal S is generated as a right ideal by some subset T, then in the statement of condition (F_J), one can insist that $t_1 \ldots t_d \in T$.

2. In particular, if the right ideal S is finitely generated, and if $t_1 \ldots t_d$ is any fixed finite set of generators for S as a right ideal, then in condition (F_J) one can insist that the integer d and the elements $t_1 \ldots t_d$ be this specific integer d and that specific set of generators $t_1 \ldots t_d$.

Proof. Clearly the set of all such families $(x_j)_{j \in J}$ is the closure of the image of $S \cdot M$ in $\widehat{M}$ for the uniform topology. And the set of all such families $(t_1 y_{1j} + \ldots + t_d y_{dj})_{j \in J}$ is $S \cdot \widehat{M}$. Therefore the condition is equivalent to:

The closure of the image of $S \cdot M$ in $\widehat{M}$ for the uniform topology is $S \cdot \widehat{M}$.
And, as we have seen in the proof of Remark 1, this is equivalent to conition ②
of Lemma 1.2.1.   □

<u>Remark 3</u>. Let the hypotheses be as in Remark 2. Let K be an infinite cardinal number such that the topological abelian group A has a neighborhood base B at zero consisting of K elements. Then, for all sets J of cardinality $\geq K$, the condition (F_J) is equivalent to (F_B).

Proof. Let J be any set of cardinality $\geq K$, and let $(x_j)_{j \in J}$ be any family of elements of S that converges to zero in A. Then there exists a subset Z of J, such that $J - Z$ has cardinality $\leq K$, and such that x_j is in the closure of zero for all $j \in Z$. (*Proof.* Let $N = \bigcup_{U \in B} \{j \in J : x_j \notin U\}$. Then N is a union of K finite sets, and therefore has cardinality at most K. Let $Z = J - N$. Then for $j \in Z$, we have $x_j \in U$, for all $U \in B$, and therefore x_j is in the closure of zero in A.) Since the family $(x_j)_{j \in J}$ converges to 0 in A, the restricted family $(x_j)_{j \in B}$ converges to 0 in A. Therefore there exists $(y_{ij})_{1 \leq i \leq d, j \in B}$ such that the conclusion of condition F_B holds. Define $y_{ij} = 0$ for $1 \leq i \leq d$, $j \in Z$. Then $(y_{ij})_{1 \leq i \leq d, j \in J}$ obeys the conclusion of the condition in F_J.   □

<u>Remark 4</u>. In particlular, under the hypotheses of Lemma 1.2.1, if the topological ring A admits a denumerable neighborhood base at zero, then for any open right ideal S in A, the equivalent conditions of Lemma 1.2.1 hold for a particular free left A-module of infinite rank iff they hold for all free left A-modules of infinite rank. (I.e., these conditions hold for all free left A-modules iff they hold for the free left A-module $A^{(\omega)}$.)

Proposition 1.2.4. *Let A be a topological ring such that the topology is given by right ideals. Let M be any abstract left A-module. Then the following conditions are equivalent:*

① *The A-adic completion $\widehat{M}$ of M is A-adically complete.*
② *The uniform and A-adic topologies on $\widehat{M}$ coincide.*
Note: Of course, by Lemma 1.2.1, another condition equivalent to the above is: "For every open right ideal S in A, the three equivalent conditions of Lemma 1.2.1 hold."

Proof. Obviously, since $\widehat{M}$ is by definition complete for the uniform topology, condition ② implies condition ①. Let $(\widehat{M}, unif)$ denote the A-module $\widehat{M}$ regarded as a topological module with the uniform topology, and let $\widehat{M}$ denote $\widehat{M}$ regarded as a topological module with the A-adic topology. Then the uniform topology is coarser than the A-adic, so the identity: $\widehat{M} \xrightarrow{id} (\widehat{M}, unif)$ is continuous. On the other hand, since $\widehat{M}$ is by condition ① A-adically complete, and since $(\widehat{M}, unif)$ is by definition the completion of M as a topological left A-module, there exists a unique continuous homomorphism: $f : (\widehat{M}, unif) \to \widehat{M}$ of topological left A-modules such that the composite: $M \to (\widehat{M}, unif) \xrightarrow{f} \widehat{M}$ is the natural map. Then the composite (1): $(\widehat{M}, unif) \xrightarrow{f} \widehat{M} \xrightarrow{id} (\widehat{M}, unif)$ is a continuous endomorphism of the topological left A-module $(\widehat{M}, unif)$ such that the composite with the natural map $M \to (\widehat{M}, unif)$ is itself. By universality of the completion, the composite (1) is the identity. Therefore f is the identity. Therefore the uniform and A-adic topologies on $\widehat{M}$ coincide. $\square$

Proof of the Note: An open base for the A-adic topology of $\widehat{M}$ is $\{S \cdot \widehat{M} : S$ is an open right ideal in $A\}$. And an open base for the uniform topology on $\widehat{M}$ is $\left\{\overline{S \cdot \widehat{M}} : S$ is an open right ideal in $A\right\}$. If condition (2) of Lemma 1.2.1 holds for all open right ideals S in A, then these two sets coincide. Conversely, if these two sets coincide, then condition ② of Proposition 1.2.4 holds. $\square$

Corollary 1.2.5. *Let A be a topological ring such that the topology is given by right ideals, and let F be a free left A-module. Let $(e_i)_{i \in I}$ be any fixed basis for the left A-module F. Then conditions ① and ② below are equivalent. And each of them implies condition ③ below:*

- *① The A-adic completion $\widehat{F}$ is A-adically complete.*

- *② The uniform and A-adic topologies on $\widehat{F}$ coincide.*

- *③ Given any family $(a_i)_{i \in I}$ indexed by the set I of elements of the topological ring $\widehat{A}$ that converges to zero for the uniform topology of $\widehat{A}$, we have that the Cauchy series for the uniform topology: $\sum_{i \in I} a_i e_i$ in the left A-module $\widehat{F}$, is convergent in $\widehat{F}$ for the A-adic topology.*

Proof. By Proposition 1.2.4 ① is equivalent to ②. Clearly ① and ② imply ③ (since in a complete topological abelian group all Cauchy series converge).
                     $\square$

Remark 5. If the free module F of Corollary 1.2.5 has a basis indexed by a set J, then condition (2) of the Corollary is equivalent to the statement that condition F_J holds, for all open right ideals S in A. Therefore, by Remark 3 above, if the topological ring A has an open neighborhood basis B of open right ideals, then the two equivalent conditions of Corollary 1.2.5 hold for all free modules F iff these two equivalent conditions hold for the free left A-module $A^{(B)}$.

Lemma 1.2.6. *Let A be a topological ring such that the topology is given by denumerably many right ideals. Then*

- *(1) If $f : M \to N$ is any homomorphism of abstract left A-modules, and if M is A-adically complete and N is A-adically Hausdorff, then* Im(f) *is A-adically complete.*

- *(2) If M is an abstract left A-module that is A-adically Hausdorff and if N_1 and N_2 are left submodules of M that are A-adically complete, then the left submodule: $N_1 + N_2$ of M is A-adically complete.*

Corollary 1.2.7. *Let A be a topological ring such that the topology is given by denumerably many right ideals, and let $f : M \to N$ be an epimorphism of abstract left A-modules. Then the induced homomorphism on the A-adic completions: $\hat{f} : \widehat{M} \to \widehat{N}$ is an epimorphism. Also, for every open right ideal S in A, the induced continuous homomorphism of abelian groups from the completion of $S \cdot M$ into the completion of $S \cdot N$ is surjective (where $S \cdot M$ and $S \cdot N$ are regarded as topological abelian groups with the induced topologies from M and N respectively).*

Note: Whether or not the topology of A admits a denumerable neighborhood base at zero, the proof also shows that the A-adic topology of N is induced from that of M by the epimorphism f, and is the quotient topology from M.

Proofs of the Note to Corollary 1.2.7 and Lemma 1.2.6, part (1). First, the note to Corollary 1.2.7 is obvious, since an open base for the neighborhood system at zero in M, respectively N, is $\{S \cdot M\colon S$ is an open right ideal in $A\}$, respectively $\{S \cdot N\colon S$ is an open right ideal in $A\}$, and since $f(S \cdot M) = S \cdot N$, for all open right ideals S in A (this holds whether or not the neighborhood system at zero in A admits a denumerable neighborhood base at zero). Therefore, under the hypotheses of Lemma 1.2.6, part (1), we have that Im(f) in its A-adic topology is a Hausdorff quotient-module of M. The conclusion of Lemma 1.2.6, part (1), then follows Corollary 1.1.2.   □

Proof of Lemma 1.2.6, part (2). $N_1 + N_2$ is the image of the homomorphism of abstract left A-modules: $f : N_1 \oplus N_2 \to M$ that has for coordinates, the inclusion: $N_1 \to M$ and $N_2 \to M$. Part (1) of Lemma 1.2.6 applied to this f completes the proof.   □

Proof of Corollary 1.2.7. By applying Corollary 1.1.3 to the continuous homomorphism of abelian groups $f : M \to N$, we have that $\hat{f} : \widehat{M} \to \widehat{N}$ is an epimorphism of topological abelian groups. Similarly, by applying Corollary 1.1.3 to the homomorphism of topological abelian groups $f|(S \cdot M) : S \cdot M \to S \cdot N$ we have that the induced homomorphism on the completions (for the topologies induced from M and N) is surjective, for every open right ideal S in A.   □

Corollary 1.2.8. *Let A be a topological ring such that the topology is given by denumerably many right ideals. Let S be any open right ideal in A. Then the following two conditions are equivalent:*
(1) For every left A-module M, the three conditions of Lemma 1.2.1 hold.

(2) For a fixed free left A-module of denumerable rank (say $M = A^{(\omega)}$), the three equivalent conditions of Lemma 1.2.1 hold for this fixed free left A-module of denumerable rank.

Proof. Clearly (1) implies (2). Assume (2). Then by Remark 4 above, we have that (1) holds for all free left A-modules. Now let M be any abstract left A-module. Then choose an epimorphism of left A-modules $F \to M$ where F is a free left A-module. Then, taking A-adic completions, we have the map $\widehat{F} \to \widehat{M}$. By Corollary 1.2.7, this homomorphism is necessarily an epimorphism.

We know that the three equivalent conditions of Lemma 1.2.1 hold for the left A-module F. Since $S \cdot F$ maps onto $S \cdot M$, we have by Corollary 1.2.7 that the completion of $S \cdot F$ maps onto the completion of $S \cdot M$. Since we know that the three equivalent conditions of Lemma 1.2.1 hold for the left A-module F, we have that the completion of $S \cdot F$ is $S \cdot \widehat{F}$. Since the completion of $S \cdot M$ is the closure $\overline{\iota(S \cdot M)}$ of the image of $S \cdot M$ in $\widehat{M}$ for the uniform topology, it follows that $S \cdot \widehat{F}$ maps onto $\overline{\iota(S \cdot M)}$. But clearly, the image of $S \cdot \widehat{F}$ in $\widehat{M}$ is $S \cdot \widehat{M}$. Therefore $S \cdot \widehat{M} = \overline{\iota(S \cdot M)}$, and in particular $S \cdot \widehat{M}$ is closed in $\widehat{M}$ for the uniform topology of $\widehat{M}$. Therefore condition ② of Lemma 1.2.1 holds for M. $\qquad\square$

The following non-Noetherian Example shows that there are complete topological rings, such that the topology is given by right ideals that do not obey the equivalent conditions of Proposition 1.2.4, and also do not obey the equivalent conditions of Corollary 1.2.5. (Notice that the topological ring in the Example has a denumerable neighborhood base at zero – in fact, it is given by the powers of an ideal in a commutative, non-Noetherian ring.)

Example 2. Let k be a field, and let $A_0 = k[(T_i)_{i \geq 1}]$, the ring of polynomials in denumerably many variables over the field k. Let S_0 be the ideal generated by $\{T_i : i \geq 1\}$, and regard A_0 as a topological ring with the topology given by the ideals $\{S_0^i : i \geq 1\}$, the positive powers of S_0. Then let A be the completion of the topological ring A_0, and let S be the closure of S_0 in A. Thus, explicitly, $A = \{ \sum_{m \in M} a_m m : a_m \in k, \text{ all } m \in M \}$, where M is the set of all monomials in the variables $T_1, T_2, T_3, \ldots$. And S is the set of elements of A with constant term zero (i.e., such that $a_1 = 0$, where 1 denotes the monomial $T_1^0 T_2^0 \cdots T_i^0 \cdots$). Then the topological rings A_0 and A each has a denumerable neighborhood base at zero, and the topology in the complete topological ring A is the S-adic topology.

Let F be a free abstract A-module of denumerable rank, say with basis $e_1, e_2, \ldots, e_i, \ldots$. Then we show that condition ③ of Corollary 1.2.5 does not hold.

In fact, the sequence: $(T_1, T_2^2, T_3^3, \ldots, T_i^i, \ldots)$ of elements of A converges A-adically to zero. Nevertheless, the Cauchy series: $\sum_{i \geq 1} T_i^i e_i$ in $\widehat{F}$ does not converge for the A-adic topology. For, if it did converge A-adically, then since the set-theoretic identity: $\widehat{F} \to (\widehat{F}, unif)$ is continuous, the limit x of this series in $\widehat{F}$ would also be the limit of this convergent series in $(\widehat{F}, unif)$. But then, by definition of convergence, given any A-adically open neighborhood of zero in $\widehat{F}$, for example $S \cdot \widehat{F}$ [1] there exists an integer $n \geq 1$ such that $x - \sum\limits_{1 \leq i \leq n} T_i^i e_i \in S \cdot \widehat{F}$.

But then, in $(\widehat{F}, unif)$, we have that (1): $\sum\limits_{i \geq n+1} T_i^i e_i \in S \cdot \widehat{F}$.

Recall that the elements of $S \cdot \widehat{F}$ are the set of finite sums: $s_1 f_1 + \cdots + s_m f_m$ such that $m \geq 1$, $s_1, \ldots, s_m \in S$, and $f_1, \ldots, f_m \in \widehat{F}$. So, if (1) holds, then it follows that there exist an integer $m \geq 1$ and elements $s_1, \ldots, s_m \in S$ such that $\sum\limits_{i \geq n+1} T_i^i e_i \in \{s_1, \ldots, s_m\} \cdot \widehat{F}$. Then, if I is the ideal in the commutative ring A generated by $\{s_1, \ldots, s_m, T_1, \ldots T_n\}$, then we have that:

(1) I is generated by $n + m$ elements

(2) $T_i^i \in I$, $i \geq 1$, and

(3) $I \subset S$.

But then, let N be an integer $> n + m$. Then consider the continuous homomorphism of complete topological rings $A \to k[[T_1, \ldots, T_N]]$ from A into the ring of formal power series in N variables that induces the identity on the subring $k[[T_1, \ldots, T_N]]$ of A, and that sends T_j into zero, for $j \geq N + 1$. Then, if J denotes the image of I under this epimorphism of rings, then (1), (2), and (3) above imply:

(1) J is generated by $n + m$ elements

(2) $T_i^i \in J, 1 \leq i \leq N$, and

(3) $J \subset (T_1, \ldots, T_N)$.

Properties (2) and (3) imply that $\mathrm{rad}(J) = (T_1, \ldots, T_N)$. But then, in the regular local ring of dimension N, $k[[T_1, \ldots, T_N]]$, we have that there is an ideal J that is generated by $n + m < N$ elements, and such that the radical of this ideal J is the maximal ideal, a contradiction. Therefore, the Cauchy series $\sum\limits_{i \geq 1} T_i^i e_i$ in $\widehat{F}$ indeed does not converge for the A-adic topology, as asserted; so that, also, condition ③ of Corollary 1.2.5 does not hold for the free A-module F.

Therefore, by Corollary 1.2.5:

① The A-adic completion $\widehat{F}$, of a denumerably generated free A-module F, is not A-adically complete (i.e., the abstract left A-module $\widehat{F}$ is not complete for its A-adic topology).

② F as above, then the uniform topology on $\widehat{F}$ is strictly coarser than the A-adic topology.

Another way of recording ① above is:

③ If $\widehat{\widehat{F}}$ denotes the A-adic completion of the abstract A-module $\widehat{F}$, then the natural homomorphism (from $\widehat{F}$ into its A-adic completion) $\widehat{F} \to \widehat{\widehat{F}}$ is not onto. (Of course, the homomorphism is, however, necessarily injective, since $\widehat{F}$ is A-adically Hausdorff (since it is even Hausdorff for its uniform topology, a coarser topology than the A-adic topology).)

---

[1] We will see that $S \cdot \widehat{F}$ is open for the A-adic topology of $\widehat{F}$, but not for the uniform topology of $\widehat{F}$. See Note 6 following the Example.

Proof. In general, a left A-module M is A-adically complete iff the natural map: $M \to \widehat{M}$ is an isomorphism of left A-modules. By ①, $\widehat{F}$ is not A-adically complete. Therefore the natural map of $\widehat{F}$ into its A-adic completion: $\widehat{F} \to \widehat{\widehat{F}}$ is not an isomorphism of abstract A-modules. Since this mapping is a monomorphism, it therefore must fail to be an epimorphism.   □

<u>Note 1</u>: An explicit element of $\widehat{\widehat{F}}$ that is not in the image of the above monomorphism $\iota : \widehat{F} \to \widehat{\widehat{F}}$ is the limit, in $(\widehat{\widehat{F}}, unif)$, (where $(\widehat{\widehat{F}}, unif)$ denotes the completion of the left A-module $\widehat{F}$ in its A-adic topology), of the infinite series $\sum_{i \geq 1} T_i^i e_i$. (The reason for this is that the A-adic topology of $\widehat{F}$ is the induced topology from $(\widehat{\widehat{F}}, unif)$. Therefore, if the indicated Cauchy series (in the complete left A-module $(\widehat{\widehat{F}}, unif)$) converges in $(\widehat{\widehat{F}}, unif)$ to the image of an element of $\widehat{F}$, then it would have to converge A-adically to that same element in $\widehat{F}$. And we have seen above that that series in $\widehat{F}$, although A-adically Cauchy, is not A-adically convergent.)

<u>Note 2</u>: There is also another, different, monomorphism: $\widehat{F} \to \widehat{\widehat{F}}$. Namely, let $k : F \to \widehat{F}$ denote the inclusion. Then, throwing through the functor "A-adic completion" we obtain a map $\widehat{k} : \widehat{F} \to \widehat{\widehat{F}}$. Then by construction, $\widehat{k}$ is continuous considered as a map of topological left A-modules: $(\widehat{F}, unif) \to (\widehat{\widehat{F}}, unif)$. In particular, the topology induced from $\widehat{\widehat{F}}$ on $\widehat{F}$ by the mapping $\widehat{k}$ is coarser than the uniform topology of $\widehat{F}$. But the topology induced on $\widehat{F}$ by $\iota : \widehat{F} \to (\widehat{\widehat{F}}, unif)$, as we have seen, is the A-adic topology of $\widehat{F}$, which, as we have seen, is strictly finer than the uniform topology of $\widehat{F}$. Therefore the mappings ι and $\widehat{k}$ are distinct. (Notice also that the element: $\sum_{i \geq 1} T_i^i e_i$ of $(\widehat{F}, unif)$ maps into the element $\sum_{i \geq 1} \iota(T_i^i e_i)$ of $(\widehat{\widehat{F}}, unif)$; so that this latter element of $\widehat{\widehat{F}}$ is in the image of $\widehat{k}$, but is not in the image of ι.)

<u>Note 3</u>: If we throw the sequence: $F \xrightarrow{k} \widehat{F} \xrightarrow{id} (\widehat{F}, unif)$ of topological left A-modules through the functor, "completion of a topological left A-module", then we obtain the commutative diagram of topological left A-modules:

$$
\begin{array}{ccccc}
(\widehat{F}, unif) & \xrightarrow{\ \widehat{k}\ } & (\widehat{\widehat{F}}, unif) & \xrightarrow{\ \rho\ } & (\widehat{F}, unif) \\
\uparrow{\scriptstyle k} & & \uparrow{\scriptstyle \iota} & & \uparrow{\scriptstyle id} \\
F & \xrightarrow{\ k\ } & \widehat{F} & \xrightarrow{\ id\ } & (\widehat{F}, unif)
\end{array}
$$

where ρ is the image if $id : \widehat{F} \to (\widehat{F}, unif)$ under the functor "completion of a topological left A-module."

From the above commutative diagram, we can read off that:

(1) The homomorphisms of left A-modules ι and $\widehat{k}$ (which are distinct by Note 2 above) from $\widehat{F} \to \widehat{\widehat{F}}$ are both split monomorphisms, with the same canonical epimorphism: $\rho : \widehat{\widehat{F}} \to \widehat{F}$ splitting them both (i.e., $\rho \circ \iota = \rho \circ \widehat{k} = $ identity of $\widehat{F}$).

(2) It follows that, although we have seen in Note 2 above that the uniform topology of $(\widehat{\widehat{F}}, unif)$ induces the uniform topology of $(\widehat{F}, unif)$ by means of the map $\widehat{k} : \widehat{F} \to \widehat{\widehat{F}}$, that nevertheless the topology induced by $\widehat{\widehat{F}}$ on $\widehat{F}$ by the A-adic topology of $\widehat{\widehat{F}}$, and by either of the monomorphisms ι or $\widehat{k} : \widehat{F} \to \widehat{\widehat{F}}$, is the A-adic topology of $\widehat{F}$.

(3) The monomorphisms ι and $\widehat{k} : \widehat{F} \to \widehat{\widehat{F}}$, although different by Note 2, agree on the subset F of $\widehat{F}$. (As we have observed in Note 2, ι and $\widehat{k}$ disagree on the element: $\sum_{i \geq 1} T_i^i e_i$ of $(\widehat{F}, unif)$.)

(4) Although the monomorphisms ι and $\widehat{k}$ from $\widehat{F}$ to $\widehat{\widehat{F}}$ are distinct, nevertheless there exists an automorphism α of the abstract left A-module $\widehat{\widehat{F}}$ such that $\alpha \circ \widehat{k} = \iota$. Indeed, there is a unique such α, that also has the property that $\rho \circ \alpha = \rho$.

Proof. From the commutative diagram, we have that:

- $\widehat{\widehat{F}}$ is the internal direct sum of $\text{Im}\,(\widehat{k})$ and $\text{Ker}\,(\rho)$, and also that

- $\widehat{\widehat{F}}$ is the internal direct sum of $\text{Im}\,(\iota)$ and $\text{Ker}\,(\rho)$.

The unique α as indicated is the automorphism of the abstract left A-module such that the restriction to $\text{Im}(\widehat{k})$ is the composite: $\text{Im}(\widehat{k}) \approx \widehat{\widehat{F}} \approx \text{Im}(\iota)$, and such that the restriction of α to $\text{Ker}(\rho)$ is the identity. $\qquad\square$

(5) ι and $\widehat{k}$ agree on F, but not on all of $\widehat{F}$. Therefore $\widehat{k} - \iota$ is a non-zero homomorphism of abstract left A-modules from $\widehat{F}$ into $\widehat{\widehat{F}}$ that is zero on F. Therefore $\widehat{k} - \iota$ defines a non-zero homomorphism from $\widehat{F}/F$ into $\widehat{\widehat{F}}$. (The significance of this will be made clear in Chapter 2, §4.)

In fact, it is easy to see that if $\overline{F}$ denotes the A-adic closure of the image of F in $\widehat{F}$, then ι and $\widehat{k}$ agree on $\overline{F}$; and in fact $\overline{F} = \text{Ker}(\widehat{k} - \iota)$. Also, since $\rho \circ \iota = \rho \circ \widehat{k} = $ identity of $\widehat{F}$, $\text{Im}(\widehat{k} - \iota) \subset \text{Ker}(\rho)$. Therefore $\widehat{k} - \iota$ defines, by passing to the quotient, a monomorphism of abstract left A-modules from the non-zero left A-module $\widehat{F}/\overline{F}$ into $\text{Ker}(\rho)$.

(6) Suppose, for every ordinal number i, that we define, by induction, left A-modules F_i as follows: $F_0 = F$. Having defined F_i, define $F_{i+1} = \widehat{F_i}$, and let $\iota_i : F_i \to F_{i+1}$ be the natural map from F_i into its A-adic completion. And, for every limit ordinal α, define $F_\alpha = \bigcup_{i < \alpha} F_i$.

Perhaps we should record some of the constructions made in Example 2 above, as they apply in greater generality. The above arguments prove:

Corollary 1.2.9. *Let A be a complete topological ring such that the topology is given by right ideals. Let F be an arbitrary abstract left A-module (not necessarily free). Let ι be the natural map from $\widehat{F}$ into its A-adic completion $\widehat{\widehat{F}}$ (where $\widehat{F}$ is given its A-adic topology). Let $k : F \to \widehat{F}$ denote the natural map. Then we have the two homomorphisms of abstract left A-modules, ι and $\widehat{k}$, from $\widehat{F}$ into $\widehat{\widehat{F}}$. The set-theoretic identity function is a continuous homomorphism of topological left modules* id $: \widehat{F} \to (\widehat{F}, unif)$. *Throwing this continuous homomorphism through the functor "$\widehat{}$" (completion of a uniform space) gives a continuous homomorphism of topological left modules* $\rho : (\widehat{\widehat{F}}, unif) \to (\widehat{\widehat{F}}, unif)$. *Then:*

(1) The homomorphisms of left A-modules ι and k from $\widehat{F}$ into $\widehat{\widehat{F}}$ are both split monomorphisms, with the same splitting ρ (i.e., $\rho \circ \iota = \rho \circ \widehat{k} =$ identity of $\widehat{F}$).

(2) The uniform topology of $(\widehat{\widehat{F}}, unif)$ induces the uniform topology of $(\widehat{F}, unif)$ by means of $\widehat{k}$; and the topology induced by $\widehat{\widehat{F}}$ on $\widehat{F}$, by the A-adic topology of of $\widehat{\widehat{F}}$, by either of the monomorphisms ι or $\widehat{k} : \widehat{F} \to \widehat{\widehat{F}}$, is the A-adic topology of $\widehat{F}$.

(3) The monomorphisms ι and $\widehat{k} : \widehat{F} \to \widehat{\widehat{F}}$ agree on the subset F of $\widehat{F}$.

(4) There exists a unique automorphism α of the abstract left A-module $\widehat{\widehat{F}}$ such that $\alpha \widehat{k} = \iota$ and such that $\rho \circ \alpha = \rho$.

(5) We have the commutative diagram with exact rows displayed in Note 3 following Example 2 above.

<u>Note</u>: As we have seen in Example 2 above, even if F is a free left A-module, $\widehat{k}$ and ι can be different. $\widehat{k} : (\widehat{F}, unif) \to (\widehat{\widehat{F}}, unif)$ is always uniformly continuous; $\iota : (\widehat{F}, unif) \to (\widehat{\widehat{F}}, unif)$ is uniformly continuous iff $\widehat{k} = \iota$.

<u>Exercise</u>. The notations being as in Example 2 above, show that:

(A) For $1 \leq i \leq j$ ordinal numbers, the mapping: $F_i \to F_j$ is a split monomorphism of left A-modules with a canonical splitting, but is never an isomorphism.

(B) Therefore the F_i, $i \geq 1$, "never stops."

(C) $\underset{i \text{ an ordinal}}{\cup} F_i$ is not a set (it is a proper class).

(D) For $i < j$, the topology induced by the A-adic topology of F_j on F_i is the A-adic topology.

(E) For every ordinal $i \geq 2$, $\widehat{F}$ is canonically a proper direct summand of F_i (in strictly more than one way. E.g., one such inclusion is $\widehat{F} = F_1 \subset F_i$. Another is $\widehat{F} \overset{k}{\to} \widehat{\widehat{F}} = F_2 \subset F_i$).

(F) For every ordinal $i \geq 2$, there is more than one endomorphism of the left A-module F_i that induces the identity of F. (E.g., the identity of F_i is one such, and a different endomorphism, that is neither injective nor surjective, is the composite: $F_i \to \widehat{F} \hookrightarrow F_i$).

(5) The proof that a free left A-module F of denumerable rank does not obey condition ③ of Corollary 1.2.5, given at the beginning of Example 2, also shows

that the condition (F_P) of Remark 3 fails for the open ideal S in the ring A, where P is the set of positive integers. Therefore, by Remark 3 and Lemma 1.2.1, it follows that the A-adically open subgroup (in fact, left submodule) $S \cdot \widehat{F}$ is not open for the uniform topology of $\widehat{F}$. Also, by Lemma 1.2.1, it follows likewise that (the A-adically closed submodule) $S \cdot \widehat{F}$ is also not closed for the uniform topology of $\widehat{F}$. It follows that the A-adically open-and-closed left submodule $S^i \cdot \widehat{F}$ is not open in $\widehat{F}$ for the uniform topology (since otherwise $S \cdot \widehat{F}$, which contains $S^i \cdot \widehat{F}$, would also have to be open for the uniform topology); and therefore by Lemma 1.2.1 that $S^i \cdot \widehat{F}$ is also not closed for the uniform topology of $\widehat{F}$, for each integer $i \geq 1$.

(6) By Remark 4 above (since the topological ring A has a denumerable neighborhood base at zero), it follows that, for every free left module F_0 of infinite (whether denumerable or not) rank over the ring A, we have the analogous results for F_0. (E.g., $M = F_0$ fails to obey the conditions of Proposition 1.2.4, and therefore we have ①, ②, and ③ of this Example for F_0, etc.)

Remark 6. One might wonder: Given an arbitrary topological ring A such that the topology is given by right ideals; and an arbitrary abstract left A-module M: Does there necessarily exist a pair (ι, N), where N is an A-adically complete left A-module and $\iota : M \to N$ is a homomorphism of abstract left A-modules; such that the pair (ι, N) is universal with these properties? The answer is:

Let A be any topological ring such that the topology is given by right ideals, and let M be any left A-module. Then the following two conditions are equivalent:

(1) There exists (N, ι) as above.

(2) The A-adic completion $\widehat{M}$ of M is A-adically complete.

When these two conditions hold, then $N = \widehat{M}$ and ι is the natural map: $M \to \widehat{M}$.

Proof. Clearly, if (2) holds, then so does (1), with $N = \widehat{M}$. Conversely, suppose that (N, ι) as in (1) exists. Let $(\widehat{M}, unif)$ denote the complete topological left A-module, the completion of M in the category of topological left A-modules. Then, by universality, there exists a unique continuous homomorphism of left A-modules $(\widehat{M}, unif) \xrightarrow{\beta} N$, compatible with the identity of M. For every open right ideal S in A, the left A-module M/SM is A-adically discrete, and therefore A-adically complete. Therefore, by universality of (N, ι), there exists a unique homomorphism: $N \to M/SM$ of left A-modules compatible with the identity of M. Passing to the inverse limit over open right ideals S in A, we obtain that there is a unique homomorphism of abstract left A-modules $N \xrightarrow{\gamma} \widehat{M}$ compatible with the identity of M. Since β and γ are continuous, the composite $(\widehat{M}, unif) \xrightarrow{\beta} N \xrightarrow{\gamma} \widehat{M} \xrightarrow{id} (\widehat{M}, unif)$ is a continuous endomorphism of the topological left A-module $(\widehat{M}, unif)$ compatible with the identity of M. Since the image of M is dense in the Hausdorff topological space $(\widehat{M}, unif)$, it follows that the composite is the identity of $\widehat{M}$. On the other hand, the composite: $N \xrightarrow{\gamma} \widehat{M} \xrightarrow{\beta} N$ is an endomorphism of the A-adically complete left A-module N compatible with the identity of M. By the universal mapping property satisfied by (N, ι), it follows that $\beta \circ \gamma$ is therefore the identity of N. Therefore $\widehat{M} \approx N$, and $\widehat{M}$ is A-adically complete, proving (2).   □

Remark 7. The pathology of Example 2 above is quite typical of modules that do not obey the equivalent conditions of Proposition 1.2.4. That is, if A is any topological ring such that the topology is given by right ideals, and if M is any abstract left A-module such that the two equivalent conditions of Proposition 1.2.4 fail for M, then: ①, ②, and ③ above all hold (with "M" replacing "F". In fact, each of conditions ①, ②, and ③ for M is equivalent to the negation of each of the two equivalent conditions of Proposition 1.2.4.) Similarly, Notes 2 and 3 above (except for the parenthetical part of Note 3; unless M is free and A is complete), and the entire Exercise following, go through for such an M. (Similarly for Note 1 if M is free and A is complete – an analogue of Note 1 can also be stated that makes sense and holds even for such an M that is not free.) (And similarly for the analogue of Note 6 above if M is free and A is complete; in fact, if A is complete and if M is free of rank K and does not obey the conditions of Proposition 1.2.4, then trivially every free left A-module of rank $\geq K$ fails to obey the conditions of Proposition 1.2.4 — e.g., a free module of larger rank than K contains an isomorphic copy of M as a direct summand. More elaborately, one can alternatively use Remark 3 above to prove this.)

Remark 8. Let A_0 be the topological ring described in Example 2 above (so that A is the completion of A_0 as a topological ring). Then it is easy to see that the A_0-adic topology on $A = \widehat{A_0}$ (considered as a module over A_0) is strictly finer than the $\widehat{A_0}$-adic topology (the one for which $A = \widehat{A_0}$ is the completion of A_0 – i.e., the uniform topology on the A_0-module $\widehat{A_0}$).

For, in fact, the open ideal $S_0 \subset A_0$ is such that $S_0 \cdot A$ is not open in A (for the uniform topology on A). To prove this, by Lemma 1.2.1 it suffices to show that $S_0 \cdot A$ is not closed in A (for the uniform topology). In fact, the element: $x = T_1 + T_2^2 + T_3^3 + \ldots + T_i^i + \ldots$ of A (the limit of the infinite series is for the uniform topology on A) is clearly in the closure of S_0 in A (for the uniform topology), but is not in $S_0 \cdot A$. For, if $x \in S_0 \cdot A$, then since S_0 is the ideal in A_0 generated by $\{T_1, T_2, \ldots, T_i, \ldots\}$, we have that $x \in \{T_1, T_2, \ldots, T_i, \ldots\} \cdot A$, and therefore there would exist an integer m such that $x \in \{T_1, \ldots, T_m\} \cdot A$. But then, if N is any integer $\geq m$, then throwing through the unique continuous homomorphism of topological rings $A \to k[[T_{m+1}, \ldots, T_N]]$ that induces the identity on $k[[T_{m+1}, \ldots, T_N]]$ and maps T_i into zero for $i \neq m+1, \ldots, N$, we would obtain that the obviously non-zero element $T_{m+1}^{m+1} + \ldots + T_N^N$ of the ring $k[[T_{m+1}, \ldots, T_N]]$ lies in the zero ideal, a contradiction.

Remark 8.1. The argument given in Example 2 above, through the observation ①, goes through to prove that: If A is the complete topological ring of Example 2, then the abstract A-module A^ω (the direct product of denumerably many copies of A) is not A-adically complete. (In fact, the argument is the same, and likewise shows that the A-adically Cauchy series: $\sum_{i \geq 1} T_i^i e_i$ cannot converge A-adically in A^ω, where $e_i = (\delta_{ij})_{0 \leq j < \infty}$, for all i, $0 \leq i < \infty$.) In fact, in the proof, one simply substitutes "A^ω" for "$\widehat{F}$" throughout, and also "$(A^\omega, prod)$" for "$(\widehat{F}, unif)$", where $(A^\omega, prod)$ denotes A^ω, regarded as a complete topological left A-module with the direct product topology (each factor A being given the A-adic topology). (Or, if one prefers, one can use alternatively the topology on A^ω given by the submodules, $\{S^\omega : S$ is an open ideal in $A\}$. This alternative, finer-than-the-product topology also makes A^ω into a complete topological left A-module.)

We conclude this section by introducing the notion of a *divisible element*, and of an *infinitely divisible element*, of a left module. Note: These concepts will be used extensively throughout this book.

Definition 1. Let A be a topological ring such that the topology is given by right ideals. Let M be an abstract left A-module. Then an element $x \in M$ is *A-divisible* iff x is in the closure of $\{0\}$ for the A-adic topology.

Remark 9. If A is a topological ring such that the topology is given by right ideals, and if M is an abstract left A-module, then clearly an element $x \in M$ is A-divisible iff $x \in I \cdot M$ (i.e., iff x can be written in the form $i_1 m_1 + \ldots + i_n m_n$ for some $n \geq 0$, $i_1, \ldots, i_n \in I$ and $m_1, \ldots, m_n \in M$), for all open right ideals $I \subset A$. Intuitively, the condition "$x \in I \cdot M$" is to be thought of as "x is divisible by the right ideal I." Therefore, the condition that "$x \in I \cdot M$ for all I" is to be thought of as "x is divisible by every open right ideal I in A for all open right ideals I in A", i.e., "x is divisible in topological ring A".

Proposition 1.2.10. *Let A and M be as in Definition 1. Then the set of all A-divisible elements of M is a left submodule of M, and is closed for the A-adic topology of M.*

Notation: The set of all A-divisible elements of M will be denoted by $\operatorname{div} M$. (The Proposition therefore asserts that $\operatorname{div} M$ is an abstract left submodule of M.)

Proof. The indicated set is, by Definition 1, the closure of the left submodule: $\{0\}$ in the topological left A-module consisting of M together with its A-adic topology. □

Remark 10. It is amusing to attempt proving that $\operatorname{div} M$ is a left submodule of M without using any topology directly, but instead using §1, Exercise 1. The proof is not difficult.

Notice, of course, that $\operatorname{div} M = \bigcap_{I \in \mathscr{R}} I \cdot M$, where $\mathscr{R}$ is the collection of open right ideals in M. Of course, if I is an open right ideal in A that is not a two-sided ideal, then in general $I \cdot M$ is only a subgroup (not, in general, a left submodule) of M; so the above equation alone does not immediately imply Proposition 1.2.10.

Remark 11. Let A and M be as in Definition 1. Then by Proposition 1.2.10, $\operatorname{div} M$ is a left submodule of M (and in fact an A-adically closed left submodule of M). Also, by definition, every element $x \in \operatorname{div} M$ is A-divisible *in the left*

A-module M. However, it is easy to give examples such that there are elements $x \in div\, M$ that are not *A*-divisible in $div\, M$. (See Exercise 5 below.)

Definition 2. Let *A* be a topological ring such that the topology is given by right ideals and let *M* be an abstract left *A*-module. Then an element $x \in M$ is *infinitely A-divisible* iff there exists an abstract left *A*-submodule *N* of *M* containing *x* and such that every element *y* of *N* is *A*-divisible *in N*, in the sense of Definition 1.

Remark 12. Clearly, if $f : M \rightarrow N$ is a homomorphism of abstract left *A*-modules, where *A* is as in Definition 1, then *f* maps *A*-divisible (resp. infinitely *A*-divisible) elements of *M* into *A*-divisible (resp. infinitely *A*-divisible) elements of *N*. Applying this observation to inclusion maps and using Remark 11 above, it follows that one can give an example of an *A*-divisible element of a left *A*-module *M* that is not infinitely *A*-divisible in *M*. (Namely, take *x*, *A* and *M* as in Remark 11. Then $x \in div\, M$, and is therefore *A*-divisible in *M*; but *x* is not infinitely *A*-divisible in *M*. For, otherwise there exists a left *A*-submodule *N* of *M*, such that every element of *N* is *A*-divisible in *N*, and such that $x \in N$. But then, applying this Remark to the inclusion: $N \rightarrow M$ we have that every element of *N* is divisible in *M*, i.e., $N \subset div\, M$. Then applying the result of this Remark to the inclusion: $N \rightarrow div\, M$, it follows that every element of *N* is divisible in $div\, M$. In particular *x* is divisible in $div\, M$, contrary to the hypotheses on *x* in Remark 11.)

On the other hand, it is also immediately clear by Remark 12 that an infinitely *A*-divisible element of *M* is always *A*-divisible.

Proof. Suppose $x \in M$ is infinitley *A*-divisible in *M*. Then let *N* be a left *A*-submodule of *M* containing *x* such that every element of *N* is *A*-divisible in *N*. Then by Remark 12 applied to the inclusion: $N \rightarrow M$, we have that every element of *N* is *A*-divisible in *M*. Therefore *x* is *A*-divisible in *M*.   □

Proposition 1.2.11. *Let A and M be as in Definition 2. Then the set of all elements of M that are infinitely A-divisible in M is an abstract left A-submodule of M, which we call the* infinitely *A-divisible part of M. The infinitely A-divisible part of M can be characterized, alternatively, as being: the largest abstract left A-submodule N of M such that every element of N is A-divisible in N.*

Note: *In general, we have the inclusion of submodules: (infinitely A-divisible part of M)* $\subset div\, M$. *The abstract left A-submodule (infinitely A-divisible part of M) is not, in general, A-adically closed in M. The A-adic closure of the infinitely A-divisible part of M in M is always* $div\, M$.

Proof. If N_1 and N_2 are left *A*-submodules of *M* such that every element of N_i is *A*-adically divisible in N_i, $i = 1, 2$, then $N_1 + N_2$ is clearly another such left *A*-module of *M*. Therefore the sum of all such left *A*-submodules is a maximum left *A*-submodule *N* of *M*, such that $x \in N$ implies *x* is *A*-adically divisible in *N*. Then by Definition 2, *N* is the set of all infinitely *A*-divisible elements of *N*.

The inclusion in the Note follows from the observation made just before this Proposition. Since $div\, M$ is the closure of $\{0\}$, that same inclusion implies that the closure of the (infinitely A-divisible part of M) is $div\,M$. The fact that the inclusion of the Note is in general strict follows from the observation made at the end of Remark 12. □

Corollary 1.2.12. *Let A be a topological ring such that the topology is given by right ideals, and let M be an abstract left A-module. Then the following four conditions are equivalent:*

① *The A-adic topology of M is the indiscrete topology (i.e., the only A-adically open subset of M is M).*

② *For every open right ideal I in A, we have that $I \cdot M = M$.*

③ *Every element of M is A-divisible in M.*

④ *Every element of M is infinitely A-divisible in M.*

Proof. Since the A-adic topology of M is given by open subgroups, this topology is indiscrete iff the only A-adically open subgroup of M is M, i.e., iff $I \cdot M = M$, for all open right ideals I in A. Therefore, ① iff ②. By Definition 1, an element $x \in M$ is I-divisible iff $x \in I \cdot M$ for all open right ideals I in A. Therefore ② iff ③. ④ implies ③ follows from the inclusion in the Note to Proposition 1.2.11. On the other hand, if ③ holds, then taking $N = M$ in Definition 2, we have that every $x \in M$ is infinitely A-divisible in M; i.e., ③ implies ④. □

Remark 13. Let A be a topological ring such that the topology is given by right ideals, and let M be an abstract left A-module. Then we say that the abstract left A-module M is A-divisible iff it obeys the four equivalent conditions of Corollary 1.2.12.

Exercise 1. In general, if M is an abstract left A-module (where A is as in Definition 1), then the closed left A-submodule $div\,M$ is not, in general, an A-divisible left A-module. However the (in general not A-adically closed) left A-submodule (infinitely A-divisible part of M) of M, is always a divisible left A-module; and can, in fact, be characterized as being the largest A-divisible left A submodule of M. Show, more precisely, that if A and M are as in Definition 1, then the following conditions are equivalent:

(1) The abstract left A-module $div\,M$ is A-divisible.

(2) The abstract left A-submodule (infinitely divisible part of M) is closed in M for the A-adic topology.

(3) An element $x \in M$ is A-divisible in M iff it is infinitely A-divisible in M.

Note: Let A be a ring and let $t \in A$ be an element of the ring A. Then we have the right ideal $t \cdot A$ in the ring A. The right ideal $t \cdot A$ is a two-sided ideal iff $A \cdot t \subset t \cdot A$. (Of course, this is the case if t is in the center of the ring A.)

Exercise 2. Let A be a ring and let $t \in A$ be an element of the ring A. Then by the *t-adic topology on A* – or, more precisely, the *right t-adic topology on A* — we mean the topology on A given by the right ideals $\{t^n A : n \geq 1\}$. A with this topology is always a topological abelian group. Show that A with the (right) t-adic topology is a topological ring iff the following condition holds: (*) For every $a \in A$, there exists an integer $n \geq 0$ (depending on the element $a \in A$) such that $at^n \in tA$.

Hint: Use §1, Exercise 0.

Note: A more general result than Excercise 2 above is §3, Exercise 3 (where a comprehensive proof is provided).

A topological ring A such that the topology is the right t-adic topology for some element $t \in A$ will be called a *right t-adic ring* – or, more briefly, a *t-adic ring*.

An example of a t-adic ring is: Let $A = \mathcal{O}$, any discrete valuation ring that is not a field, and take t to be any fixed non-zero element of the maximal ideal of $\mathcal{O}$. Then the t-adic topology of $\mathcal{O}$ is the usual one (so that an open neighborhood base at zero in $\mathcal{O}$ is the set of positive powers of the maximal ideal of $\mathcal{O}$).

Exercise 3. Let A and t be as in Exercise 2, and such that condition (*) of Exercise 2 holds. Let M be any abstract left A-module. Then show that:

(1) An element $x \in M$ is A-divisible in the sense of Definition 1 iff for every integer n, there exists $y \in M$ such that $x = t^n \cdot y$ in M.

(2) An element $x \in M$ is infinitely A-divisible iff there exists a sequence x_i, $i \geq 0$, of elements of M, such that $x_0 = x$, and such that $t \cdot x_{i+1} = x_i$, for all integers $i \geq 0$.

Conclude that (3): An element $x \in M$ is A-divisible, respectively: infinitely A-divisible, in M in the sense of Definition 1, respectively: 2, above, iff x is t-divisible, respectively: infinitely t-divisible, in the sense of [COC], p. 407. Henceforth, in the situation of this Exercise, we use the term "A-divisible" and "t-divisible" interchangeably; and similarly for "infinitely A-divisible" and "infinitely t-divisible."

Exercise 4. Let $\mathcal{O}$ be a discrete valuation ring not a field and let t be a generator for the maximal ideal of $\mathcal{O}$. Then construct an $\mathcal{O}$ module M having no non-zero infinitely t-divisible elements, and yet such that there exists a non-zero t-divisible element in M.

(Hint: There are plenty of such examples in [COC].)

We conclude this section with a technical lemma that will be needed in Chapter 2.

Lemma 1.2.13. *Let A be a topological ring such that the topology is given by denumerably many right ideals. Let M and N be topological left A-modules, such that M is A-adically complete and N is A-adically Hausdorff, and such that the topologies of the topological left A-modules M and N are coarser than the A-adic topology. Let $f : M \to N$ be a continuous homomorphism of topological left A-modules. Suppose that the image $\mathrm{Im}(\mathbf{f})$ of f is A-adically dense in N. Then f is surjective.*

Proof. First, let $A = I_0 \supset I_1 \supset I_2 \supset I_3 \supset \ldots$ be a complete system of open neighborhoods of zero in A each of which is a right ideal in A. Let $x \in N$. Then, by induction on the integer $n \geq 0$, we construct sequences y_i, $i \geq 0$, and x_i, $i \geq 0$, such that $y_i \in I_i \cdot M$ and $x_i \in I_i \cdot N$, for all $i \geq 0$, and such that: (1) $x = f(y_0) + f(y_1) + \ldots + f(y_{n-1}) + x_n$.

Case 1. $n = 0$. Then define $x_0 = x$.

Case 2. $n > 0$. Then, by the inductive assumption, we have elements $x_0, \ldots, x_{n-1}, y_0, \ldots, y_{n-2}$ such that $y_i \in I_i \cdot M$, $0 \leq i \leq n - 2$, $x_i \in I_i \cdot N$, $0 \leq i \leq n - 1$, and such that: (2) $x = f(y_0) + \ldots + f(y_{n-2}) + x_{n-1}$. Since $x_{n-1} \in$

$I_{n-1} \cdot N$, there exists an integer $d \geq 0$, elements $\iota_1, \ldots, \iota_d \in I_{n-1}$ and elements $z_1, \ldots, z_d \in N$ such that (3): $x_{n-1} = \iota_1 z_1 + \ldots + \iota_d z_d$. Since A is a topological ring such that the topology is given by the right ideals $\{I_i : i \geq 0\}$, by §1, Exercise 0, there exists an integer $m \geq 0$ such that (4): $\iota_j I_m \subset I_n$, all integers j, $1 \leq j \leq d$. Also, since the image $\mathtt{Im(f)}$ of f is A-adically dense in N, for each j, $1 \leq j \leq d$, we have that every A-adic neighborhood of z_j contains an element of $\mathtt{Im(f)}$. In particular there exist elements $y_{n-1}^{(1)}, \ldots, y_{n-1}^{(d)} \in M$ such that, for every $1 \leq j \leq d$, we have that $z_j - f(y_{n-1}^{(j)}) \in I_m \cdot N$. Let $x_n^{(j)}$ denote this element of $I_m \cdot N$, $1 \leq j \leq d$. Then

(5): $z_j = f(y_{n-1}^{(j)}) + x_n^{(j)}$, $1 \leq j \leq d$, and

(6): $x_n^{(j)} \in I_m \cdot N$, $1 \leq j \leq d$. Let

(7): $y_{n-1} = \iota_1 y_{n-1}^{(1)} + \ldots + \iota_d y_{n-1}^{(d)}$, and let

(8): $x_n = \iota_1 x_n^{(1)} + \ldots + \iota_d x_n^{(d)}$. Then since $\iota_1, \ldots, \iota_d \in I_{n-1}$, we have that:

(9): $y_{n-1} \in I_{n-1} \cdot M$. Also by equation (6), $\iota_j x_n^{(j)} \in \iota_j \cdot I_m \cdot M$, which by equation (4) is contained in $I_n \cdot M, 1 \leq j \leq d$. Therefore by equation (8),

(10): $x_n \in I_n \cdot M$. Substituting equation (5) into equation (3) and using equations (7) and (8), we have that:

$$x_{n-1} = \iota_1 z_1 + \ldots + \iota_d z_d$$

$$= (\iota_1 f(y_{n-1}^{(1)}) + \iota_1 x_n^{(1)}) + \ldots + (\iota_d f(y_{n-1}^{(d)}) + \iota_d x_n^{(d)}) = f(y_{n-1}) + x_n.$$

That is, that

(11): $x_{n-1} = f(y_{n-1}) + x_n$. Equations (2) and (11) imply equation (1) which together with equations (9) and (10) complete the construction by induction.

Therefore, we indeed have elements $x_i \in N$ and $y_i \in M$, $i \geq 0$, as described above. But then, since $y_i \in I_i \cdot M$, for all $i \geq 0$, we have that the infinite series

$$(12) : \sum_{i \geq 0} y_i$$

in the left A-module M is a Cauchy series in M for the A-adic topology. Since the topology of M is coarser than the A-adic, and since M is complete, this series has a unique limit y in M. Then, since $f : M \to N$ is a continuous homomorphism of left topological A-modules, it follows that $f(y)$ is a limit of the series

$$(13) : \sum_{i \geq 0} f(y_i)$$

in the topological left A module N. Since the topological left A-module N is by hypotheses Hausdorff, it follows that $f(y)$ is the unique limit of the infinite series (13) in N. But, by equation (1), and the fact that $x_n \in I_n \cdot N$, for all integers $n \geq 0$, it follows that the infinite series (13) converges to x in N for the A-adic topology. Since the topology of N is coarser than the A-adic topology, the infinite series (13) also converges to x in N. Therefore $f(y) = x$. Since $x \in N$ is an arbitrary element, it follows that f is surjective. $\square$

Remark 14. The hypotheses in Lemma 1.2.13 that "M is complete and N is Hausdorff," can be replaced by the weaker hypotheses: "M/(the closure of zero) is complete, and Im(f) contains the closure of $\{0\}$." The proof is the same. However, the assertion of this Remark is not really stronger than that of the Lemma; since the Remark, for the map $f : M \to N$, it is also an easy Corollary of the Lemma, applied to the continuous homomorphism of Hausdorff topological left A-modules induced by f,

$$M/(\text{closure of } \{0\} \text{ in } M) \to N/(\text{closure of } \{0\} \text{ in } N).$$

Remark 15. (1) In the statement of Lemma 1.2.1, suppose we weaken the hypotheseis, "S is an open right ideal in A" to "S is a closed right ideal in A." Then the proof of Lemma 1.2.1 (where we use §1, Exercise 0, part 4, instead of §1, Exercise 0, part 3) shows that, conditions ① and ② of that Lemma continue to be equivalent (but not ③). And that the last paragraph of that Lemma continues to hold as well. (The Note to the Lemma continues to hold as well — the parenthetical part being modified to, "A complete system of open neighborhoods of zero in $S \cdot M$ is $\{(S' \cdot M) \cap (S \cdot M) : S'$ is an open right ideal in $A\}$.) Let us call this resulting true statement "Lemma 1.2.1 (modified)".

(2) Similarly, in Corollary 1.2.2 suppose that we modify the hypotheses on S to read, that "S is a closed (instead of open) right ideal, such that there exists a finite set $x_1, \ldots, x_d$ of generators for S as a right ideal, such that, for every open right ideal S' in A, we have that the right ideal $x_1 S' + \ldots + x_d S'$ is open in S." Then once again, the proof of Corollary 1.2.2 then shows that the two equivalent conditions of Lemma 1.2.1 (modified) hold.

Let us call this generalization of Corollary 1.2.2 "Corollary 1.2.2 (modified)."

(3) Once again, suppose that we have the hypotheses of the Note to Theorem 1.2.3. Then let S be any closed finitely generated right ideal in A, such that, the right ideal $x_{i1} S + \ldots + x_{id_i} S$ is open in S, for all integers $i \geq 1$. Then the proof of Theorem 1.2.3 shows that: For every left A-module M, the subgroup $S \cdot \widehat{M}$ of $\widehat{M}$ is closed (for the A-adic, or equivalently uniform, topology of $\widehat{M}$) in $\widehat{M}$. And $S \cdot \widehat{M}$ is the closure of the image of $S \cdot M$ in $\widehat{M}$. Also, $S \cdot \widehat{M}$ is the completion, as a topological abelian group, of $S \cdot M$ (for the induced topology from the A-adic topology of $S \cdot M$). A complete system of neighborhoods of zero in $S \cdot M$ is $\{(x_1 \cdot S' + \ldots + x_d \cdot S') \cdot M : S'$ is an open right ideal in $A\}$, where $\{x_1, \ldots, x_d\}$ is a set of generators for S as a right ideal, such that $x_1 \cdot S' + \ldots + x_d \cdot S'$ is open in S, for all open right ideals S' in A. (Another open base at zero is: $\{S'' \cdot M : S''$ is a right ideal in $A, S'' \subset S$, and S'' is open in $S\}$.) Let us call this result "Theorem 1.2.3, Note (modified)." A special case of "Theorem 1.2.3, Note (modified)" is: Under the hypotheses of Theorem 1.2.3, let S be any closed two-sided ideal in A, such that S is right finitely generated, and such that for every open two-sided ideal S', we have that $S' \cdot S$ is open in S. Then, the conclusion of "Theorem 1.2.3, Note (modified)" hold. Let us call this latter result, "Theorem 1.2.3 (modified)."

1.3 Admissible and Strongly Admissible Topological Rings

After considering Proposition 1.2.4 and Corollary 1.2.5 of §2, and also the (lengthy) Example 2 of §2, one is led to pose the following

Definition 1. Let A be a topological ring. Then A is *admissible* (or more precisely *right admissible*) iff the topology of A is given by right ideals, and in addition any one of the following several equivalent conditions hold:

[1] For every free left A-module F, the A-adic completion $\widehat{F}$ of F is complete in its A-adic topology.

[1'] Let K be any fixed cardinal number such that the topological ring A has a base for the neighborhood system at zero of $\leq K$ elements. Then the free left A-module $A^{(K)}$ of rank K is such that $\widehat{A^{(K)}}$ is A-adically complete.

[1''] K a fixed cardinal number as in [1'], then for every open ideal S in A, condition (F_K) of Remark 2 of §2 holds.

[2] For every free left A-module F, the A-adic topology on the A-adic completion $\widehat{F}$ of F coincides with the uniform topology of $\widehat{F}$.

[3] For every free left A-module F, the A-adic topology of $\widehat{F}$ is such that, $\widehat{F}$, together with the A-adic topology, is a completion of F (together with its A-adic topology) in the category of topological left A-modules (or, equivalently, in the category of topological abelian groups).

[4] Let F be a free left A-module. Let $\widehat{F}$ be the A-adic completion of F and let $\widehat{\widehat{F}}$ be the A-adic completion of $\widehat{F}$. Then the natural monomorphism $\iota : \widehat{F} \to \widehat{\widehat{F}}$ of $\widehat{F}$ into its A-adic completion is an isomorphism of left A-modules.

[5] F is as in [4] above, we have the image $\widehat{k} : \widehat{F} \to \widehat{\widehat{F}}$ under the functor "A-adic completion" of the natural map $k : F \to \widehat{F}$. Then $\widehat{k}$ (which is always a canonically split monomorphism), is an isomorphism of left A-modules.

[6] F as in [4] above, ι and $\widehat{k}$ as in [4] and [5] above, then we have that $\iota = \widehat{k}$.

[7] F as in [4] above, for every open right ideal S in the topological ring A, we have that the two equivalent conditions of Lemma 1.2.1 of §2 hold for the left A-module F.

[2'], resp. [3'], resp. [4'], resp. [5'], resp. [6'], resp. [7'] Let K be any fixed cardinal number as in condition [1'], F a free left A-module of rank K. Then F obeys condition [2], resp. [3], resp. [4], resp. [5], resp. [6], resp. [7].

[8] Let F be a free abstract left A-module and let I be an open right ideal in the ring A. Then $I \cdot \widehat{F}$ is open in $\widehat{F}$ for the uniform topology of $\widehat{F}$ (i.e., for the topology on $\widehat{F}$ such that $\widehat{F}$ is the completion as a topological left A-module F).

[9] Let F be a free abstract left A-module and let I be an open right ideal in A. Then $I \cdot \widehat{F}$ is closed in $\widehat{F}$ for the uniform topology of $\widehat{F}$.

[10] Let F be a free abstract left A-module and let I be an open right ideal in A. Then the natural map: $F/(I \cdot F) \to \widehat{F}/(I \cdot \widehat{F})$ is bijective.

[11] Let S be a set and let I be an open right ideal in A. Then $I \cdot \widehat{A^{(S)}} = \{(a_i)_{i \in S} \in \widehat{A^{(S)}} : a_i \in I \text{ for all } i \in S\}$.

Proof that the above many conditions are equivalent. Equivalence of [2] and [3] is obvious. Equivalence of [1] and [2] follows from Corollary 1.2.5 of §2. Equivalence of [1] and [4] follows as in the Proof of the analagous assertion (number ③) of the (lengthy) Example 2 of §2. Equivalence of [5] and [6] with these conditions follows readily from the commutative diagram in Note 2 of the (lengthy) Example 2 of §2. Equivalence of conditions [1] and [2] with condition [7], as observed in the Note to Proposition 1.2.4 of §2, follows from Lemma 1.2.1 of §2. Equivalence of [1] and [1'] follows from Remarks 2 and 3 of §2. Finally, having established that [1] iff [i] (for any given fixed free abstract left A-module F), and that [1] iff [1'], it follows that [1] iff [i'], for each integer $i, 2 \le i \le 7$.

In general: If G is a topological group such that the topology is given by open subgroups, and if U is an open subgroup of G, if $\widehat{G}$ is the completion of G and if $\overline{U}$ is the closure of U in $\widehat{G}$, then $\overline{U}$ is an open subgroup of $\widehat{G}$, and the induced function on the left coset spaces: $G/U \to \widehat{G}/\overline{U}$ is a bijective function of the discrete topological spaces.

If F is a free abstract left A-module, then the uniform topology on $\widehat{F}$ is coarser than the A-adic. Therefore, [2] holds above iff the A-adic topology is coarser than the uniform topology. I.e., iff condition [8] holds. Therefore [2] iff [8].

If F is a free abstract left A-module, then taking G to be the topological group (F with its A-adic topology) and $U = I \cdot F$, where I is an open right ideal in A, then (image of U) $\subset I \cdot \widehat{F} \subset \overline{U}$. Therefore $I \cdot \widehat{F}$ is closed in $\widehat{F}$ for the uniform topology iff $I \cdot \widehat{F} = \overline{U}$. Since $\overline{U}$ is an open subgroup of $(\widehat{F}, unif)$, it follows that the subgroup $I \cdot \widehat{F}$ of $\widehat{F}$ is closed in $(\widehat{F}, unif)$ iff $I \cdot F = \overline{U}$ iff $I \cdot \widehat{F}$ is open in $(\widehat{F}, unif)$. Therefore [8] iff [9].

Since $G/U \to G/\overline{U}$ is bijective, we have that the natural map: $F/(I \cdot F) \to \widehat{F}/\overline{U}$ is bijective. Since $I \cdot \widehat{F} \subset \overline{U}$, it follows that condition [10] holds for I iff $I \cdot \widehat{F} = \overline{U}$. Since also (image of U) $\subset I \cdot \widehat{F} \subset \overline{U}$, so that $I \cdot \widehat{F}$ is dense in $\overline{U}$, we have that $I \cdot \widehat{F} = \overline{U}$. iff $I \cdot \widehat{F}$ is closed in $(\widehat{F}, unif)$. Therefore [10] iff [8].

Finally, if S is a set, then $A^{(S)}/(I \cdot A^{(S)}) = (A/I)^{(S)}$ for all open right ideals I in A. Therefore if $F = A^{(S)}$, then $F/(I \cdot F) = (A/I)^{(S)}$. The image of the map $\widehat{F}/(I \cdot \widehat{F}) \to A^{(S)}/I^{(S)}$ is also $(A/I)^{(S)}$. Therefore [10] holds for the fixed open right ideal I iff the natural map: $\widehat{F}/(I \cdot \widehat{F}) \to A^S/I^S$ is injective. I.e., iff $I^S \cap \widehat{F} = I \cdot \widehat{F}$. That is, iff $I^S \cap \widehat{A^{(S)}} = I \cdot \widehat{A^{(S)}}$. But this is the displayed equation in [11]. Therefore [10] holds iff [11] holds.   □

An apparently slightly more stringent condition on a topological ring is:

<u>Definition 2</u>. Let A be a topological ring. Then A is *strongly admissible* (or, more precisely, *strongly right admissible*) iff the topology of A is given by right ideals, and any one of the following equivalent conditions holds:

[1] For every left A-module M, the A-adic completion $\widehat{M}$ of A is complete in its A-adic topology.

[2] For every left A-module M, the A-adic topology on the A-adic completion $\widehat{M}$ of M, is such that, $\widehat{M}$ together with the A-adic topology is a completion of M (together with its A-adic topology) in the category of topological left A-modules (or, equivalently, in the category of topological abelian groups).

[3] For every left A-module M and every open right ideal S in A, we have that the condition of Remark 1 of §2 holds.

[4] Let M be a left A-module, let $\widehat{M}$ be the A-adic completion of M and let $\widehat{\widehat{M}}$ be the A-adic completion of $\widehat{M}$. Then the natural monomorphism $\iota : \widehat{M} \to \widehat{\widehat{M}}$ of $\widehat{M}$ into its A-adic completion is an isomorphism of left A-modules.

[5] If M is any abstract left A-module, then we have that the image $\widehat{k}$: $\widehat{M} \to \widehat{\widehat{M}}$ under the functor "A-adic completion" of the natural map $k : M \to \widehat{M}$ ($\widehat{k}$ is always necessarily a canonically split monomorphism) is an isomorphism of left A-modules.

[6] If M is any left A-module, and if ι and $\widehat{k} : \widehat{M} \to \widehat{\widehat{M}}$ are defined as in [4] and [5] above, then $\iota = \widehat{k}$.

[7] If M is any left A-module, and if S is any open right ideal in the topological ring A, then we have that the three equivalent conditions of Lemma 1.2.1 of §2 hold.

[8] For every left A-module M, and every open right ideal I in the ring A, we have that $I \cdot \widehat{M}$ is open in $\widehat{M}$ for the uniform topology of $\widehat{M}$ (i.e., the topology on $\widehat{M}$ such that $\widehat{M}$ is the completion as a topological left A-module of M).

[9] For every left A-module M, we have that $I \cdot \widehat{M}$ is closed in $\widehat{M}$ for the uniform topology of $\widehat{M}$.

[10] For every left A-module M, and every open right ideal I in the ring A, we have that the natural map: $M/(I \cdot M) \to \widehat{M}/(I \cdot \widehat{M})$ is bijective.

Proof of the equivalence of the above ten conditions in Definition 2. Entirely similar to the corresponding Proof of the equivalence of the conditions of Definition 1 — with Proposition 1.2.4 of §2 replacing Corollary 1.2.5 of §2; and Lemma 1.2.1 of §2 replacing Remark 2 of §2. □

Proposition 1.3.1. *Let A be a topological ring such that the topology is given by right ideals. Then the following two conditions are equivalent:*

[1] A is admissible

[2] Both conditions [2a] and [2b] below hold:

[2a] The completion $\widehat{A}$ of A is such that the uniform and A-adic topologies coincide.

and [2b] The complete topological ring $\widehat{A}$ is admissible.

Proof. If A is admissible, then condition [2] of Definition 1, applied to the free abstract left A-module A, is equivalent to [2a] above. Thus, we can assume that the topological ring A obeys condition [2a].

Let S be any set. Then the A-adic completion of the free abstract left A-module $A^{(S)}$ and the $\widehat{A}$-adic completion of the free abstract left $\widehat{A}$-module $\widehat{A}^{(S)}$, are the same as abstract left $\widehat{A}$-modules. (E.g, by Lemma 1.3.4 later in this section, we have that $\widehat{A^{(S)}} \subset \widehat{A}^S$, and $\widehat{\widehat{A}^{(S)}} \subset \left(\widehat{\widehat{A}}\right)^S = \widehat{A}^S$ — where in both formulae, "$\widehat{A}$" and "$\widehat{\widehat{A}}$" mean "completion for the uniform topology". And $\widehat{A^{(S)}}$ (respectively: $\widehat{\widehat{A}^{(S)}}$) is the closure of the indicated subset in $\widehat{A}^S$; namely, both of these are equal to $\{(a_i)_{i \in S} \in \widehat{A}^S : (a_i)_{i \in S}$ converges to zero in $\widehat{A}\}$. Call this $\widehat{A}$-module $\widehat{F}$. By [2a] the uniform and A-adic topologies coincide in $\widehat{A}$. Therefore, for any abstract left $\widehat{A}$-module M, we have that the A-adic and $\widehat{A}$-adic topologies of M coincide. In particular, this is so for $\widehat{F}$. Therefore $\widehat{F}$ is complete for the A-adic topology iff $\widehat{F}$ is complete for the $\widehat{A}$-adic topology. I.e., condition [1] of Definition 1 holds for the abstract left A-module $A^{(S)}$ iff condition [1] of Definition 1 holds for the abstract left $\widehat{A}$-module $\widehat{A}^{(S)}$. Therefore A is admissible iff $\widehat{A}$ is admissible.   □

Proposition 1.3.2. *Let A be a topological ring such that the topology is given by right ideals. Then the following two conditions are equivalent:*

[1] A is strongly admissible

[2] Both conditions [2a] and [2b] below hold:

[2a] The completion $\widehat{A}$ of A is such that the uniform and A-adic topologies coincide and

[2b] The complete topological ring $\widehat{A}$ is strongly admissible.

Proof. Entirely similar to that of Proposition 1.3.1.   □

Remark 0.3. Let A be a (not necessarily complete) topological ring such that the topology is given by right ideals. Then (whether or not the topological ring A is admissible or strongly admissible) condition [2a] of Proposition 1.3.1, and of Proposition 1.3.2 is equivalent to the ten conditions, [1] – [10], of Definition 2 in the special case $M = A$ (A regarded as a left-module over itself). (And these are also equivalent to each of the conditions [1], [2], [4], [5], [6], [7], [8], [9], [10] of Definition 1 in the special case that the free module F is A (A regarded as an abstract left module over itself). Again, the proof is the same.

(It should be noted that condition [2a] of Proposition 1.3.1 and Proposition 1.3.2 does not always hold. E.g., in Remark 8 of §2, (following Example 2 of §2), we have shown that the topological ring A_0 of Example 2 does not obey this condition.)

Remark 1. Let A be a topological ring such that there exists a denumerable neighborhood base at zero. Then A is strongly admissible iff A is admissible.

Proof. In fact, by Corollary 1.2.8 of §2, condition [7] of Definition 2 above is then equivalent to condition [7'] of Definition 1 above.   □

Example 1. Let A be a topological ring, such that the topology is given by denumerably many two-sided ideals, each of which is finitely generated as a right ideal; and such that for every open two-sided ideal S in A, we have that the ideal S^2 is open. Then the topological ring A is strongly admissible.

Proof. Theorem 1.2.3 □

Example 2. Let A be a topological ring such that the topology of A is given by denumerably many open right ideals $S_i, i \geq 1$, such that,

(1): For every $i \geq 1$, there exists a finite set $x_{i1}, \ldots, x_{id_i}$ of elements of S_i that generate S_i as a right ideal, and such that for every open right ideal S' in A, we have that the right ideal $x_{i1}S' + \ldots + x_{id_i}S'$ is open in A. Then the topological ring is strongly admissible.

Proof. The Note to Theorem 1.2.3 of §2. □

Remark: Example 2 is a priori more general than Example 1. However, if the topological ring A admits an open neighborhood base at zero consisting of *two-sided ideals*, then it is easy to see that the conditions of Examples 1 and 2 are identical.

Example 3. Let A be an abstract ring, and let $F_i, i \geq 1$ be an arbitrary sequence of finite subsets of the ring A. Then let
$$\underline{J} = \{F_{j_1}^{i_1} \cdots F_{j_n}^{i_n} \cdot A : n \geq 1, i_1, \ldots, i_n, j_1, \ldots, j_n \text{ integers} \geq 1\}.$$ Then $\underline{J}$ is a denumerable set of right ideals in the ring A. Then using §1, Exercise 1, it is not difficult to show that: Necessary and sufficient conditions for the denumerable collection $\underline{J}$ of right ideals to be a base for the neighborhood system at zero for a (necessarily uniquely determined) topology on the set A that makes A into a *topological ring*, is that the following hold:

(*): For every finite subset S of A and every integer $i \geq 1$, there exists an integer $m \geq 0$ and positive integers $i_1, \ldots, i_m$ and $j_1, \ldots, j_m$, such that $SF_{i_1}^{j_1} \cdots F_{i_m}^{j_m} \subset F_i \cdot A$. [2]

Proof. If multiplication is continuous at $(c, 0)$ for all $c \in S$, then since $F_i \cdot A$ is an open neighborhood of zero, there exists a base open neighborhood, say $F_{i_1}^{j_1} \cdots F_{i_m}^{j_m} \cdot A$, of zero, such that $c \cdot F_{i_1}^{j_1} \cdots F_{i_m}^{j_m} \cdot A \subset F_i \cdot A$, for all $c \in S$; it is equivalent to say: such that (1): $S \cdot F_{i_1}^{j_1} \cdots F_{i_m}^{j_m} \subset F_i \cdot A$. Thus, the condition (*) is necessary. Conversely, suppose that the condition (*) holds. Then, by induction on $k_1 + \ldots + k_n$, we show that, for every integer $n \geq 0$, and every pair of sequences $k_1, \ldots, k_n, r_1, \ldots, r_n$ of integers ≥ 1, that there exists an integer m and sequence $i_1, \ldots, i_m, j_1, \ldots, j_m \geq 1$ such that

(2): $S \cdot F_{i_1}^{j_1} \cdots F_{i_m}^{j_m} \subset F_{r_1}^{k_1} \cdots F_{r_n}^{k_n} \cdot A$.

In fact, for $k_1 + \cdots + k_n = 0$, this is trivial. For $k_1 + \ldots + k_n = 1$, this is condition (*). Suppose that $k_1 + \ldots k_n > 1$ and that the assertion is established for smaller values of $k_1 + \ldots + k_n$. Then by the inductive hypothesis, there exists an integer $m \geq 0$, and sequences $i_1, \ldots, i_m, j_1, \ldots, j_m \geq 1$, such that

---

[2] In this displayed formula, by "$SF_{i_1}^{j_1} \cdots F_{i_m}^{j_m}$", we prefer, in this case, to mean the *set product* (a finite set), rather than the "set of finite sums of the set product". (However, the condition (*) remains identical under either interpretation.)

$S \cdot F_{i_1}^{j_1} \cdots F_{i_m}^{j_m} \subset F_{r_1}^{k_1} \cdots F_{r_n}^{k_n - 1} \cdot A$. (If $k_n = 1$, then if w is the largest integer $\leq n - 1$ such that $k_w \neq 0$, then this is equivalent to the assertion: $S \cdot F_{i_1}^{j_1} \cdots F_{i_m}^{j_m} \subset F_{r_1}^{k_1} \cdots F_{r_w}^{k_w} \cdot A$, which is true by the inductive hypothesis, since $k_1 + \cdots k_w = k_1 + \cdots k_n - 1$.) Then, since $S \cdot F_{i_1}^{j_1} \cdots F_{i_m}^{j_m}$ is a finite set, there exists a finite subset T of A such that

(3): $S \cdot F_{i_1}^{j_1} \cdots F_{i_m}^{j_m} \subset F_{r_1}^{k_1} \cdots F_{r_n}^{k_n - 1} \cdot T$. Then, applying condition (*) to the finite set T (in lieu of S) and the integer r_n (in lieu of i), we have that there exist an integer $v \geq 0$ and sequences $a_1, \ldots, a_v, b_1, \ldots, b_v \geq 1$ such that

(4): $T \cdot F_{a_1}^{b_1} \cdots F_{a_v}^{b_v} \subset F_{r_n} \cdot A$. But then, multiplying equation (3) on the right by $F_{a_1}^{b_1} \cdots F_{a_v}^{b_v}$, we obtain that:

$$S \cdot F_{i_1}^{j_1} \cdots F_{i_m}^{j_m} \cdot F_{a_1}^{b_1} \cdots F_{a_v}^{b_v} \subset F_{r_1}^{k_1} \cdots F_{r_n}^{k_n - 1} \cdot T \cdot F_{a_1}^{b_1} \cdots F_{a_v}^{b_v} \subset$$

$$F_{r_1}^{k_1} \cdots F_{r_n}^{k_n - 1} \cdot F_{r_n} \cdot A = F_{r_1}^{k_1} \cdots F_{r_n}^{k_n} \cdot A,$$

proving equation (2).

By §1, Exercise 0, to complete the proof that the collection of right ideals $\underline{J}$ defines a topology on A such that A is a topological ring, it is necessary and sufficient to show that:

(5): $I, I' \in \underline{J}$ imply there exists $I'' \in \underline{J}$ such that $I'' \subset I$, $I'' \subset I'$, and

(6): $I \in \underline{J}$, $c \in A$ imply there exists $I' \in \underline{J}$ such that $c \cdot I' \subset I$.

In fact, suppose that $I, I' \in \underline{J}$, say $I = F_{a_1}^{b_1} \cdots F_{a_v}^{b_v} \cdot A$ and $I' = F_{r_1}^{k_1} \cdots F_{r_n}^{k_n} \cdot A$. Then by equation (2) applied to the finite set $S = F_{a_1}^{b_1} \cdots F_{a_v}^{b_v}$, there exist an integer $m \geq 0$ and a sequence $i_1, \ldots, i_m, j_1, \ldots, j_m$ such that

$$F_{a_1}^{b_1} \cdots F_{a_v}^{b_v} F_{i_1}^{j_1} \cdots F_{i_m}^{j_m} \subset F_{r_1}^{k_1} \cdots F_{r_n}^{k_n} \cdot A.$$

Since obviously

$$F_{a_1}^{b_1} \cdots F_{a_v}^{b_v} \cdot F_{i_1}^{j_1} \cdots F_{i_m}^{j_m} \subset F_{a_1}^{b_1} \cdots F_{a_v}^{b_v} \cdot A,$$

we therefore have that the element

$$I'' = F_{a_1}^{b_1} \cdots F_{a_v}^{b_v} \cdot F_{i_1}^{j_1} \cdots F_{i_m}^{j_m} \cdot A$$

of $\underline{J}$ is contained in both I and I', proving (5). On the other hand, given I and c as in (6), say $I = F_{r_1}^{k_1} \cdots F_{r_n}^{k_n} \cdot A$, then by condition (*) applied to the finite set $S = \{c\}$, we have that there exists an integer $m \geq 0$ and integers $i_1, \cdots, i_m, j_1, \cdots, j_m \geq 1$ such that

$$c \cdot F_{i_1}^{j_1} \cdots F_{i_m}^{j_m} \subset F_{r_1}^{k_1} \cdots F_{r_n}^{k_n} \cdot A.$$

But then the element $I' = F_{i_1}^{j_1} \cdots F_{i_m}^{j_m} \cdot A$ of $\underline{J}$ satisfies the conclusion of condition (6). □

Given the abstract ring A, and a denumerable collection $\mathscr{F} = \{F_i : i \geq 1\}$ of finite subsets of A, such that condition (*) of Example 3 holds, the topology described in Example 3 will be called the *right topology determined by the finite* sets F_i, $i \geq 1$. Then given an abstract ring A and a denumerable collection F_i, $i \geq 1$, of finite subsets of A that obey condition (*), the ring A together with the right topology determined by the finite sets F_i, $i \geq 1$, is strongly admissible, by the Note to Theorem 1.2.3 of §2.

Example 3 is in fact identical in generality to Example 2.

Example 4. Let A be an abstract ring and let $T_1, T_2, \ldots$ be a sequence of finite subsets of the ring A, such that $A \cdot T_i \subset T_i \cdot A$, for all $i \geq 1$. Let $\mathscr{I}$ be the set of all right ideals (= two sided ideals) generated by all finite products of the sets $T_1, T_2, \ldots$. Thus, explicitly, $\mathscr{I} = \{T_{i_1} \cdots T_{i_n} \cdot A : i_1, \ldots, i_n \geq 1, n \geq 1\}$, and let τ be the topology defined by the collection of right (in fact, two-sided) ideals $\mathscr{I}$. Then (A, τ) is a topological ring, and is strongly admissible. Example 4 is of course a special case of Example 3.

Example 5. Let A be an abstract ring and let F be a finite subset of A. Suppose that (*): For every $b \in A$, there exists a positive integer n (depending on b) such that $b \cdot F^n \subset F \cdot A$. Let τ be the topology defined by the set $\mathscr{I}$ of right ideals generated by positive powers of the set F : $\mathscr{I} = \{F^i \cdot A : i \geq 1\}$. Then (A, τ) is a topological ring, and is strongly admissible. Example 5 is a special case of Example 3, in which $F_i = F$ for all $i \geq 1$.

Example 5.1. If A is an abstract ring and t is an element of A, then in Exercise 2 above, we have defined the *right t-adic topology* on A; it is the topology that makes A into a topological abelian group, such that a complete system of open neighborhoods of zero is the set of right ideals, $t^n A$, $n \geq 0$. As noted in Exercise 2, the right t-adic topology on A makes A into a *topological ring* iff, for every element $b \in A$, there exists an integer n (depending on b) such that $bt^n \in tA$. We will call such a topological ring a *right t-adic* topological ring. As a special case of Example 5, every right t-adic topological ring is strongly admissible.

Example 6. Let A be an abstract ring and let I be a two-sided ideal in A. Then by the *I-adic topology in A* we mean the topology defined by the positive powers $\mathscr{I} = \{I^n : n \geq 1\}$ of the ideal I. If the ideal I is right finitely generated, then the topological ring A with the I-adic topology is strongly admissible.

Example 6 is a special case of Example 2 — and is also a special case of Example 5. Example 6 includes most of the cases usually considered in algebraic geometry.

Example 7. Let $A = k[(T_i)_{i \geq 1}]$, the polynomial ring in denumerably many variables, over a non-zero ring k. Then the ring A is not right (or left) Noetherian. Let S be the right ideal (= the left ideal = the two-sided ideal) in A generated by the central elements $\{T_i : i \geq 1\}$. Then the ring A together with the S-adic topology is not admissible.

This was proved in (the lengthy) Example 2, near the end of §2.

Remark 2. Of course, the hypothesis in Example 6, that fails in Example 7, is that the ideal S is not right finitely generated.

Remark 3. All of the above Examples involve topological rings such that there exists a denumerable neighborhood system of zero. This is, of course, the most important case for algebraic-geometric and commutative-algebraic applications. It might nevertheless be interesting to see some examples in the case in which there is not a denumerable neighborhood base at zero.

We now develop certain conditions on a topological ring that are related to the condition of strong admissibility.

Theorem 1.3.3. *Let A be a topological ring such that the topology is given by some set of right finitely generated right ideals. Then if I is any set, and if M_i is an A-adically complete abstract left A-module for all $i \in I$, then the direct product abstract left A-module:*

$$\prod_{i \in I} M_i$$

is also A-adically complete.

Note 1: Theorem 1.3.3 makes no assumptions about "denumerable neighborhood bases."

Note 2: The proof of the Theorem shows also that, if A is any topological ring as in the Theorem, and if I is any set and N_i is an arbitrary abstract left A-module for every $i \in I$, then the natural homomorphism:

$$\widehat{\prod_{i \in I} N_i} \to \prod_{i \in I} \widehat{N_i}$$

is an isomorphism of left A-modules.

Proof. Let S be any right finitely generated right ideal in the ring A. Then I claim that, if I is any set, and if for each $i \in I$, M_i is any abstract left A_i-module, then the submodules,

$$(1) : S \cdot \prod_{i \in I} M_i = \prod_{i \in I} (S \cdot M_i)$$

of $\prod_{i \in I} M_i$, are equal.

In fact let $t_1, \ldots, t_d$ be any finite set of right generators for the right ideal S. Then $S = t_1 A + \ldots + t_d A$, and

$$S \cdot \prod_{i \in I} M_i = t_1 \cdot \prod_{i \in I} M_i + \ldots + t_d \cdot \prod_{i \in I} M_i = \prod_{i \in I} (t_1 \cdot M_i) + \ldots + \prod_{i \in I} (t_d \cdot M_i)$$

$$= \prod_{i \in I} (t_1 \cdot M_i + \ldots + t_d \cdot M_i) = \prod_{i \in I} (S \cdot M_i),$$

verifying equation (1).

Therefore, by equation (1), for every open right finitely generated right ideal S in A, we have that:

$$\left(\prod_{i \in I} M_i \right) \Big/ \left(S \cdot \prod_{i \in I} M_i \right) = \left(\prod_{i \in I} M_i \right) \Big/ \left(\prod_{i \in I} S \cdot M_i \right) \approx \prod_{i \in I} M_i / (S \cdot M_i).$$

Therefore, passing to the inverse limit for all such open, right finitely generated, right ideals S, we have that:

$$\varprojlim_{S} \left(\left(\prod_{i \in I} M_i \right) \Big/ \left(S \cdot \prod_{i \in I} M_i \right) \right) \approx \varprojlim_{S} \prod_{i \in I} (M_i/(S \cdot M_i)) \approx \prod_{i \in I} \varprojlim_{S} (M_i/(S \cdot M_i)).$$

That is,

$$(2): \widehat{\prod_{i \in I} M_i} \approx \prod_{i \in I} \widehat{M_i}.$$

This proves Note 2 to the theorem. If now M_i is A-adically complete for all $i \in I$, then the natural map: $M_i \to \widehat{M_i}$ is an isomorphism of abstract left A-modules. Therefore, in this case, equation (2) is equivalent to the assertion that $\prod_{i \in I} M_i$ is A-adically complete. $\qquad\square$

The following elementary Lemma will be used in Chapter 5; it is related to Note 2 above, but requires only that A be a standard topological ring.

Lemma 1.3.4. *Let A be an arbitrary topological ring such that the topology is given by right ideals. If I is a set, and if M_i is an abstract left A-module for all $i \in I$, then the natural map is a monomorphism:*

$$(1): \widehat{\underset{i \in I}{\oplus} M_i} \subset \prod_{i \in I} \widehat{M_i}.$$

Proof. For every abstract left A-module N, we have that

$$(2): \widehat{N} = \varprojlim_{J \in \mathscr{S}} N/JN,$$

where $\mathscr{S}$ is the set of all open right ideals in the topological ring A. (Notice also that $N/JN = (A/J) \otimes_A N$.) Hence the left side of equation (1) is

$$(3): \varprojlim_{J \in \mathscr{S}} \left((A/J) \underset{A}{\otimes} \underset{i \in I}{\oplus} M_i \right).$$

Since tensor product commutes with arbitrary direct sums, this can be rewritten as

$$(4): \varprojlim_{J \in \mathscr{S}} \underset{i \in I}{\oplus} (M_i/J \cdot M_i).$$

An element $x = (x_J)_{J \in \mathscr{S}}$ of the left A-module in (4) maps into zero in the right side of equation (1) *iff* the image of x in $\widehat{M_i}$ is zero, for all $i \in I$ — and equation (2) with $N = M_i$ tells us that this latter happens *iff* the image of x in $M_i/J \cdot M_i$ is zero for all $i \in I$ and all $J \in \mathscr{S}$. So to complete the proof we must show that: If an element x of the group (4) is such that the image of x in M_i is in $J \cdot M_i$ for all $J \in \mathscr{S}$ and all $i \in I$, then $x = 0$.

Every element x of the group (4) is of the form $x = (x_J)_{J \in \mathscr{S}}$, where $x_J \in \underset{i \in I}{\oplus} (M_i/J \cdot M_i)$. The image of x in $M_i/J \cdot M_i$ is simply the ith coordinate of

$x_J \in \underset{i\in I}{\oplus} (M_i/J \cdot M_i)$. If this is zero for all $i \in I$, then $x_J = 0$. This being true for all $J \in \mathscr{S}$, we have that $x = (x_J)_{J\in\mathscr{S}} = 0$.   □

Remark 4. Examples 1 - 6 above all obey the hypotheses of Theorem 1.3.3.

Theorem 1.3.5. *Let A be a topological ring such that the topology is given by right ideals. Then the following three condions are equivalent.*

[1] Let M be an A-adically complete left A-module, and let N be an A-adically closed submodule. Then N is A-adically complete.

[2] Let $f : M \to M'$ be a homomorphism of abstract left A-modules, and let $K = \mathtt{Ker}(f)$. If M is A-adically complete and M' is A-adically Hausdorff, then K is A-adically complete.

[3] Let

be a diagram of abstract left A-modules, such that M and N are A-adically complete, and L is A-adically Hausdorff. Then the fiber product abstract left A-module $N \underset{L}{\times} M$ is A-adically complete.

Proof. [1] implies [2]. Since f is a homomorphism of abstract left A-modules, it follows that f is continuous for the A-adic topologies of M and M'. Since M' is A-adically Hausdorff, $\{0\}$ is A-adically closed in M'. Therefore $\mathtt{Ker}(f) = f^{-1}(\{0\})$ is A-adically closed in M. But then by [1] K is A-adically complete.

[2] implies [1]. Let M' be the quotient A-module $M' = M/N$. Then the natural map into the quotient: $f : M \to M'$ is an epimorphism of abstract left A-modules. By the note to Corollary 1.2.7 of §2, it follows that the A-adic topology on M' is the quotient topology from the A-adic topology of M. Since the kernel of f is N, which by hypothesis is closed in M for the A-adic topology, it follows that M' is Hausdorff for the quotient topology from M, that is for the A-adic topology of M'. Therefore by [2], $N = \mathtt{Ker}(f)$ is A-adically complete.

[2] implies [3]. The fiber product $N \underset{L}{\times} M$ is the kernel of the homomorphism of abstract left A-modules $f : N \times M \to L$ such that $f(n, m) = \phi(n) - \psi(m)$, (where $\phi : N \to L$ and $\psi : M \to L$ are the maps in the diagram in [3]).

[3] implies [2]. The kernel $K = \mathtt{Ker}(f)$ of f is isomorphic as an abstract left A-module to the fiber product of the diagram:

$$
\begin{array}{ccc}
 & & M \\
 & & \downarrow \Gamma_f \\
M \times \{0\} & \xrightarrow{\ inclusion\ } & M \times M'
\end{array}
$$

where Γ_f is the graph of f (i.e., the homomorphism such that $\Gamma_f(m) = (m, f(m))$ for all $m \in M$).   □

Corollary 1.3.6. *Let A be a topological ring such that the topology is given by right ideals. Then the following three conditions are equivalent:*

[1] Let $f : M \to M'$ be a homomorphism of left A-modules, and let $K = \text{Ker}(f)$. If M and M' are A-adically complete then so is K.

[2] Let

be a diagram of left A-modules, such that M, N, and L are A-adically complete. Then the fiber product abstract left A-module $N \underset{L}{\times} M$ is A-adically complete.

[3] Let $\mathscr{C}$ be a finite category (i.e., one with only finitely many objects and only finitely many maps), anld let $M : \mathscr{C}^{\circ} \rightsquigarrow$ (the category of abstract left A-modules) be an arbitrary contravariant functor. If, for every object $i \in \mathscr{C}$, we have that the abstract left A-module $M(i)$ is A-adically complete, then the abstract left A-module

$$\varprojlim_{i \in \mathscr{C}} M(i)$$

is A-adically complete

Proof. The proof that [1] iff [2] is the same as in the proof of Theorem 1.3.5, that ([2] of Theorem 1.3.5 iff [3] of Theorem 1.3.5).

[3] implies ([2] and [1]), since kernels and finite fiber products are special cases of inverse limits over a finite category, and [1] implies [3], since the $\varprojlim_{i \in \mathscr{C}} M(i)$ can be interpreted as the kernel of a map from a finite product of several of the $M(i)$'s (with a few repeats) into another finite such direct product of several of the $M(i)$'s, in general with some repeats. $\square$

(Remark: The reader not too familiar with general finite inverse limits can ignore condition [3] of the last Corollary.)

Theorem 1.3.7. *Let A be a topological ring such that the topology is given by right ideals. Then the following five conditions are equivalent:*

[1] The topological ring A is strongly admissible.

[2] Conditions [2a] and [2b] below both hold:

[2a] The three equivalent conditions of Theorem 1.3.5 hold, and

[2b] The direct product of an arbitrary indexed family of A-adically complete left A-modules is A-adically complete.

[3] Conditions [3a] and [3b] below both hold:

[3a] The three equivalent conditions of Corollary 1.3.6 hold.

[3b] Condition [2b] above holds.

[4] Let D be a directed set, and let $(M_i, \alpha_{ij})_{i,j \in D}$ be an arbitrary inverse system of abstract left A-modules. If M_i is A-adically complete for all $i \in D$, then the inverse limit abstract left A-module

$$\varprojlim_{i \in D} M_i$$

is A-adically complete.

[5] *Let $\mathscr{C}$ be an arbitrary category such that the objects form a set, and let*

$$M : \mathscr{C}^o \rightsquigarrow \text{(the category of abstract left A-modules)}$$

be an arbitrary contravariant functor. If, for every object $i \in \mathscr{C}$, we have that the abstract left A-module $M(i)$ is A-adically complete, then the abstract left A-module

$$\varprojlim_{i \in \mathscr{C}} M(i)$$

is A-adically complete.

Proof. [5] implies [4] is obvious.

[4] implies [1]: Let M be an abstract left A-module. Then for every open right ideal S in A, M/SM is such that the A-adic topology is discrete. Therefore M/SM is A-adically complete. But then by condition [4], the inverse limit

$$\widehat{M} = \varprojlim_{S \in \mathscr{R}} M/SM$$

(where $\mathscr{R}$ denotes the collection of open right ideals of A), is A-adically complete. M being arbitrary, it follows that A is strongly admissible.

[1] implies [2a]. If [1] holds, we show that Theorem 1.3.5, condition [1] holds. In fact, let M be A-adically complete and let N be an A-adically closed left submodule of M. Let $\widehat{N}$ be the A-adic completion of N. Then, by condition [1], the topological ring A is strongly admissible, so that $\widehat{N}$ is A-adically complete; and $\widehat{N}$ with its A-adic topology is the completion of N as a topological left A-module. Therefore, there exists a unique (continuous) homomorphism of left A-modules $\beta : \widehat{N} \to M$ such that the composite:

$$N \to \widehat{N} \xrightarrow{\beta} M$$

is the inclusion $N \hookrightarrow M$. (Of course, every homomorphism of left A-modules is A-adically continuous.) Since $\widehat{N}$ with its A-adic topology is the completion of N as a topological left A-module, we have that the image of N is A-adically dense in $\widehat{N}$. Throwing through the A-adically continuous function β, it follows that N is A-adically dense in $\text{Im}(\beta)$. However, since N is closed in M, we have that N is closed in $\text{Im}(\beta)$. Therefore $N = \text{Im}(\beta)$. But then the composite:

$$N \to \widehat{N} \xrightarrow{\beta} N$$

is the identity of N. Also, the composite:

$$\widehat{N} \xrightarrow{\beta} N \to \widehat{N}$$

is an endomorphism of $\widehat{N}$ compatible with the identity of N. Since A is strongly admissible, we have that $\widehat{N}$ with its A-adic topology is the completion of N as a topological left A-module. Therefore the composite $\widehat{N} \overset{\beta}{\to} \widehat{N}$ is the identity of $\widehat{N}$. Therefore the natural mapping: $N \to \widehat{N}$ is an isomorphism; i.e., N is A-adically complete.

[1] implies [2b]: Let I be a set, and suppose that M_i is a complete left A-module, for all $i \in I$. Let $\widehat{\prod_{i \in I} M_i}$ denote the A-adic completion of the abstract left A-module $\prod_{i \in I} M_i$. Let $(\prod_{i \in I} M_i, product)$ denote the topological left A-module, with abstract left A-module $\prod_{i \in I} M_i$, and with the direct product topology. This is also the direct product of the topological left A-modules M_i (with their A-adic topologies) in the category of all topological left A-modules. Therefore, $(\prod_{i \in I} M_i, product)$ is a complete topological left A-module, and the identitiy is a continuous homomorphism of topological left A-modules: $\prod_{i \in I} M_i \to (\prod_{i \in I} M_i, product)$.

Since the topological ring A is strongly admissible, we have that $\widehat{\prod_{i \in I} M_i}$ with its A-adic topology is the completion as a topological left A-module of $\prod_{i \in I} M_i$, with its A-adic topology. Therefore, by universality, there exisits a unique continuous homomorphism of topological left A-modules:

$$\beta : \widehat{\prod_{i \in I} M_i} \to \left(\prod_{i \in I} M_i, product \right)$$

such that the composite:

$$\prod_{i \in I} M_i \to \widehat{\prod_{i \in I} M_i} \overset{\beta}{\to} \left(\prod_{i \in I} M_i, product \right)$$

is the set-theoretic identity of $\prod_{i \in I} M_i$. Then the composite of the homomorphisms of left A-modules:

$$\widehat{\prod_{i \in I} M_i} \overset{\beta}{\to} \prod_{i \in I} M_i \to \widehat{\prod_{i \in I} M_i}$$

is an endomorphism of the abstract left A-module $\widehat{\prod_{i \in I} M_i}$ (and is therefore continuous for the A-adic topology) that induces the identity on $\prod_{i \in I} M_i$. Since A is strongly admissible, $\widehat{\prod_{i \in I} M_i}$ together with its A-adic topology is the completion of $\prod_{i \in I} M_i$ (with its A-adic topology) in the category of topological left A-modules. Therefore the composite:

$$\widehat{\prod_{i \in I} M_i} \overset{\beta}{\to} \prod_{i \in I} M_i \to \widehat{\prod_{i \in I} M_i}$$

is the identity. Therefore the natural map: $\prod\limits_{i \in I} M_i \to \widehat{\prod\limits_{i \in I} M_i}$ is an isomorphism; i.e., $\prod\limits_{i \in I} M_i$ is A-adically complete.

[2] implies [3]: Clearly, Theorem 1.3.5, condition [2] implies Corollary 1.3.6, condition [1].

[3] implies [5]: Let E be the set of all objects in the category $\mathscr{C}$, and let $E' = \{(i, j, f) : i, j \in E,$ and $f : i \to j$ is a map in the category $\mathscr{C}\}$. Then, for every $(i, j, f) \in E'$, we have the composite homomorphism:

$$\prod_{h \in E} M(h) \overset{p_j}{\to} M(j) \overset{M(f)}{\to} M(i),$$

and also the ith projection:

$$\prod_{h \in E} M(h) \overset{p_i}{\to} M(i),$$

and therefore also the difference:

$$(1): \qquad \prod_{h \in E} M(h) \xrightarrow{\ p_i - M(f) \circ p_j\ } M(i).$$

The maps (1), for all $(i, j, f) \in E'$, define a map:

$$(2): \prod_{h \in E} M(h) \to \prod_{(i,j,f) \in E'} M(i)$$

and $\varprojlim\limits_{i \in \mathscr{C}} M(i)$ is the kernel of the map (2).

But then, by condition [3b], since $M(h)$ is A-adically complete for all $h \in E$, we have that $\prod\limits_{h \in E} M(h)$ and $\prod\limits_{(i,j,f) \in E'} M(i)$ are both A-adically complete. Therefore by condition [3a] applied to the map (2), we have that the kernel $\varprojlim\limits_{i \in \mathscr{C}} M(i)$ of the map (2) is A-adically complete.   □

Remark: The reader who does not understand condition [5] of the last Theorem can ignore this condition. (In the above proof, then simply delete the first part: "[5] implies [4]"; and modify the last part, "[3] implies [5]," to read "[3] implies [4]"; and modify the last part of the proof, by taking $E =$ the underlying set of the directed set D, and $E' = \{(i, j) \in E \times E : i \leq j$ in $D\}$.)

Remark 5. After Theorem 1.3.3, and the equivalence of conditions [1] and [2] of Theorem 1.3.7, the statement (proved in Example 2) that: "Given a topological ring A obeying the hypotheses of Example 2, then A is strongly admissible", is equivalent to: "Given a topological ring A obeying the hypotheses of Example 2, then A obeys condition [2a] of Theorem 1.3.7." In fact, it is not difficult to give a direct, naive, proof of this latter assertion. After Theorem 1.3.3, and the equivalence of conditions [1] and [2] of Theorem 1.3.7, such a direct proof can be interpreted as being an alternative proof to §2, Note to Theorem 1.2.3 (and therefore also of §2, Theorem 1.2.3, and of Example 2 above). However, §2, Corollary 1.2.2, would not immediately follow from such a proof; so the procedure for establishing Example 2 above that we have adopted is perhaps superior.

Remark 6. It appears possible that condition [2a] of Theorem 1.3.7 alone might be enough to prove strong admissibility. (In fact, in Remark 5 above we have observed that a naive attempt to prove condition [3a] can be successfully concluded directly, if one assumes that the complete topological ring A obeys the hypotheses of Example 2 above. However, a similar naive attempt without the hypotheses of Example 2, appears to run into difficulties.) This tends to support the belief that [2a], and perhaps very little else, might already be equivalent to strong admissibility.

In the next chapter, we will see another, far more significant, condition that is equivalent to admissibility.

Chapter 2

The C-completion of an Abstract Module over a Topological Ring

2.1 Some Preliminaries

In this chapter, A denotes a fixed topological ring such that the topology is given by right ideals. *We make no other assumptions on the topological ring A.* (Although later, we will show that admissibility of the topological ring A would be equivalent to certain deep results.)

Definition. Let A be a topological ring such that the topology is given by right ideals. Let I be any set, and let $(a_i)_{i \in I}$ be an indexed family of elements in the ring A indexed by the set I. Then we say that the family $(a_i)_{i \in I}$ of elements in the topological ring A is *zero-bound* iff, either the set I is finite, or else the family $(a_i)_{i \in I}$ converges to zero in A. That is, iff for every open right ideal S in A, we have that $a_i \in S$ for all but at most finitely many $i \in I$.

Notice, therefore, that given a topological ring A such that the topology is given by right ideals; a set I; and an indexed family $(a_i)_{i \in I}$ of elements of A; that the following conditions are equivalent:

(1) $(a_i)_{i \in I}$ is zero-bound.

(2) The infinite series $\sum_{i \in I} a_i$ in the topological ring A is a Cauchy series.
(That is: The net in A: $F \to \sum_{i \in F} a_i$ is a Cauchy net, where F runs through the elements of the set consisting of all finite subsets of I, directed by inclusion.)

(3) For every left A-module M, and every family $(x_i)_{i \in I}$ of elements of M, we have that the infinite series:

$$\sum_{i \in I} a_i \cdot x_i$$

45

in the left A-module M, is a Cauchy series, for the A-adic topology of M. (That is: The net in M, $F \to \sum\limits_{i \in F} a_i \cdot x_i$, is a Cauchy net; where F runs through the set of all finite subsets of I directed by inclusion; and the topology on M is the A-adic topology.)

Remark 1. The proof of the equivalence of the above three conditions is immediate; the easiest chain of implications is (3) implies (2) implies (1) implies (3).

Remark 2. Let A be a topological ring such that the topology is given by $\leq K$ right ideals, where K is an infinite cardinal number. If I is any set, and if $(a_i)_{i \in I}$ is any family of elements in the ring A, then necessary and sufficient conditions for the family $(a_i)_{i \in I}$ to be zero-bound, is that:

There exists a subset J of I, such that J is of cardinality $\leq K$; such that, for every $i \in I$ with i not in J, we have that a_i is in the closure of zero in A; and such that the family $(a_i)_{i \in J}$ is zero-bound.

Proof. In fact, necessity of the above assertion is the same as the assertion made in Chapter 1, §2, Remark 3 in the first sentence of the proof of that Remark, and is proved there. Sufficiency is immediate.   □

Remark 2.1. Let A be a topological ring such that the topology is given by right ideals. Let I be a set and let $(a_i)_{i \in I}$ be an indexed family of elements of A. Then necessary and sufficient conditions that $(a_i)_{i \in I}$ be zero-bound is that, for every denumerable subset D of I, the family $(a_i)_{i \in D}$ is zero-bound.

Proof. If $(a_i)_{i \in I}$ is zero-bound, then clearly $(a_i)_{i \in J}$ is zero-bound for every subset J of I. On the other hand, if $(a_i)_{i \in I}$ is not zero-bound, then there exists an open right ideal S in A such that $a_i \notin S$ for infinitely many $i \in S$. Therefore there exists a denumerable subset D of I such that $a_i \notin S$ for all $i \in D$. And then the family $(a_i)_{i \in D}$ is not zero-bound.   □

Lemma 2.1.1. *Let A be a complete topological ring such that the topology is given by right ideals. Let I and J be sets, and let $(a_{ij})_{(i,j) \in I \times J}$ and $(c_i)_{i \in I}$ be indexed families of elements of A, such that, for every $i \in I$, the family $(a_{ij})_{j \in J}$ is zero-bound, and such that the family $(c_i)_{i \in I}$ is zero-bound. Then the family of elements of A, indexed by the set J,*

$$\left(\sum_{i \in I} c_i a_{ij} \right)_{j \in J}$$

is zero-bound.

Note. Of course, in the above statement, for every $j \in J$, the (perhaps infinite) sum

$$\sum_{i \in I} c_i a_{ij}$$

means the unique limit of this Cauchy series in the complete topological ring A.

Proof. In fact, let S be any open right ideal in the ring A. Then, since $(c_i)_{i \in I}$ converges to zero in A, there exists a finite subset F of I such that $c_i \in S$ for all $i \in I$ with i not in F. Since S is an open subgroup of A, and therefore also closed,

$$(1): \sum_{i \in I - F} c_i a_{ij} \in S,$$

for all $j \in J$. On the other hand, since for every $i \in I$, the family $(a_{ij})_{j \in J}$ is zero-bound in A for $i \in I$, therefore the family $(c_i \cdot a_{ij})_{j \in J}$ is likewise zero-bound in A. Hence there exists a finite subset G_i of J such that $c_i \cdot a_{ij} \in S$ for all $j \in J - G_i$. Then if $G = \underset{i \in F}{\cup} G_i$, then G is a finite subset of J, and for all $j \in J$ with $j \notin G$, we have that $c_i \cdot a_{ij} \in S$ for all $i \in F$, whence

$$(2): \sum_{i \in F} c_i a_{ij} \in S.$$

But then for all $j \in J$ with $j \notin G$, by equations (1) and (2), we have that

$$\sum_{i \in I} c_i a_{ij} = \sum_{i \in I - F} c_i a_{ij} + \sum_{i \in F} c_i a_{ij} \in S$$

S being an arbitrary open right ideal in A, it follows that the family of elements of A, indexed by the set J,

$$\left(\sum_{i \in I} c_i a_{ij} \right)_{j \in J}$$

is zero-bound in A. □

Let A be any topological ring such that the topology is given by open right ideals. Then $\widehat{A}$ denotes the completion of A *as a topological ring.* Thus, $\widehat{A}$ is a complete topological ring, and a complete system of neighborhoods of zero in the topological ring $\widehat{A}$ is the set of right ideals, $\{\widehat{S} : S$ is an open right ideal in $A\}$.

<u>Remark 3</u>. In the above formula, for every open right ideal S in A (or, more generally, for any open additive subgroup S of A), $\widehat{S}$ denotes, as usual, the completion of S as a topological abelian group. This is canonically isomorphic, as a topological abelian group, to its image in $\widehat{A}$ with the induced topology from $\widehat{A}$ (where in this chapter our convention is that the topology that we take on $\widehat{A}$ is the one such that $\widehat{A}$ is the completion of A as a topological ring, or equivalently as topological group; or equivalently as uniform space. In the terminology of Chapter 1, this is the *uniform* topology on $\widehat{A}$, as opposed to, e.g., the A-adic topology on $\widehat{A}$. Cf. Chapter 1, §1, Exercise 3, for a proof. Therefore for every open right ideal S in A, we have that $\widehat{S}$ is an open right ideal in $\widehat{A}$ (for the only topology that we ever discuss on $\widehat{A}$ in this Chapter)).

(Notice that $\widehat{S}$ cannot in general be defined to be "the A-adic completion" of S, since if the right ideal S is not a two-sided ideal, then S is not a left A-module, and the terminology "A-adic completion" has been restricted to left

A-modules. Another correct description of $\widehat{S}$ for any open right ideal (or even for any open subgroup) S in A, is $\widehat{S} = \overline{\iota(S)}$, where $\iota : A \to \widehat{A}$ is the natural homomorphism from the topological ring A into its completion $\widehat{A}$, and $\overline{\iota(S)}$ denotes the closure of $\iota(S)$ for the topology of the complete topological ring $\widehat{A}$ — cf. again Chapter 1, §1, Example 3).

Remark 4. It should be noted that, if the ring A does not obey condition [3'a] of the Note following Remark 5 of Chapter 1, §2, then *a priori* there is no reason to believe that the topology of $\widehat{A}$ is not strictly coarser than the A-adic topology on $\widehat{A}$. (This latter, finer, topology on $\widehat{A}$ is the one such that an open neighborhood base at zero is $\{S \cdot \widehat{A} : S$ is an open right ideal in $A\}$.) See Chapter 1, §2, Remark 8 for a counterexample. The only topology on $\widehat{A}$ that we will always be interested in, however, is the one such that $\widehat{A}$ with this topology is the completion, as a topological ring, of A.

If this detail (two topologies on $\widehat{A}$) should be bothersome to the reader, the reader can, if he wishes, simply assume that the fixed topological ring A is complete, so that $A = \widehat{A}$. It is not difficult to see that this does not really reduce the generality very much – since we shall observe later that, the basic construction of this section — the C-completion of an abstract left A-module — is such that: If M is any abstract left A-module, then the C-completion of M *qua* left A-module, coincides with the C-completion of $\widehat{A} \underset{A}{\otimes} M$ *qua* abstract left $\widehat{A}$-module.

Example. Let A be a topological ring, and let $\widehat{A}$ be the completion of A (as always in this Chapter, with the topology such that $\widehat{A}$ is the completion as a topological ring of A. I.e., with the topology given by the right ideals $\{\widehat{S} : S$ is an open right ideal in $A\}$). Let I be any set and let $(a_i)_{i \in I}$ be an indexed family of elements of $\widehat{A}$. Then the following conditions are equivalent:

(1) $(a_i)_{i \in I}$ is zero-bound in $\widehat{A}$.

(2) For every open right ideal S in A, we have that $a_i \in \widehat{S}$ for all but finitely many $i \in I$.

(3) The element $(a_i)_{i \in I}$ of the Cartesian product set $\widehat{A}^I$, lies in the subset $\widehat{A^{(I)}}$, (where $A^{(I)}$ denotes the free left A-module, and $\widehat{A^{(I)}}$ denotes the A-adic completion of the abstract left A-module $A^{(I)}$).

The proof of this example is easy — (once one notes, that, the natural map: $\widehat{A^{(I)}} \to A^I$ is indeed injective — see Chapter 1, §3, Lemma 1.3.4).

2.2   C-complete Left A-modules

Definition 1. Let A be a topological ring such that the topology is given by right ideals, and let M be an abstract left A-module. Then an *infinite sum structure* on M is:

(S) The giving — for every set I, for every indexed family $(x_i)_{i \in I}$ of elements of M, and every family $(a_i)_{i \in I}$ of elements of $\widehat{A}$ [1] that is zero-bound in $\widehat{A}$ [2] of an element, denoted

(4) $\sum\limits_{i \in I} a_i x_i$

of the abstract left A-module M, such that the following axioms hold:

(IS1): For all sets I and all indexed families $(x_i)_{i \in I}$ of elements of M, and for all $j \in I$,

$$\sum_{i \in I} \delta_{ji} x_i = x_j$$

(IS2):

$$\sum_{i \in I} (a_i + b_i) x_i = \sum_{i \in I} a_i x_i + \sum_{i \in I} b_i x_i,$$

whenever x_i, $i \in I$, are elements of M, and $(a_i)_{i \in I}$ and $(b_i)_{i \in I}$ are families of elements that are zero-bound in $\widehat{A}$.

(IS3):

$$\sum_{i \in I} (c a_i) \cdot x_i = c \sum_{i \in I} a_i \cdot x_i$$

for all $c \in A$, all sets I, all $(a_i)_{i \in I}$ families indexed by I of elements of $\widehat{A}$ that are zero-bound, and all $(x_i)_{i \in I}$ families of elements of M, and

(IS4): If I and J are sets, and if $(a_{ij})_{(i,j) \in I \times J}$ and $(c_i)_{i \in I}$ are indexed families of elements of $\widehat{A}$ such that, for every $i \in I$, the family $(a_{ij})_{j \in J}$ is zero-bound in $\widehat{A}$; and such that the family $(c_i)_{i \in I}$ is zero-bound in $\widehat{A}$; then for every family $(x_i)_{i \in I}$ of elements of M indexed by the set I, we have that

$$\sum_{j \in J} \left(\sum_{i \in I} c_i a_{ij} \right) x_j = \sum_{i \in I} c_i \left(\sum_{j \in J} a_{ij} x_j \right).$$

Notes 1. In $(IS4)$, the infinite sum: $\sum\limits_{i \in I} c_i a_{ij}$ is of course understood to mean the limit of this convergent series in the topological ring $\widehat{A}$.

2. Note also that, by §1, Lemma 2.1.1, applied to the complete topological ring

---

[1]Notice that $a_i, i \in I$, are elements of $\widehat{A}$, the completion of A – and are not necessarily assumed to be in the image of A in $\widehat{A}$.

[2]This condition on the indexed family $(a_i)_{i \in I}$ is equivalent to:

For every open right ideal S in A, there exists a finite subset F of I such that $a_i \in \widehat{S}$ for all $i \in I$ with $i \notin F$

Another equivalent formulation is:

$$(a_i)_{i \in I} \in \widehat{A^{(I)}}$$

In the latter formulation, of course, $\widehat{A^{(I)}}$ denotes the A-adic completion of the free left A-module $A^{(I)}$.

See the Example at the end of §1.

$\widehat{A}$, we have that, in the above displayed equation, the left side makes sense — since by §1, Lemma 2.1.1, the family: $(\sum_{i \in I} c_i a_{ij})_{j \in J}$ is zero-bound in $\widehat{A}$.

<u>Remark 1</u>. Given datum (S), obeying Axioms $(IS1)$ and $(IS4)$, then Axiom $(IS2)$ is equivalent to the following, apparently weaker, condition:

(IS2$_{weak}$): Let $I = \{1, 2\}$. Let $(a_1, a_2) = (1, 0)$ and $(b_1, b_2) = (0, 1)$. Then for every $(x_1, x_2) \in M^I$, we have that

$$\sum_{i \in I}(a_i + b_i) \cdot x_i = \sum_{i \in I} a_i x_i + \sum_{i \in I} b_i x_i.$$

Proof. Obviously, $(IS2)$ implies $(IS2_{weak})$. Assume that we have $(IS2_{weak})$, and let J be any set. Let $(c_j)_{j \in J}$ and $(d_j)_{j \in J}$ be families of elements that are zero-bound in $\widehat{A}$, and let $(x_j)_{j \in J}$ be any family of elements in M, indexed by the set J. Then let $I = \{1, 2\}$ and let $e_{1j} = c_j$, $e_{2j} = d_j$, for all $j \in J$. Then the doubly-indexed family $(e_{ij})_{(i,j) \in I \times J}$ and the singly indexed family $(f_1, f_2) = (1, 1) \in A^I$ obey the hypotheses of Axiom $(IS4)$. Therefore,

$$(1) : \sum_{j \in J}\left(\sum_{i \in I} f_j e_{ij}\right) x_j = \sum_{i \in I} f_j \left(\sum_{j \in J} e_{ij} x_j\right).$$

In this equation, by Note 1 to $(IS4)$, we have that

$$(2) : \sum_{i \in I} f_i e_{ij} = e_{1j} + e_{2j} = c_j + d_j$$

for all $j \in J$. Since also $f_i = a_i + b_i$, for all $i \in I = \{1, 2\}$, by Axiom $(IS2_{weak})$ we have that

$$(3) : \sum_{i \in I} f_i \left(\sum_{j \in J} e_{ij} x_j\right) = \sum_{i \in I} a_i \left(\sum_{j \in J} e_{ij} x_j\right) + \sum_{i \in I} b_i \left(\sum_{j \in J} e_{ij} x_j\right).$$

But, by Axiom $(IS1)$ applied to the set $I = \{0, 1\}$, we also have that:

$$(4) : \sum_{i \in I} a_i \left(\sum_{j \in J} e_{ij} x_j\right) = \sum_{j \in J} e_{1j} x_j = \sum_{j \in J} c_j x_j, \text{ and}$$

$$(5) : \sum_{i \in I} b_i \left(\sum_{j \in J} e_{ij} x_j\right) = \sum_{j \in J} e_{2j} x_j = \sum_{j \in J} d_j x_j.$$

Substituting equations (2), (3), (4), and (5) into equation (1), we have that

$$\sum_{j \in J}(c_j + d_j) x_j = \sum_{j \in J} c_j x_j + \sum_{j \in J} d_j x_j$$

J being an arbitrary set, and $(c_j)_{j \in J}$, $(d_j)_{j \in J}$ any pair of zero-bound families indexed by J of elements of $\widehat{A}$, and $(x_j)_{j \in J}$ being any family of elements of M, this proves $(IS2)$. $\square$

Remark 2. The proof of Remark 1 also shows that: If A is any topological ring such that the topology is given by right ideals, if M is any abstract left A-module, and if we have a datum as in (S) above for the left A-module M, such that Axiom $(IS4)$ holds, then Axiom $(IS2)$ holds iff

(*): If $I = \{1, 2\}$, and if $(a_1, a_2) = (1, 1)$, then for every $(x_1, x_2) \in M^I$, we have that

$$\sum_{i \in I} a_i x_i = x_1 + x_2.$$

The explicit proof of axiom $(IS2)$, from (*); tracing through the last Remark, is as follows: Given J a set, $x_j \in M$, all $j \in J$, and $(a_j)_{j \in J}$, $(b_j)_{j \in J}$ zero-bound in $\widehat{A}$; then define $I = \{1, 2\}, c_1 = c_2 = 1, a_{1j} = a_j; a_{2j} = b_j$, all $j \in J$. Then Axiom $(IS4)$ applies to $I, J, (c_i)_{i \in I}, (a_{ij})_{(i,j) \in I \times J}$, and $(x_j)_{j \in J}$. Substituting (*) into the identity of $(IS4)$ gives $(IS2)$.

Remark 3. Given datum (S), obeying Axioms $(IS1)$ and $(IS4)$, then Axiom $(IS3)$ is equivalent to the following, apparently weaker condition:

($IS3_{weak}$) Let $c \in \widehat{A}$ and let $x \in M$. Then define $I = \{0\}$, $a_0 = c$, and $x_0 = x$. Then

$$\sum_{i \in I} a_i x_i = c \cdot x.$$

(where the left side means the infinite sum structure applied to the set I, the zero bound family $(a_i)_{i \in I}$ of elements of $\widehat{A}$, and the family $(x_i)_{i \in I}$ of elements of M).

Proof. First, notice that an argument similar to that of the proof of Remark 1, shows that Axiom $(IS3)$ is equivalent to:

(1) Let $c \in A$, $a \in \widehat{A}$, and $x \in M$. Let $I = \{0\}$, $b_0 = a$, and $x_0 = x$. Then

$$\sum_{i \in I} (c \cdot b_i) \cdot x_i = c \cdot \left(\sum_{i \in I} b_i \cdot x_i \right)$$

(where, the left side means the infinite sum structure applied to $(c \cdot b_i)_{i \in I}$ and $(x_i)_{i \in I}$, and the right side involves the infinite sum structure applied to $(b_i)_{i \in I}$ and $(x_i)_{i \in I}$, and the product "$c\cdot$" involves the A-module structure on M).

In condition (1), if we consider the special case in which the element $a \in \widehat{A}$ is the element $1 \in \widehat{A}$, then by $(IS1)$ applied to $I = \{0\}$,

$$\sum_{i \in I} b_i \cdot x_i = x.$$

Also, in the ring $\widehat{A}$,

$$c \cdot b_i = c, i \in I = \{0\}.$$

Therefore, equation (1) becomes $(IS3_{weak})$. Therefore, $(IS3_{weak})$ is necessary for $(IS3)$.

On the other hand, suppose that we have equation $(IS3_{weak})$. Then, if $c \in A$, $a \in \widehat{A}$, and $x \in M$, let $I = J = \{0\}$, $c_0 = c$, $a_{00} = a$, and $x_0 = x$. Then by axiom $(IS4)$,

$$(2) : \sum_{j \in J} \left(\sum_{i \in I} c_i a_{ij} \right) \cdot x_j = \sum_{i \in I} c_i \left(\sum_{j \in J} a_{ij} x_j \right).$$

If we let $b_0 = a$, then

$$\sum_{i \in I} c_i a_{ij} = c \cdot a = c \cdot b_j \text{ in } \widehat{A}, \text{ all } j \in I = \{0\},$$

so that the left side of equation (2) is:

$$(3) : \sum_{j \in J} (c b_j) \cdot x_j.$$

By $(IS3_{weak})$, the right side of equation (2) is

$$(4) : c \cdot \sum_{j \in J} a_{ij} x_j = c \cdot \sum_{j \in J} b_j \cdot x_j.$$

Substituting (3) and (4) into (2), we obtain condition (1) (with J replacing I). Therefore in the presence of $(IS4)$ and $(IS1)$, $(IS3_{weak})$ implies condition (1), and therefore implies $(IS3)$.   □

Remark 4. Let A be a topological ring such that the topology is given by right ideals and let M be an abstract left A-module together with an infinite sum structure as defined in Definition 1. Then, if n is any integer ≥ 0, if $a_1, \ldots, a_n \in A$, $x_1, \ldots, x_n \in M$, and if $I = \{1, \ldots, n\}$, then necessarily:

$$(1) : \sum_{i \in I} a_i x_i = a_1 x_1 + \ldots + a_n x_n$$

— where the sum on the left side of equation (1) is the one given by the datum (S) of Definition 1; and where the sum and scalar products on the right side of equation (1) are the ones deduced by virtue of M being an abstract left A-module.

(In fact, more generally, equation (1) holds, when $a_1, \ldots, a_n \in \widehat{A}$ rather than A; where the sum and scalar product on the right side of equation (1) are then taken to be the ones deduced from the abstract left $\widehat{A}$-module structure of M described in Corollary 2.2.2 below. The proof is the same, with $(IS3')$ of Corollary 2.2.3 below replacing $(IS3)$ of Definition 1.)

Proof. By axioms $(IS2)$ and $(IS3)$ applied to the set $I = \{1, \ldots, n\}$, it suffices to prove the assertion in the case that there is a j, $1 \leq j \leq n$, such that $a_i = \delta_{ij}$, $1 \leq i \leq n$. But then equation (1) is Axiom $(IS1)$ for the set $I = \{1, \ldots, n\}$.   □

Remark 5. Let A be a topological ring such that the topology is given by $< K$ right ideals, where K is any infinite cardinal number. Let M be an abstract left A-module. Then, to give an infinite sum structure on the abstract left A-module M, in the sense of Definition 1, it is sufficient to give datum (S) as in Definition 1, *only for sets I of cardinality $< K$*, such that Axioms $(IS1)$, $(IS2)$, $(IS3)$, and $(IS4)$ all hold (for all sets I and J of cardinality $< K$).

In fact, given such an (apparently weaker) structure on M, then for any set H of arbitrary cardinality, and for any family $(a_h)_{h \in H}$ of elements $\widehat{A}$ that is zero-bound, and any family $(x_h)_{h \in H}$ of elements of M, first note that, by §1, Remark 2, we have that there exists a subset I of H such that $\mathrm{card}(I) < K$ and such that $a_h = 0$ for all $h \in H$, $h \notin I$. Then define

$$\sum_{h \in H} a_h \cdot x_h = \sum_{h \in I} a_h \cdot x_h.$$

Then, using that $(IS1)$, $(IS2)$, $(IS3)$, and $(IS4)$ hold for sets of cardinality $< K$, it is not difficult to show that:

(1) The above definition of $\sum\limits_{h \in H} a_h \cdot x_h$ is independent of the choice of the subset I of H (such that $\mathrm{card}(I) < K$, and such that $a_h = 0$ for all $h \in H$ with $h \notin I$).

(2) The resulting datum (S) for the abstract left A-module M is an infinite sum structure for M.

(3) The infinite sum structure (S) thus defined on M is the only infinite sum structure on M such that, the restriction to sets T of cardinality $< K$ is the given restricted datum.

Remark 6. Let A and K be as in Remark 5 above. Let I be any fixed set such that the neighborhood system of A has a base of cardinality $\leq$ that of the cardinality of I. Then, taking the successor of the cardinal of I and using Remark 5 above, it is not difficult to show that, to give an infinite sum structure on an abstract left A-module M, it is equivalent to giving datum (S) of Definition 1, *but only for the given fixed set I* (i.e., only for those families of elements of A and of M *that are indexed by the given fixed set I*), such that Axioms $(IS1)$, $(IS2)$, $(IS3)$, and $(IS4)$ hold for the fixed set I (where, in Axiom $(IS4)$, we insist that $I = J =$ the given fixed set I).

However, Definition 1 above is perhaps a bit more elegant to work with.

Remark 7. Another definition of an *infinite sum structure*, equivalent to that of Definition 1, is:

Definition 1′. Let A be a topological ring such that the topology is given by right ideals and let M be an abstract left A-module. Then an *infinite sum structure on M* is, the following data:

(S') The giving, for every set I and every family $(x_i)_{i \in I}$ indexed by I of elements of M, of a homomorphism of abstract left A-modules:

$$\sigma_{(x_i)_{i \in I}} : \widehat{A^{(I)}} \to M,$$

such that the following two axioms hold:

(IS1′) For all $j \in I$, the image, under $\sigma_{(x_i)_{i \in I}}$ composed with the natural map $\iota : A^{(I)} \to \widehat{A^{(I)}}$ of the jth canonical basis element $e_j = (\delta_{ij})_{i \in I}$ of the free A-module $A^{(I)}$, is x_j:

$$\sigma_{(x_i)_{i \in I}}(\iota(e_j)) = x_j, \text{ for all } j \in J.$$

(IS4′) Let I and J be sets, let $\alpha : \widehat{A^{(I)}} \to \widehat{A^{(J)}}$ be a homomorphism of left $\widehat{A}$-modules that is continuous for the uniform topologies on $\widehat{A^{(I)}}$ and $\widehat{A^{(J)}}$ (— to give an $\widehat{A}$-homomorphism that obeys this continuity condition is equivalent to giving a homomorphism : $A^{(I)} \to \widehat{A^{(J)}}$ of abstract left A-modules). Let $(x_j)_{j \in J}$ be any family indexed by J of elements of M. For every $i \in I$, let $y_i = \sigma_{(x_j)_{j \in J}}(e_i)$, where e_i is the ith canonical basis element of the free left A-module $A^{(I)}$. Then the triangle:

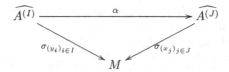

is commutative.

Proof that Definition 1 and Definition 1′ are equivalent. First, notice that, by the Example at the end of §1, the families $(a_i)_{i \in I}$ indexed by I of elements of $\widehat{A}$ that are zero-bound are the same as the elements of the A-adic completion $\widehat{A^{(I)}}$ of the free left A-module $A^{(I)}$. If M is an abstract left A-module, then to give data as in (S) of Definition 1, is to give, for every set I, a function:

$$\widehat{A^{(I)}} \times M^I \to M$$

obeying the axioms. Using the canonical bijection between

$$M^{\widehat{A^{(I)}} \times M^I} \text{ and } \left(M^{\widehat{A^{(I)}}} \right)^{M^I},$$

it is equivalent to give, for every set I and every element $(x_i)_{i \in I} \in M^I$, a function $\sigma_{(x_i)_{i \in I}} : \widehat{A^{(I)}} \to M^I$. Then, in this notation, Axioms $(IS2)$ and $(IS3)$ together are equivalent to the single condition: For every set I and every $(x_i)_{i \in I} \in M^I$, $\sigma_{(x_i)_{i \in I}}$ is a homomorphism of abstract left A-modules, $(IS1)$ and $(IS1′)$ are then immediately seen to be equivalent.

Finally, $(IS4)$ and $(IS4′)$ are equivalent. To prove this, first note that to give α as in $(IS4′)$ is equivalent to giving $(a_{ij})_{(i,j) \in I \times J}$ as in $(IS4)$. And, that in the notation of Definition 1, for every $i \in I$, y_i as in $(IS4′)$ becomes $\sum_{j \in J} a_{ij} \cdot x_j$. Therefore, commutativity of the diagram in $(IS4′)$ at the typical element $(c_i)_{i \in I}$ is equivalent in the notation of Definition 1, to the displayed formula in $(IS4)$. □

The proof of the next proposition will be given later.

Proposition 2.2.1. *Let A be a topological ring such that the topology is given by right ideals, and let M be an abstract left A-module. Suppose we have an infinite sum structure on the abstract left A-module M. Then*

$(IS5)$ $\sum_{i \in I} a_i \cdot (x_i + y_i) = \sum_{i \in I} a_i \cdot x_i + \sum_{i \in I} a_i y_i$, *for all sets I, all $(a_i)_{i \in I}$ families*

of elements of $\widehat{A}$ indexed by I that are zero-bound and for all pairs $(x_i)_{i \in I}$ and $(y_i)_{i \in I}$ of families indexed by I of elements of M.

$(IS6)$ $\sum_{i \in I} (a_i \cdot c_i) \cdot x_i = \sum_{i \in I} a_i \cdot (c_i \cdot x_i)$, *for all sets I, all $c_i \in \widehat{A}$, $i \in I$, all*

$x_i \in M$, $i \in I$, *and for all families $(a_i)_{i \in I}$ of elements of $\widehat{A}$ indexed by I that are zero-bound.*

<u>Note</u>. In the equation $(IS6)$, the left side means: datum (S) applied to the zero-bound family $(a_i c_i)_{i \in I}$ of elements of $\widehat{A}$, and to the family $(x_i)_{i \in I}$ of elements of M. The right side means: datum (S) applied to the zero-bound family $(a_i)_{i \in I}$ of elements of $\widehat{A}$, and to the family $(c_i \cdot x_i)_{i \in I}$ of elements of M.

Corollary 2.2.2. *Let A be a topological ring such that the topology is given by right ideals, and let M be an abstract left A-module, together with an infinite sum structure as in Definition 1. Then the scalar multiplication: $A \times M \to M$ naturally extends to a scalar multiplication: $\widehat{A} \times M \to M$ such that M becomes an abstract left $\widehat{A}$-module.*

Proof. If $a \in \widehat{A}$ and $x \in M$, then let I be the set consisting of a single element 0. Let $a_0 = a$, and $x_0 = x$. Then $(a_i)_{i \in I}$ is a zero-bound family of elements of $\widehat{A}$ and $(x_i)_{i \in I}$ is a family of elements of M as in datum (S), so that from the infinite sum structure on M, we have the element $\sum_{i \in I} a_i x_i$ of M. Define $a \cdot x = \sum_{i \in I} a_i x_i$. Then by Axioms $(IS1)$ and $(IS3)$ with $I = \{0\}$, (i.e., Axioms $(IS1)$ and $(IS3_{weak})$), this definition of a function $\widehat{A} \times M \to M$ extends the scalar product: $A \times M \to M$ of the abstract left A-module M. And Axioms $(IS1)$, $(IS2)$, and $(IS4)$ (with $I = J = \{0\}$), and $(IS5)$ of Proposition 2.2.1 above (in the special case $I = \{0\}$) then imply that the above definition of a product: $\widehat{A} \times M \to M$ does indeed make the abstract abelian group M into an abstract left $\widehat{A}$-module, such that the scalar product extends the scalar product of M as an abstract left A-module. $\qquad\square$

Corollary 2.2.3. *Let A and M be as in Proposition 2.2.1. Then*

$$(IS3') : \sum_{i \in I} (c \cdot a_i) \cdot x_i = c \cdot \sum_{i \in I} a_i x_i,$$

whenever $c \in \widehat{A}$, $(a_i)_{i \in I}$ is a zero-bound family in $\widehat{A}$ and x_i, $i \in I$, are elements of M (where the left side of $(IS3')$ is interpreted as in $(IS3)$; and where the *scalar product of c and $\sum_{i\in I} a_i x_i$ on the right side of $(IS3')$ is the scalar product for the structure of M as a $\hat{A}$-module as defined in Corollary 2.2.2).*

The proof of Corollary 2.2.3 is delayed.

<u>Definition 2</u>. If A is a topological ring such that the topology is given by right ideals, then a *C-complete left A-module* is an abstract left A-module M together with an infinite sum structure on the left A-module M in the sense of Definition 1 above.

By Corollary 2.2.2 above, if M is a C-complete left A-module, then M has a natural structure as an abstract left $\hat{A}$-module, such that the scalar multiplication extends that of M as an abstract left A-module.

<u>Example 1</u>. Let M be any abstract left A-module and let $\widehat{M}$ be the A-adic completion of M. Then there is induced an infinite sum structure on the A-module $\widehat{M}$.

Proof. First note that $\widehat{M}$ is a complete topological left $\hat{A}$ module for the uniform topology on $\widehat{M}$ (cf. Chapter 1, §2). Then given any set I, any family $(a_i)_{i\in I}$ of elements of $\hat{A}$ that is zero-bound in $\hat{A}$ and any indexed family $(x_i)_{i\in I}$ of elements of M, define $\sum_{i\in I} a_i x_i$ to be the limit of this convergent series in the complete topological $\hat{A}$-module $\widehat{M}$ for the uniform topology. Then it is trivial to note that Axioms $(IS1)$ - $(IS4)$ all hold. [3]

Given an abstract left A-module M, we will often use $\widehat{M}$ to denote the C-complete left A-module, in the sense of Definition 2 that is thus obtained.

□

<u>Remark</u>. As we shall see, Example 1 above has a special significance. Nevertheless, it can be generalized as follows:

<u>Example 1.1</u>. Let M be any complete topological left A-module, such that the topology of M is coarser than the A-adic topology. Then M has a natural structure as a C-complete left A-module.

Proof. Let $(M, A\text{-adic})$ denote the topological left A-module consisting of the abstract A-module of M together with the A-adic topology. Then by hypothesis, the identity of M is continuous: $id_M : (M, A\text{-adic}) \to M$. Since M is by hypothesis complete, by the universal mapping property definition of "completion" we have that id_M extends to a continuous function of topological A-modules: $(\widehat{M}, unif) \xrightarrow{f} M$. Then $M \xrightarrow{\iota_M} \widehat{M} \xrightarrow{f} M$ is the identity of M (so that both ι_M and f split as maps of abstract left A-modules).

Given $x \in M$, the function: $a \to a \cdot x$ from A into M is continuous (since by hypothesis M is a topological left A-module). Since M is complete, this

---

[3] As noted in Chapter 1, §2, the uniform topology on $\widehat{M}$ is, in general, coarser than the A-adic topology on $\widehat{M}$. Intermediate between these topologies is the $\hat{A}$-adic topology. The Example at the end of Chapter 1, §2, shows that these three topologies on $\widehat{M}$ can in general be distinct.

function extends uniquely to a function $\mu_x : \widehat{A} \to M$. For $a \in \widehat{A}$ and $x \in M$, define $a \cdot x = \mu_x(a)$. Then we have defined a function: $\mu : \widehat{A} \times M \to M$.

If the same method of definition were applied to $\widehat{M}$ for the uniform topology, then (since $\widehat{M}$ is a topological left $\widehat{A}$-module, and therefore for any $x \in \widehat{M}$, the scalar multiplication, $a \to a \cdot x$, for $a \in \widehat{A}$, $\widehat{A} \to \widehat{M}$, is the unique continuous extension of $a \to a \cdot x$, for $a \in A$, $A \to M$), one obtains the usual scalar multiplication: $\widehat{A} \times \widehat{M} \to \widehat{M}$ of the left $\widehat{A}$-module $\widehat{M}$. Therefore the diagram:

$$
\begin{array}{ccc}
\widehat{A} \times \widehat{M} & \xrightarrow[\text{multiplication}]{\text{scalar}} & \widehat{M} \\
{\scriptstyle \widehat{A} \times f} \downarrow & & \downarrow {\scriptstyle f} \\
\widehat{A} \times M & \xrightarrow{\mu} & M
\end{array}
$$

is commutative. Therefore the map μ is induced from the scalar product: $\widehat{A} \times \widehat{M} \to \widehat{M}$ by passing to the quotient. Therefore the quotient left A-module M of $\widehat{M}$, is a quotient left $\widehat{A}$-module. In particular, we have that: There exists a unique way of introducing a scalar product: $\widehat{A} \times M \to M$, extending the given scalar product: $A \times M \to M$, such that M becomes an abstract left $\widehat{A}$-module, and such that $f : \widehat{M} \to M$ is a homomorphism of abstract left $\widehat{A}$-modules.

Suppose now that I is a set, $(a_i)_{i \in I}$ is a zero-bound family of elements of $\widehat{A}$ and $(x_i)_{i \in I}$ is an arbitrary family of elements of M. Then I claim that the infinite series

$$(1) : \sum_{i \in I} (a_i \cdot x_i)$$

in the complete topological group M is convergent.

For, since $(a_i)_{i \in I}$ is zero-bound, and since $\widehat{M}$ is a topological $\widehat{A}$-module, we have that the infinite series:

$$(2) : \sum_{i \in I} (a_i \cdot \iota_M(x_i))$$

is convergent in $\widehat{M}$ for the uniform topology. Since $f : \widehat{M} \to M$ is $\widehat{A}$-linear, and since $f \circ \iota_M =$ identity of M, the image of the infinite series (2) in $\widehat{M}$ under f is the infinite series (1) in M. But since (2) is convergent in $\widehat{M}$, and f is continuous, it follows that (1) is convergent in M.

If now $(a_i)_{i \in I}$ is zero-bound in $\widehat{A}$ and $(x_i)_{i \in I}$ is an arbitrary family of elements in M, define the infinite sum $\sum_{i \in I} a_i \cdot x_i$ to be the limit of the infinite series (1) in the complete topological abelian group M. Then the function $f : \widehat{M} \to M$ preserves infinite sums (for the infinite sum structure of $\widehat{M}$ defined in Example 1). Since the infinite sum structure in $\widehat{M}$ obeys the axioms $(IS1)$ – $(IS4)$, it folows that these axioms likewise hold in M for the datum just defined. Therefore M possesses a natural infinite sum structure, as asserted. $\qquad\square$

Remark. Under the hypotheses of Example 1.1, notice that

(1) The homomorphism $f : \widehat{M} \to M$ is infinitely linear (in the sense of Definition 1 of §3 below) (and in particular is also $\widehat{A}$-linear).

(2) For every $x \in M$, the formula: $a \to a \cdot x$, from $\widehat{A}$ into M, is continuous (for the $\widehat{A}$-module structure of M that comes from the infinite sum structure defined in Example 1.1), and

(3) If $(a_i)_{i \in I}$ is zero-bound in $\widehat{A}$, and $(x_i)_{i \in I}$ is an arbitrary family in M, then the infinite sum: $\sum_{i \in I} a_i x_i$ is the limit, for the topology of M, of the infinite series $\sum_{i \in I} (a_i \cdot x_i)$.

Remark. Given any abstract left A-module M, then knowledge of the infinite sum structure on $\widehat{M}$ is equivalent to knowing (1) the structure of $\widehat{M}$ as a left $\widehat{A}$-module; and also (2) the limits of all of the Cauchy series:

$$\sum_{i \in I} a_i x_i,$$

for all sets I, for all families $(a_i)_{i \in I}$ indexed by the set I of elements of $\widehat{A}$ that converge to zero in $\widehat{A}$, and for all families $(x_i)_{i \in I}$ indexed by the set I of elements of M^I, for the *uniform* (*not*, in general, either the A-adic or the $\widehat{A}$-adic) topology on $\widehat{M}$ (i.e., the topology on $\widehat{M}$ such that $\widehat{M}$, together with this topology, is the completion in the category of topological abelian groups (or, equivalently, topological rings), of M together with its A-adic topology). If the topological ring A has a denumerable neighborhood base at zero and if the complete topological ring $\widehat{A}$ is strongly admissible, then knowledge of these limits is easily seen to be enough to retrieve the uniform (or, since $\widehat{A}$ is strongly admissible, equivalently the $\widehat{A}$-adic) topology on $\widehat{M}$ — since then every Cauchy filter in $\widehat{M}$ admits a denumerable base, and therefore comes from a denumerable Cauchy series. And every Cauchy series for the $\widehat{A}$-adic topology in any A-module M is easily seen to be of the form of one in datum (S) of Definition 1.

However, if the topology of A *does not* admit a denumerable neighborhood base at zero, or if the complete topological ring $\widehat{A}$ is not strongly admissible, then this is not in general clear (since, e.g., if the neighborhood system at zero in A does not admit a denumerable base, then it is not clear whether or not every Cauchy filter in $\widehat{M}$ comes from a Cauchy series indexed by some set (of arbitrary cardinality)). Therefore, in this case, the infinite sum structure on $\widehat{M}$ may contain *strictly less* information than knowledge of the uniform topology on $\widehat{M}$, and of the $\widehat{A}$-module structure of $\widehat{M}$.

Note. The topological considerations of the last Remark, if confusing to the reader, can be totally ignored, as they are unnecessary to the understanding of the most important constructions of this Chapter.

Remark 8. As noted in Remark 4 above, if M is a C-complete left A-module, then the structure of M as an abstract left A-module can be determined from the infinite sum structure. This suggests giving an alternative, equivalent, definition of "C-complete left A-module" entirely in terms of infinite sums. Here is such an equivalent definition (in the fullest generality).

Definition 2′. Let A be a topological ring such that the topology is given by right ideals. Then a *C-complete left A-module* is a set M, together with

(S) The giving — for every set I, for every indexed family $(x_i)_{i \in I}$ of elements of M, and for every indexed family $(a_i)_{i \in I}$ of elements of $\hat{A}$ that is zero-bound in $\hat{A}$, of an element, denoted

$$\sum_{i \in I} a_i \cdot x_i$$

of the set M, such that Axioms $(IS1)$ and $(IS4)$, as stated in Definition 2, above, hold.

Proof that Definition 2′ is equivalent to Definition 2. Clearly, if M is as in Definition 2, then M is as in Definition 2′. It remains to show that, given a set M and datum (S) obeying Axioms $(IS1)$ and $(IS4)$; then there is a unique abstract left A-module structure on M such that Axioms $(IS2)$ and $(IS3)$ also hold.

In fact, if $x_1, x_2 \in M$, then let $a_1 = a_2 = 1$, and define

$$(1) : x_1 + x_2 = \sum_{i \in I} a_i x_i$$

(where the sum on the right side is the one defined in datum (S) above). Also, if $x \in M$ and $a \in A$, then let $J = \{0\}$, $a_0 = a$, and $x_0 = x$, and define

$$a \cdot x = \sum_{j \in J} a_j x_j$$

(where the sum on the right side is the one defined by datum (S) above).

Notice that, by condition (*) of Remark 2 and condition $(IS3_{weak})$ of Remark 3, respectively, if there is any abstract left A-module structure on M such that Axioms $(IS2)$ and $(IS3)$ hold, then that left A-module structure must be such that, respectively, equations (1) and (2) above, hold. Therefore, to complete the proof of the equivalence of Definitions 2 and 2′, it is necessary and sufficient to show that:

(3) With the sum and scalar product defined in equations (1) and (2), M becomes an abstract left A-module; and that

(4) Axioms $(IS2)$ and $(IS3)$ then hold.

In fact, if $x \in M$, then (5) $1 \cdot x = x$ — this is by Axiom $(IS1)$ in the case $I = \{0\}$. Let $a_1, a_2 \in A$ and $x \in M$. Then let $I = \{1, 2\}$, $J = \{0\}$, and let $x_0 = x$, $c_1 = c_2 = 1$, $a_{10} = a_1$, $a_{12} = a_2$. Then, by Axiom $(IS4)$ applied to the sets I, J, and the families $(a_{ij})_{(i,j) \in I \times J}$, $(c_i)_{i \in I}$, and $(x_j)_{j \in J}$, we have that

$$(6) : \sum_{j \in J} \left(\sum_{i \in I} c_i a_{ij} \right) \cdot x_j = \sum_{i \in I} c_i \left(\sum_{j \in J} a_{ij} \cdot x_j \right)$$

(where the expression $\sum_{i \in I} c_i a_{ij}$ utilizes the sum and product in the ring A; and all other summations use the structure in (S)). Then $\sum_{i \in I} c_i a_{ij} = a_1 + a_2$, and by the definition of "scalar product" in equation (2), the left side of equation (6) is:

$$(7) : (a_1 + a_2) \cdot x.$$

Also, by the definition of "scalar product" in equation (2), we have that

$$\sum_{j \in J} a_{ij} x_j = a_i \cdot x \text{ , for all } i \in I.$$

Then, by the definition of "sum of two elements in M" in equation (1), it follows that the right side of equation (6) is

$$(8) : a_1 \cdot x + a_2 \cdot x.$$

Equations (6), (7), and (8) imply

$$(9) : (a_1 + a_2) \cdot x = a_1 \cdot x + a_2 \cdot x$$

for all $a_1, a_2 \in A$ and all $x \in M$.

Next, if $c, a \in A$ and $x \in M$, then let $I = J = \{0\}$, $c_0 = c$, $a_{00} = a$, and $x_0 = x$. Then by Axiom $(IS4)$ applied to the families $(a_{ij})_{(i,j) \in I \times J}$, $(c_i)_{i \in I}$ and $(x_j)_{j \in J}$, we have that equation (6) above holds. But then, by definition of the "scalar product in M", equation (2) above, we have that

$$\sum_{j \in J} a_{ij} \cdot x_j = a \cdot x,$$

and therefore by another use of equation (2),

$$(10) : \sum_{i \in I} c_i \left(\sum_{j \in J} a_{ij} \cdot x_j \right) = c \cdot (a \cdot x).$$

In the ring A, we have that

$$\sum_{i \in I} c_i a_{ij} = c \cdot a.$$

Therefore, by an application of equation (2),

$$(11) : \sum_{j \in J} \left(\sum_{i \in I} c_i \cdot a_{ij} \right) \cdot x_j = (c \cdot a) \cdot x.$$

Equations (6), (10), and (11) imply that

$$(12) : (c \cdot a) \cdot x = c \cdot (a \cdot x)$$

for all $c, a \in A$, and all $x \in M$.

Next, let $c \in A$ and $x_1, x_2 \in M$. Let $j_0 = 1$ or 2. Then, by Axiom $(IS4)$ applied with $J = \{1,2\}$, $I = \{0\}$, $c_0 = c$, $a_{0j} = \delta_{j,j_0}$, $j = 1,2$, we have that equation (6) above holds. But then

$$(13) : \sum_{i \in I} c_i a_{ij} = c\delta_{j,j_0} \text{ in the ring } A,$$

and by Axiom $(IS1)$ applied to the set J and the element j_0,

$$(14) : \sum_{j \in J} a_{ij} \cdot x_j = x_{j_0} \ j_0 = 1,2.$$

Therefore, applying equation (2) to the right side of equation (6), we have that

$$(15) : \sum_{j \in \{1,2\}} (c \cdot \delta_{j,j_0}) \cdot x_j = c \cdot x_{j_0} \text{ for all } c \in A, \text{ for } j_0 = 1,2, \text{ and all } x_1, x_2 \in M.$$

Next, if $c \in A$ and $x_1, x_2 \in M$, then applying Axiom $(IS4)$ in the case that $I = J = \{1,2\}$, $c_1 = c_2 = 1$, $a_{ij} = c \cdot \delta_{ij}$, $1 \leq i$, $j \leq 2$, we obtain equation (6) above. But, by equation (15) above,

$$\sum_{j \in J} a_{ij} x_j = c \cdot x_i, \ i = 1,2.$$

And, by the definition of "sum of two elements of M" in equation (1),

$$(17) : \sum_{i \in I} c_i (c \cdot x_i) = c \cdot x_1 + c \cdot x_2.$$

Also, in the ring A, we have that

$$(18) : \sum_{i \in I} c_i a_{ij} = c, \text{ for } i = 1,2.$$

Substituting equations (16), (17), and (18) into equation (6), we obtain that

$$(19) : \sum_{j \in \{1,2\}} c \cdot x_j = c \cdot x_1 + c \cdot x_2, \text{ for all } c \in A, \text{ all } x_1, x_2 \in M.$$

(In equation (19), the expression " $\sum_{j \in \{1,2\}} c \cdot x_j$ " of course means " $\sum_{i \in I} a_i x_i$ as in datum (S), where $I = \{1,2\}$, and $a_1 = a_2 = c$".)

Next, given $c \in A$ and $x_1, x_2 \in M$, apply Axiom $(IS4)$ in the case that $I = \{0\}$, $J = \{1,2\}$, $a_{01} = a_{02} = 1$, and $c_0 = c$. We obtain equation (6) above. But then, by definition of "sum of two elements of M," equation (1) above, we have that

$$(20) : \sum_{j \in J} a_{0j} \cdot x_j = x_1 + x_2.$$

And, therefore by definition of "scalar product in M," equation (2) above, we have that

$$(21): \sum_{i \in I} c_i \left(\sum_{j \in J} a_{ij} \cdot x_j \right) = c \cdot (x_1 + x_2).$$

On the other hand, since in the ring A

$$(22): \sum_{i \in I} c_i \cdot a_{ij} = c,, \text{ for } j = 1, 2,$$

substituting equations (21) and (22) into equation (6) gives

$$(23): \sum_{j \in \{1,2\}} c \cdot x_j = c \cdot (x_1 + x_2), \text{ for all } c \in A, \text{ for all } x_1, x_2 \in M.$$

Equations (19) and (23) imply that, for all $c \in A$ and all $x_1, x_2 \in M$, we have that

$$c \cdot (x_1 + x_2) = c \cdot x_1 + c \cdot x_2.$$

Equations (5), (7), (12), and (24) imply that, for the definition of "sum" and "scalar product" in equations (1) and (2) above, M does indeed become an abstract left A-module. That is, we have verified assertion (3) above. To complete proving the equivalence of Definitions 2 and 2', it therefore suffices to verify assertion (4) — i.e., to prove that Axioms $(IS2)$ and $(IS3)$ hold. But, by the definition of "sum of two elements" given in equation (1) above, we have that condition (*) of Remark 2 holds. Since by hypothesis Axiom $(IS4)$ holds, by Remark 2 it follows that $(IS2)$ holds. Also, by the definition of the "scalar product of an element of A by an element of M" given in equation (2) above, we have that condition $(IS3_{weak})$ of Remark 3 holds. Since by hypothesis $(IS1)$ and $(IS4)$ hold, from Remark 3 it follows that $(IS3)$ holds.   □

Remark 9. The definition, Definition 2', given in Remark 8 above, of a "C-complete left A-module," is perhaps the most elegant way to define a C-complete left A-module.

Given datum (S) as in Definition 2' of Remark 8, one *could* introduce the following intuitive notation. Given a set I, a zero-bound family $(a_i)_{i \in I}$ of elements of $\widehat{A}$, and a family $(x_i)_{i \in I}$ of elements of M, then interpret x as being a "column vector" of length I

$$\mathbf{x} \uparrow = [(x_i)_{i \in I}]$$

and $(a_i)_{i \in I}$ as being a "row vector" of length I

$$\overrightarrow{\mathbf{a}} = [(a_i)_{i \in I}].$$

Then write (1): $\overrightarrow{\mathbf{a}} \cdot \mathbf{x} \uparrow$ in lieu of $\sum_{i \in I} a_i x_i$. Then, also, if I and J are arbitrary sets, and if $(a_{ij})_{(i,j) \in I \times J}$ is any family of elements of $\widehat{A}$ indexed by the set $I \times J$, then introduce the matrix

$$\overrightarrow{\mathbf{a}} \uparrow = (a_{ij})_{(i,j) \in I \times J}.$$

Given such a matrix, with the property that the ith row $\vec{\mathbf{a}_i} = (a_{ij})_{j \in J}$ is zero-bound for all $i \in I$, if $\vec{\mathbf{x}} = (x_j)_{j \in J}$ is a row vector in M indexed by the set J, then introduce the notation

$$(2) : \vec{\mathbf{a}} \uparrow \cdot \vec{\mathbf{x}}$$

to stand for the "column vector" of elements of M indexed by the set I:

$$[(\vec{\mathbf{a}_i} \cdot \vec{\mathbf{x}})_{i \in I}]$$

— i.e., for the column vector

$$\left[\left(\sum_{j \in J} a_{ij} \cdot x_j \right)_{i \in I} \right].$$

Then, in this matrix notation, Axioms $(IS1)$ and $(IS4)$ become:

$(IS1)$. Let $\mathscr{I}_I$ be the identity matrix indexed by any set I, $\mathscr{I}_I = (\delta_{ij})_{(i,j) \in I \times I}$. Then for every column vector

$$\mathbf{x} \uparrow = (x_i)_{i \in I}$$

of elements of M indexed by I, we have

$$\mathscr{I}_I \cdot (\mathbf{x} \uparrow) = \mathbf{x} \uparrow.$$

$(IS4)$ Let I and J be sets, and let

$$\vec{\mathbf{a}} \uparrow = (a_{ij})_{(i,j) \in I \times J}$$

be an arbitrary matrix in the ring $\widehat{A}$, indexed by $I \times J$, such that for every $i \in I$, the ith row $\vec{\mathbf{a}_i} = (a_{ij})_{j \in J}$ is zero-bound in $\widehat{A}$. Let

$$\vec{\mathbf{c}} = (c_i)_{i \in I}$$

be any row vector of elements of $\widehat{A}$ indexed by I that is zero-bound, and let

$$\mathbf{x} \uparrow = (x_j)_{j \in J}$$

be any column vector indexed by J of elements of M. Then we have

$$(\vec{\mathbf{c}} \cdot \vec{\mathbf{a}} \uparrow) \cdot \mathbf{x} \uparrow = \vec{\mathbf{c}} \cdot (\vec{\mathbf{a}} \cdot \vec{\mathbf{x}}).$$

(In the above equation, $(\vec{\mathbf{c}} \cdot \vec{\mathbf{a}} \uparrow)$ on the left-hand side is the product of these matrixes of elements of the ring $\widehat{A}$, using the sum, product and topology of the complete topological ring $\widehat{A}$. The other three "dot products" are as defined in equations (1) and (2) of this remark, using only datum (S) of Definition 2′ of Remark 8.)

Perhaps the above motivates calling Axiom $(IS1)$ the *identity axiom*, and Axiom $(IS4)$ the *matrix axiom*.

Notice also, that in the above notations, Axioms $(IS2)$ and $(IS3)$ are:

$$(IS2): \; (\overrightarrow{\mathbf{a}} + \overrightarrow{\mathbf{b}}) \cdot \mathbf{x} \uparrow = \overrightarrow{\mathbf{a}} \cdot \mathbf{x} \uparrow + \overrightarrow{\mathbf{b}} \cdot \mathbf{x} \uparrow$$

whenever $\overrightarrow{\mathbf{a}}$ and $\overrightarrow{\mathbf{b}}$ are zero-bound row vectors in $\widehat{A}$ indexed by the same set I, and $\mathbf{x} \uparrow$ is any column vector in M, (also indexed by the same set I)

$$(IS3): (c \cdot \overrightarrow{\mathbf{a}}) \cdot \mathbf{x} \uparrow = c \cdot (\overrightarrow{\mathbf{a}} \cdot \mathbf{x} \uparrow),$$

whenever $c \in A$, $\overrightarrow{\mathbf{a}}$ is a zero-bound row vector in $\widehat{A}$ indexed by a set I, and $\mathbf{x} \uparrow$ is any column vector in M indexed by the same set I. (Here, $c \cdot \overrightarrow{\mathbf{a}}$ is the row-vector obtained by multiplying all entries of $\overrightarrow{\mathbf{a}}$ by the element $c \in A$.) Also, in these notations, condition $(IS2_{weak})$ of Remark 1 becomes

$$(IS2_{weak}): \; ([0,1] + [1,0]) \cdot \begin{bmatrix} x_1 \\ x_2 \end{bmatrix} = [1,1] \cdot \begin{bmatrix} x_1 \\ x_2 \end{bmatrix}, \text{ for all } x_1, x_2 \in M.$$

And $(IS3_{weak})$ of Remark 3 becomes

$$(IS3_{weak}): \; [\overrightarrow{\mathbf{c}}] \cdot [\mathbf{x}] \uparrow = [\mathbf{c} \cdot \mathbf{x}] \uparrow \text{ for all } c \in A, \text{ and } x \in M$$

where $[\overrightarrow{\mathbf{c}}]$, resp.: $[\mathbf{x}] \uparrow$, resp.: $[\mathbf{c} \cdot \mathbf{x}] \uparrow$ is the row, resp.: column, resp.: row vector of elements of $\widehat{A}$, resp.: M, resp.: M, indexed by $I = \{0\}$; $c \cdot x \in M$ is the scalar product for the abstract left A-module M, and $[\overrightarrow{\mathbf{c}}] \cdot [\mathbf{x}] \uparrow$ is the product of row and column vectors in (1) of this Remark.

Also, condition (*) of Remark 2 becomes

$$(*): \; [1,1] \cdot \begin{bmatrix} x_1 \\ x_2 \end{bmatrix} = x_1 + x_2, \text{ for all } x_1, x_2 \in M.$$

We have avoided introducing such notations, since the whole structure of C-complete left A-modules will be made explicit by a theorem in §4 below, making such notation superfluous; except perhaps for the isolated use in this section.

Remark 10. One might call Axioms $(IS2)$ and $(IS3)$ of Definition 1, respectively, the *additive axiom* and the *scalar axiom*. The two axioms together might be called the *linear axioms*. Then, by Remark 8 above, if we have all of the hypotheses of Definition 1 except possibly the linear axioms $(IS2)$ and $(IS3)$, then there is induced a unique structure of abstract left A-module on the set M (possibly different from the given abstract left A-module structure on M) such that with respect to this new induced structure (possibly different from the original given) abstract left A-module structure on M, all of the conditions of Definition 1 hold. The linear axioms, $(IS2)$ and $(IS3)$, are equivalent to asserting that: "The original given abstract left A-module structure on M, and this new, induced abstract left A-module structure, coincide."

Remark 11. In the proof of the equivalence of Definitions 2 and 2′ given in Remark 8, the longest part of the proof consisted in verifying that: If $a \in A$ and $x, y \in M$, then $a \cdot (x + y) = a \cdot x + a \cdot y$. Essentially the same, somewhat long, argument can be used to show that, if I is any set, $(a_i)_{i \in I}$ any indexed family of elements of $\widehat{A}$ that is zero-bound, and $(x_i)_{i \in I}$, $(y_i)_{i \in I}$ any indexed families of M, then equation $(IS5)$ of Proposition 2.2.1 above holds. That is, the conclusion, equation $(IS5)$ of Proposition 2.2.1, (and therefore also Corollary 1.1), can be proved directly from, e.g., Definition 1, by this method. One *could* also "struggle out" proofs of $(IS6)$ of Proposition 2.2.1, and of Corollary 2.2.3, by similar "matrix methods" at this point. (However, it is perhaps easier to delay giving these proofs until after establishing the structure theorem of §4 below after which all of these delayed proofs become trivial.)

Remark 12. Let A be any ring. Then we can regard A as being a topological ring by giving A the discrete topology. Then the topology of A is given by the single two-sided ideal, the zero ideal $\{0\}$. Of course, A is complete for this topology, $A = \widehat{A}$. Then by Remark 4 above, and Remark 5 above, in the case $K = \aleph_0$, we have that every abstract left A-module possesses a unique infinite sum structure in the sense of Definition 1. (Notice, in this case that, as has been observed in e.g. Remark 5 above, — and as is also obvious immediately, directly from the definitions — an indexed family $(a_i)_{i \in I}$ of elements of A, indexed by an arbitrary set I, is zero-bound in A for the discrete topology iff $a_i = 0$ for all but finitely many $i \in I$.) Therefore, if A is a discrete topological ring, then the notions of "abstract left A-module" and "C-complete left A-module" coincide.

Note. Of course, all of the observations of Remark 12 go through to any topological ring such that there exists a smallest neighborhood J of zero (such a neighborhood J is then easily seen, from the definition of "topological ring," to be necessarily a two-sided ideal, and also to be the closure of $\{0\}$) — the exceptions being, that the statement, "A is complete," must then be replaced by "$A/J = \widehat{A}$ is complete, where J is the closure of $\{0\}$. And $\widehat{A}$ is discrete." And similarly, the parenthetical observation in Remark 12 about zero-bound families should read: "An indexed family $(a_i)_{i \in I}$ of elements of A/J (one writes A/J rather than A), ..., is zero bound in A/J (again, A/J rather than A), ..., etc.

Remark 13. In Definition 1 (and similarly in Definition 2'), the datum (S) is defined such that, "For every indexed family $(a_i)_{i \in I}$ *of elements of* $\widehat{A}$, the *completion of* A, etc. ...." The reader might wonder why we have taken $(a_i)_{i \in I}$ such that $a_i \in \widehat{A}$, the completion of A, rather than merely $a_i \in A$ itself? Suppose that we only had required this weaker data. Then, if, e.g., the topology of A is Hausdorff and is given by denumerably many right ideals, then it is easy to show that, for every $a \in \widehat{A}$, there exist elements $a_i \in A$, $i \geq 1$, such that $(a_i)_{i \geq 1}$ converges to zero in A, and such that

$$a = \sum_{i \geq 1} a_i \in \widehat{A}.$$

Then, given any set J and elements $a_j \in \widehat{A}$ such that $(a_j)_{j \in J}$ is zero-bound; then for every $j \in J$, choose elements $a_{ij} \in A$, $j \geq 1$, such that

$$a_j = \sum_{i \geq 1} a_{ij} \in \widehat{A}.$$

Moreover, we can insist that the above choice be such that; if $I_1 \supset I_2 \supset \dots$ is any fixed open neighborhood base at zero for the topology of A, then, for every $k \geq 1$, if F_k is the (necessarily finite) subset of J consisting of all j such that $a_j \notin I_k$, then $a_{ij} \in I_k$ for all $j \in J - I_k$, and all $i \geq 1$. But then the indexed family $(a_{ij})_{(i,j) \in P \times J}$ of elements of A is zero-bound in A, where P denotes the set of positive integers. Now, given any elements $x_j \in M$, for all $j \in J$, define

$$\sum_{j \in J} a_j \cdot x_j = \sum_{(i,j) \in P \times J} a_{ij} \cdot x_j.$$

In this fashion, we see that an "infinite sum structure defined only for zero-bound families of A (rather than of $\widehat{A}$)," would be equivalent, after all, to an infinite sum structure as defined in Definition 1 above — at least in the case that A is Hausdorff and is such that the topology is given by denumerable many right ideals. (The Hausdorff assumption on A is not really needed. One would simply add the axiom: "$\sum_{i \in I} a_i \cdot x_i = 0$, whenever $a_i \in$ the closure of zero in A".)

However, this works only if the topology of A is given by *denumerably* many right ideals – which of course is not always the case.

Remark 14. Let A be any topological ring such that the topology is given by right ideals. Then Definition $2'$ above, of a C-complete left A-module, depends only on $\widehat{A}$. Therefore, clearly, the notions of "C-complete left A-module" and of "C-complete left $\widehat{A}$-module" coincide.

From this, it is of course immediate (via the equivalence of Definitions 2 and $2'$, for both of the topological rings A and $\widehat{A}$), that every C-complete left A-module M possesses a natural structure as an $\widehat{A}$-module. This can be characterized as being the unique structure of abstract left $\widehat{A}$-module, such that the datum (S) in Definition $2'$ obeys the equation $(IS3')$ of Corollary 2.2.3. (It follows, by Corollary 2.2.3, that the structure of an $\widehat{A}$-module thus defined must be identical to the structure of an abstract left $\widehat{A}$-module constructed in Corollary 2.2.2.) And, this observation gives an alternate, direct construction of the natural $\widehat{A}$-module structure of a C-complete left A-module M, that does not use the structure theorem of §4.

2.3 Infinitely Linear Functions

Definition 1. Let A be an arbitrary topological ring such that the topology is given by right ideals. Let M and N be C-complete left A-modules over A. Let $f : M \to N$ be a function of sets. Then we say that f is *infinitely linear* (or, more precisely, *infinitely A-linear*), if whenever I is any set, $(a_i)_{i \in I}$ is any family of elements indexed by the set I of $\widehat{A}$ that is zero-bound in $\widehat{A}$, and $(x_i)_{i \in I}$ is any family of elements indexed by the set I of M, we have that

$$(1): \quad f\left(\sum_{i \in I} a_i x_i\right) = \sum_{i \in I} a_i f(x_i).$$

Lemma 2.3.1. *If A is a topological ring such that the topology is given by right ideals, and if $\widehat{A}$ is the completion of the topological ring A, and if M and N are any two C-complete left A-modules, then M and N are also C-complete left $\widehat{A}$-modules. If $f : M \to N$ is any function of sets, then f is infinitely A-linear iff f is infintely $\widehat{A}$-linear.*

Proof. Follows immediately from Definition 1 above. □

We have observed, in Remark 14, at the end of §2, that if A is any topological ring such that the topology is given by right ideals, then C-complete left A-modules are the same as C-complete left $\widehat{A}$-modules. Combining with Lemma 2.3.1 above, we have that

Corollary 2.3.2. *If A is any topological ring such that the topology is given by right ideals, then the category of all C-complete left A-modules coincides with the category of C-complete left $\widehat{A}$-modules.*

Proposition 2.3.3. *Let $f : M \to N$ be an infinitely linear function of C-complete left A-modules. Then f is also a homomorphism of abstract left A-modules.*

Note. And in fact, f is even a homomorphism of abstract left $\widehat{A}$-modules, for the abstract left $\widehat{A}$-module structures on M and N described in §2, Corollary 2.2.2.

Proof. Let $I = \{1, 2\}$. Then by §2, Remark 5, we have that, whenever $a_1, a_2 \in A$ and $x_1, x_2 \in M$ (or N), then $\sum_{i \in I} a_i \cdot x_i$, using the infinite sum structure of M (or N), coincides with $a_1 x_1 + a_2 x_2$, for the abstract left A-module structure of M (or N). Therefore, equation (1) of Definition 1, in the special case of $I = \{1, 2\}$, implies that

$$f(a_1 x_1 + a_2 x_2) = a_1 f(x_1) + a_2 f(x_2), \text{ for all } a_1, a_2 \in A \text{ and all } x_1, x_2 \in M.$$

And this, of course, implies that f is a homomorphism of abstract left A-modules from M into N.

The proof of the Note is identical (with $\widehat{A}$ replacing A). □

Given C-complete left A-modules M and N, by a *homomorphism of C-complete left A-modules* we will mean a function, $f : M \to N$, that is infinitely linear in the sense of Definition 1.

Remark 1. Let A be a topological ring such that the topology is given by right ideals, let M and N be C-complete left A-modules, and let $f : M \to N$ be a homomorphism of abstract left A-modules. Then is f necessarily infinitely linear? Example 1 following Remark 2 below shows that this is *definitely not true* in general, even if f is $\widehat{A}$-linear, if the topological ring A is such that the completion $\widehat{A}$ is not admissible (see Chapter 1, §3, Definition 1).

A related question is: Let A be a topological ring such that the topology is given by right ideals, and let M be an abstract left A-module such that there exists an infinite sum structure on M. Then is the structure necessarily unique? Again, in Example 1 following Remark 2 below, this is shown to be false in general, if the completion $\widehat{A}$ of A is not admissible. (In fact, it is shown that there can even be more than one infinite sum structure on M that induces by means of §2, Corollary 2.2.2, the same abstract $\widehat{A}$-module structure on M.)

Also, in Example 2 below, it is shown that also: There exists a commutative topological ring A such that the topology is given by denumerably many ideals, and an abstract A-module N, such that we have two distinct structures of C-complete left A-module on N; and such that the two structures as $\widehat{A}$-module induced by these two different infinite sum structures on N, by virtue of §2, Corollary 2.2.2, are *distinct* abstract $\widehat{A}$-module structures on N (but necessarily inducing the same underlying A-module structure on N).

Remark 2. However, as we shall see in §5, the pathology of Remark 1 cannot occur for those A such that $\widehat{A}$ is admissible (see §5 below).

I should note that Examples 1 and 2 immediately following are a bit technical and topological. The reader who finds them confusing can skip these Examples, without losing too much comprehension; since the major positive theorems and constructions of this chapter do not make use of these Examples.

Example 1. Let A be a topological ring such that the topology is given by right ideals, such that

(0). There exists an abstract free left A-module F, together with a basis $(e_i)_{i \in I}$, such that there exists a family $(a_i)_{i \in I}$ indexed by the same set I of elements of $\widehat{A}$ that converges to zero in $\widehat{A}$, and such that the infinite series in the $\widehat{A}$-module $\widehat{F}$:

$$(1): \quad \sum_{i \in I} a_i e_i,$$

is not convergent in $\widehat{F}$ for the $\widehat{A}$-adic topology.

(E.g., the topological ring A_0 constructed in Chapter 1, §2, Example 2, and its A_0-adic completion $\widehat{A_0}$ both have these properties.) (Of course, by Chapter 1, Definition 1 and Chapter 1, §2, Corollary 1.2.5, such a topological ring A is not admissible.) As in §2, Example 2, let $(\widehat{\widehat{F}}, unif)$ denote the completion of $\widehat{F}$ (= $\widehat{F}$ with its $\widehat{A}$-adic topology as a topological left $\widehat{A}$-module), and let $\iota : \widehat{F} \to \widehat{\widehat{F}}$ be the natural map into the completion. Then $\widehat{F}$ has the induced topology from $(\widehat{\widehat{F}}, unif)$. Since the series

$$\sum_{i \in I} a_i e_i$$

is $\widehat{A}$-adically Cauchy in $\widehat{F}$, its image under ι

$$(2): \quad \sum_{i \in I} a_i \iota(e_i)$$

has a limit, call it x, in $(\widehat{\widehat{F}}, unif)$. Then

$$(3): x \notin \mathrm{Im}(\iota)$$

for, otherwise, say if $x = \iota(x_0)$, then since $\widehat{F}$ has the induced topology from $(\widehat{\widehat{F}}, unif)$, it would follow that the series (1) would converge $\widehat{A}$-adically to x_0 in $\widehat{F}$ (and we have constructed (1) so that it does not have a limit in $\widehat{F}$). If, now, we let $(\widehat{F}, unif)$ denote the complete topological left $\widehat{A}$-module that is the completion of F (= F with its A-adic topology) as a topological left $\widehat{A}$-module (so that $\widehat{F}$ and $(\widehat{F}, unif)$ have the same underlying abstract left $\widehat{A}$-module, but different topologies), then in $(\widehat{F}, unif)$, the series (1) of course is convergent (since it is $\widehat{A}$-adically Cauchy, and therefore also Cauchy for the uniform topology, which is coarser). Let y be the limit of (1) in $(\widehat{F}, unif)$.

Let $(\widehat{F}, C\text{-}complete)$, resp.: $(\widehat{\widehat{F}}, C\text{-}complete)$ denote the C-complete left A-module, deduced by virtue of $\widehat{F}$, respectively: $\widehat{\widehat{F}}$, being the A-adic completion of F, respectively: of $\widehat{F}$, using the fact, §2, Example 1, that the A-adic completion $\widehat{M}$ of any abstract left A-module M, has a natural structure as a C-complete left A-module. Then the homomorphism of left $\widehat{A}$-modules:

$$\iota : \widehat{F} \to \widehat{\widehat{F}}$$

cannot be infinitely linear.

Proof. Suppose ι were infinitely linear. Then, since

$$y = \sum_{i \in I} a_i e_i$$

in $(\widehat{F}, unif)$, if ι were infinitely linear we would deduce that

$$\iota(y) = \sum_{i \in I} a_i \cdot \iota(e_i) = x$$

in $(\widehat{\widehat{F}}, unif)$. (Recall §2, Example 1, that the infinite sum structure in $\widehat{M}$ is defined so that the infinite sums are simply the limits for the uniform topology of $\widehat{M}$; for all abstract left A-modules M, and in particular for $M = F$ or $\widehat{F}$.) But then, the element $x \in \widehat{\widehat{F}}$ would be in the image of ι, and we have shown, equation (3), that $x \notin \text{Im}(\iota)$. This contradiction shows that the homomorphism of abstract left $\widehat{A}$-modules

$$\iota : \widehat{F} \to \widehat{\widehat{F}}$$

is not infinitely linear, where $\widehat{F}$, respectively, $\widehat{\widehat{F}}$ is given the structure as C-complete left A-module by virtue of being the A-adic completion of F, respectively of $\widehat{F}$, as in §2, Example 1. $\square$

On the other hand, if $k : F \to \widehat{F}$ denotes the natural map into the completion, then

$$\widehat{k} : \widehat{F} \to \widehat{\widehat{F}}$$

is continuous from $(\widehat{F}, unif)$ into $(\widehat{\widehat{F}}, unif)$, and therefore $\widehat{k}$ is infinitely linear, considered as a function from $(\widehat{F}, C\text{-}complete)$ into $(\widehat{\widehat{F}}, C\text{-}complete)$. Let α be the automorphism of the abstract left $\widehat{A}$-module $\widehat{\widehat{F}}$ constructed in §2, Example 2. Then, by §2, Example 2, we have that $\alpha \circ \iota = \widehat{k}$. Therefore, the automorphism α of the left $\widehat{A}$-module $\widehat{\widehat{F}}$ *cannot be infinitely linear*, from $(\widehat{F}, C\text{-}complete)$ into itself. It follows that *there are at least two distinct* infinite sum structures on the left $\widehat{A}$-module $\widehat{F}$, both compatible with the A-module (and even both inducing the same $\widehat{A}$-module) structure on $\widehat{\widehat{F}}$ — namely, the one is the infinite sum structure that we have been working with (such that $\sum_{i \in I} a_i \cdot x_i$ is simply the limit in $(\widehat{\widehat{F}}, unif)$; and another, different, infinite sum structure, namely the image of that structure under α (i.e., such that

$$\sum_{i \in I} a_i \cdot x_i \text{ for the second infinite sum structure is } \alpha^{-1}\left(\sum_{i \in I} a_i \cdot \alpha(x_i)\right)$$

where the latter infinite sum is the limit in $(\widehat{\widehat{F}}, unif)$)).

Therefore, if $\widehat{A}$ obeys condition ⓪ above (and therefore is not admissible), then there exists an abstract free left A-module F such that there is more than one infinite sum structure on the abstract left A-module $\widehat{F}$ — all of which even induce, (by §2, Corollary 2.2.2), the same abstract left $\widehat{A}$-module structure on $\widehat{\widehat{F}}$.

Example 2. Let A be a topological ring such that the topology is given by right ideals, and such that, for the A-adic topology, $\widehat{A}$ is not complete. (Of course, $\widehat{A}$ is always complete with respect to the $\widehat{A}$-adic (= the uniform) topology. Therefore for such an A, the $\widehat{A}$-adic topology on $\widehat{A}$ is strictly coarser than the A-adic topology.) For example, the commutative ring A_0 constructed in Chapter 1, §2, Example 2, together with the S_0-adic topology, is such a topological ring. (This is proved in Chapter 1, §2, Remark 8.) Then, since $\widehat{A}$ is not A-adically complete, if $\widehat{\widehat{A}}$ denotes the A-adic (*not* the $\widehat{A}$-adic) completion of $\widehat{A}$, then the natural map $\iota : \widehat{A} \to \widehat{\widehat{A}}$ (from $\widehat{A}$ with its A-adic topology, into its completion as a topological A-module), is not an isomorphism. Also, if $k : A \to \widehat{A}$ denotes the natural map from A into its A-adic completion, then the image of k under the functor "A-adic completion" is a homomorphism of abstract left A-modules $\widehat{k} : \widehat{A} \to \widehat{\widehat{A}}$. Then, by Chapter 1, §2, Remark 7, applied to the abstract left A-module $M = \widehat{A}$, we have that $\iota \neq \widehat{k}$, although ι and k agree on A; and that there exists an automorphism α of $\widehat{\widehat{A}}$ as an abstract left A-module, such that $\alpha \circ \iota = \widehat{k}$. Then, as in Exercise 1, let us endow $\widehat{\widehat{A}}$ with the structure of C-complete left A-module, by giving $\widehat{\widehat{A}}$ the infinite sum structure by virtue of being the A-adic completion of the abstract left A-module $\widehat{A}$ (see

§2, Example 1, with $M = \widehat{A}$). Then, as in Example 1, the map

$$\widehat{k} : \widehat{A} \to \widehat{\widehat{A}}$$

is continuous for the uniform topologies, and therefore preserves infinite sums. Since $\widehat{A}$ is C-complete, by §2, Corollary 2.2.2, we have an induced structure of abstract left $\widehat{A}$-module on $\widehat{\widehat{A}}$. By the proof of Corollary 2.2.2, the $\widehat{A}$-module structure is: $a \in \widehat{A}$, $x \in \widehat{\widehat{A}}$, then $a \cdot x \in \widehat{\widehat{A}}$ is the element given by the infinite sum structure on $\widehat{\widehat{A}}$. But by §2, Example 1, this latter is the element of $\widehat{\widehat{A}} = \widehat{M}$ (where $M = \widehat{A}$) by virtue of $\widehat{M}$ being the A-adic completion of M. Since $\widehat{k}$ preserves infinite sums, in particular we have that, for $a \in \widehat{A}$,

$$(1): \ a \cdot 1 = \widehat{k}(a)$$

for the structure of $\widehat{\widehat{A}}$ as an $\widehat{A}$-module, deduced from the infinite sum structure of $\widehat{\widehat{A}}$. But, since α is an automorphism of $\widehat{\widehat{A}}$ as a left A-module, $\alpha \circ \iota = \widehat{k}$, it follows that if we "throw the infinite sum structure of $\widehat{\widehat{A}}$ through α" as in Example 1, then for the thus obtained, new infinite sum structure on $\widehat{\widehat{A}}$, we have that

$$(2): \ a \cdot 1 = \iota(a), \text{ for all } a \in \widehat{A}.$$

Since ι and $\widehat{k}$ agree on A, but do not agree on all of $\widehat{A}$, from equations (1) and (2) we deduce that the two infinite sum structures on the abstract left A-module $\widehat{\widehat{A}}$ thus defined, are such that *the induced structures as $\widehat{A}$ modules are distinct*. Thus, there exists an abstract left A-module N (namely, $N = \widehat{\widehat{A}}$), and two different structures of C-complete left A-module on N, such that the two thereby induced structures as abstract left $\widehat{A}$-modules on N are also different.

We resume our general discussion of infinitely linear functions.

Let A be a topological ring such that the topology is given by right ideals, and let M be a C-complete left A-module. Then a *C-complete left A-submodule of M* is a subset N of M, such that, whenever I is any set, $(a_i)_{i \in I}$ a zero-bound family of elements of $\widehat{A}$ and $(x_i)_{i \in I}$ a family of elements in N, then the infinite sum:

$$\sum_{i \in I} a_i x_i$$

in the C-complete left A-module M is an element of the subset N. Clearly, if N is a C-complete left A-submodule of the C-complete left A-module M, then the infinite sum structure of M induces, by restriction, an infinite sum structure on N, so that N becomes a C-complete left A-module.

If H is any subset of a C-complete left A-module M, then the intersection of all C-complete left A-submodules of M that contain H is the smallest C-complete left A-submodule of M that contains H. We call this *the C-complete left A-submodule of M generated by H*. The subset H of M is called a *set* *of generators* of M as a C-complete left A-module iff the C-complete left A-submodule of M generated by H is M.

Proposition 2.3.4. *Let M be a C-complete left A-module, let I be a set, and let $(x_i)_{i \in I}$ be an indexed family of elements of M. Then the C-complete left submodule of M generated by $\{x_i : i \in I\}$ is*

$$\{\sum_{i \in I} a_i x_i : (a_i)_{i \in I} \text{ is a zero-bound family of elements of } \widehat{A} \text{ indexed by the set } I\}.$$

Proof. From the definition of "C-complete left A-submodule," it is clear that the displayed set N is contained in any C-complete left A-submodule of M that contains $\{x_i : i \in I\}$. Therefore, it suffices to show that N is a C-complete left A-submodule of M. We give two proofs of this.   □

Proof 1. (Using Axiom $(IS4)$) In fact, let J be any set, let $(y_j)_{j \in J}$ be any indexed family of elements of N and let $(c_j)_{j \in J}$ be any zero-bound family of elements of $\widehat{A}$, indexed by the same set J. Then we must show that

$$\sum_{j \in J} c_j y_j \in N.$$

Since $y_j \in N$, by the definition of N, there exists a zero-bound family $(a_{ji})_{i \in I}$ of elements of $\widehat{A}$ indexed by the set I, such that

$$y_j = \sum_{i \in I} a_{ji} x_i.$$

But then, by Axiom $(IS4)$ we have that

$$\sum_{j \in J} c_j y_j = \sum_{i \in I} \left(\sum_{j \in J} c_j a_{ji} \right) x_i,$$

and therefore is an element of N.   □

Proof 2. This is given immediately after Remark 3 following Corollary 2.3.7 below.   □

Theorem 2.3.5. *Let A be a topological ring such that the topology is given by right ideals, and let M be an arbitrary C-complete left A-module. Then there exists an abstract free left A-module F, such that, if $\widehat{F}$ denotes the A-adic completion of F regarded as a C-complete left A-module (as defined in §2 Example 1, with $M = F$), then there exists an infinitely linear function $f : \widehat{F} \to M$ that is set-theoretically surjective.*

Proof. Let $(x_j)_{j \in J}$ be any indexed family of elements of M, such that $M = \{\sum_{j \in J} a_j x_j : (a_j)_{j \in J}$ is a zero-bound family of elements of $\widehat{A}$ indexed by the set $J\}^4$ (E.g., by §2, Remark 4, any set of generators of M as an abstract left

---

[4]Once one knows Proposition 2.3.3, an equivalent statement is: "Let $(x_j)_{j \in J}$ be a family of elements of M such that $\{x_j : j \in J\}$ is a set of generators of M as a C-complete left A-module."

A-module has this property). Then, define

$$f : \widehat{A^{(J)}} \to M$$

to be the function that to every $(a_j)_{j \in J} \in \widehat{A^{(J)}}$ — that is, to every zero-bound family $(a_j)_{j \in J}$ of elements of $\widehat{A}$ — associates the value

$$f((a_j)_{j \in J}) = \sum_{j \in J} a_j \cdot x_j.$$

Then by the hypothesis on the family $(x_j)_{j \in J}$, the function f is surjective. Regard $\widehat{A^{(J)}}$ as being a C-complete left *A*-module as in §2, Example 1 (with M = the free left *A*-module $A^{(J)}$). Then we must show that f is infinitely linear.

In fact, let I be any set, and let $(z_i)_{i \in I}$ be any indexed family of elements of $\widehat{A^{(I)}}$. Then let

$$\alpha : \widehat{A^{(I)}} \to \widehat{A^{(J)}}$$

be the unique homomorphism of left *A*-modules that is continuous for the uniform topologies on $\widehat{A^{(I)}}$ and $\widehat{A^{(J)}}$ (i.e., the topologies together with which $\widehat{A^{(I)}}$ and $\widehat{A^{(J)}}$ are respectively the completion of $A^{(I)}$ and $A^{(J)}$ as topological abelian groups (or equivalently as topological left *A*-modules), where $A^{(I)}$ and $A^{(J)}$ are given their *A*-adic topologies), such that the image $\iota(e_i)$ of the ith canonical basis vector in $A^{(I)}$ (where $\iota : A^{(I)} \to \widehat{A^{(I)}}$ is the natural map) under α is z_i:

$$\alpha(\iota(e_i)) = z_i, \text{ all } i \in I.$$

(More explicitly,

$$\alpha((a_i)_{i \in I}) = \sum_{i \in I} a_i \cdot z_i, \text{ all } (a_i)_{i \in I} \in \widehat{A^{(I)}}$$

where the infinite sum on the right side is the one using the infinite sum structure in the C-complete left *A*-module $\widehat{A^{(J)}}$.) Then, in the notations of §2, Remark 7, we have that

$$f = \sigma_{(x_j)_{j \in J}}.$$

Also, by Axiom *(IS4')* of Definition 1' of Remark 7, if $y_i = f(z_i)$ for all $i \in I$, then we have that the diagram:

$$
\begin{array}{ccc}
\widehat{A^{(I)}} & \xrightarrow{\ \alpha\ } & \widehat{A^{(J)}} \\
{\scriptstyle \sigma_{(y_i)_{i \in I}}}\Big\downarrow & \swarrow {\scriptstyle \sigma_{(x_j)_{j \in J}}} & \\
M & &
\end{array}
$$

is commutative. Where, in §2, Remark 7, $\sigma_{(y_i)_{i \in I}}$ is defined by

$$\sigma_{(y_i)_{i \in I}}((a_i)_{i \in I}) = \sum_{i \in I} a_i \cdot y_i$$

(the infinite sum on the right, using the given infinite sum structure in the C-complete left A-module M). But therefore, if $(a_i)_{i \in I}$ is any zero-bound family of elements of $\widehat{A}$, then commutativity of the above diagram, evaluated at the element $(a_i)_{i \in I} \in \widehat{A^{(I)}}$, tells us that:

$$f\left(\sum_{i \in I} a_i \cdot z_i\right) = f(\alpha((a_i)_{i \in I})) = (\sigma_{(x_i)_{i \in I}} \circ \alpha)((a_i)_{i \in I}) = \sigma_{(y_i)_{i \in I}}((a_i)_{i \in I})$$

$$= \sum_{i \in I} a_i \cdot y_i = \sum_{i \in I} a_i \cdot f(z_i).$$

The set I, the indexed family $(z_i)_{i \in I}$ indexed by the set I of elements of $\widehat{A^{(J)}}$, and the zero-bound indexed family $(a_i)_{i \in I}$ indexed by the set I of elements of $\widehat{A}$, being arbitrary, this last equation tells us that f is infinitely linear. $\square$

Proof of §2, Proposition 2.2.1: If M is any C-complete left A-module, then by Theorem 2.3.5 choose an abstract free left A-module F, and a surjective infinitely linear function

$$f : \widehat{F} \to M$$

where $\widehat{F}$ is given its structure as a C-complete left A-module as in §2, Example 1. Then, first note that the indicated identities $(IS5)$ and $(IS6)$ hold in the C-complete left A-module $\widehat{F}$ — this is because, if we regard $\widehat{F}$ as being a topological left A-module, $(\widehat{F}, unif)$, with the topology such that it is the completion of F (in its A-adic topology) as a topological group (or, equivalently, as a topological left A-module), then the infinite sum structure on $\widehat{F}$ is by definition the one such that

$$\sum_{i \in I} a_i \cdot x_i$$

= sum of this convergent infinite series in the complete topological left $\widehat{A}$-module $(\widehat{F}, unif)$. And the identities $(IS5)$ and $(IS6)$ in $\widehat{F}$ therefore follow from the corresponding identities, extended over finite subsets F of I, by passing to the limit as F runs through the directed set of all finite subsets of I.

But then, since the function f is infinitely linear and surjective, it follows that the identities $(IS5)$ and $(IS6)$ likewise hold in the C-complete left A-module M.

(E.g., to prove $(IS5)$ in M: Let $x_i', y_i' \in \widehat{F}$ be elements such that $f(x_i') = x_i$, $f(y_i') = y_i$. Then by $(IS5)$ in $\widehat{F}$, we have

$$\sum_{i \in I} a_i(x_i' + y_i') = \sum_{i \in I} a_i x_i' + \sum_{i \in I} a_i y_i'.$$

By Proposition 2.3.3, f also preserves sums of two elements for the A-module structures of $\widehat{F}$ and of M. Therefore, throwing this equation through f, we obtain that

$$\sum_{i \in I} a_i(x_i + y_i) = \sum_{i \in I} a_i x_i + \sum_{i \in I} a_i y_i,$$

proving $(IS5)$ in M. $(IS6)$ in M is proved similarly.) □

Having finally completed the proof of §2, Proposition 1, we have that the §2, Corollary 2.2.2 is true. Therefore, every C-complete left A-module has now been shown, by §2, Corollary 2.2.2, to have a natural structure as an abstract left $\widehat{A}$-module (for the specific $\widehat{A}$-module structure constructed in §2, Corollary 2.2.2). In particular, we now have that the Note to Proposition 2.3.3 of this section is true.

Proof of §2, Corollary 2.2.3: If M is any C-complete left A-module, then (as in the proof of §2, Proposition 2.2.1), let F be an abstract free left A-module and let $f : \widehat{F} \to M$ be a surjective, infinitely linear function. Then, by the Note to Proposition 2.3.3 of this section, f is also a homomorphism of abstract left $\widehat{A}$-modules (for the structures of abstract left $\widehat{A}$-modules of $\widehat{F}$ and M defined in §2, Corollary 2.2.2). But therefore, the identity $(IS3')$ in the C-complete left A-module M follows from the corresponding identity in $\widehat{F}$ — and this identity for the C-complete left A-module $\widehat{F}$ follows as in the proof of §2, Proposition 2.2.1. □

It is clear that the *composite* of infinitely linear functions of C-complete left A-modules is infinitely linear. Therefore, if A is any topological ring such that the topology is given by right ideals, then we have the *category* of all C-complete left A-modules, with infinitely linear functions for maps. We will denote this category by $\mathscr{C}_A$.

If $f, g : M \to N$ are infinitely linear functions of C-complete left A-modules, then define $f + g : M \to N$ to be the function such that

$$(f + g)(x) = f(x) + g(x) \text{ for all } x \in M.$$

Then $f + g$ is infinitely linear.

Proof. If $(a_i)_{i \in I}$ are zero-bound in $\widehat{A}$ and if x_i, $i \in I$ are elements of M, then

$$(f + g)\left(\sum_{i \in I} a_i x_i\right) = f\left(\sum_{i \in I} a_i x_i\right) + g\left(\sum_{i \in I} a_i x_i\right)$$

$$= \sum_{i \in I} a_i f(x_i) + \sum_{i \in I} a_i g(x_i) = \text{(by } (IS5) \text{ of §2, Proposition 2.2.1)}$$

$$\sum_{i \in I} a_i (f(x_i) + g(x_i)) = \sum_{i \in I} a_i ((f + g)(x_i)).$$

□

Therefore, for every topological ring A such that the topology is given by right ideals, it follows that the category $\mathscr{C}_A$ of all C-complete left A-modules and all infinitely linear functions is an *additive* category. (See, e.g., [COC], Introduction, Chapter 1 for the definition of an *additive category*.)

Theorem 2.3.6. *Let A be a topological ring such that the topology is given by right ideals, let M and N be C-complete left A-modules, and let*

$$f : M \to N$$

be an infinitely linear function. Then the set-theoretic kernel $\mathrm{Ker}(f)$ *of f is a C-complete left A-submodule of M, and the set-theoretic image* $\mathrm{Im}(f)$ *of f is a C-complete left A-submodule of N.*

Proof. Let I be a set, let $(a_i)_{i \in I}$ be a zero-bound family of elements of $\widehat{A}$ and let $(x_i)_{i \in I}$ be a family of elements of $\mathrm{Ker}(f)$. Then

$$f\left(\sum_{i \in I} a_i \cdot x_i\right) = \sum_{i \in I} a_i \cdot f(x_i) = \sum_{i \in I} a_i \cdot 0 = 0$$

(the last equality holds, since by §2, Proposition 2.2.1, equation $(IS5)$, we have that the function $(y_i)_{i \in I} \to \sum_{i \in I} a_i y_i$ from N^I into N is a homomorphism of abelian groups, and therefore maps the zero element $(0)_{i \in I}$ of N^I into zero). Therefore $\sum_{i \in I} a_i x_i \in \mathrm{Ker}(f)$. Thus $\mathrm{Ker}(f)$ is a C-complete left A-submodule of M.

On the other hand, we have that

$$\sum_{i \in I} a_i \cdot f(x_i) = f\left(\sum_{i \in I} a_i x_i\right);$$

so the set-theoretic image $\mathrm{Im}(f)$ of f is a C-complete left A-submodule of N. □

Corollary 2.3.7. *Let A be a topological ring such that the topology is given by right ideals, let M and N be C-complete left A-modules and let $f_i : M \to N$ be two infinitely linear functions, $i = 1, 2$. Let S be a set of generators for M as a C-complete left A-module. If $f_1|S = f_2|S$, then $f_1 = f_2$.*

Proof. Let $K = \mathrm{Ker}(f_1 - f_2)$. Then by Theorem 2.3.6, K is a C-complete left A-submodule of M. Since $f_1|S = f_2|S$, we have that $S \subset K$. Since S is a set of generators for M as a C-complete left A-module, by definition the only C-complete left A-submodule of M that contains S is the whole module M. Therefore, $K = M$, and $f_1 = f_2$. □

Remark 3. The analogue of Theorem 2.3.6, for "complete left A-modules," instead of "C-complete left A-modules," is probably not true in the fullest generality. Of course, the image of a map of complete left A-modules is always complete; but considering Chapter 1, §3, Theorem 1.3.7, it appears likely that, for certain topological rings A such that the topology is given by right ideals, but such that the completion $\widehat{A}$ is not strongly admissible, that there might be an example of a map of complete left A-modules such that the kernel is not A-adically complete. (In fact, by Chapter 1, §3, Theorem 1.3.3, this is the case for any topological ring A such that $\widehat{A}$ is not strongly admissible, but such that the topology of A is given by some (not necessarily denumerable) set of *right finitely generated* right ideals. It seems likely that such an example might exist).

Although, of course, by Chapter 1, §3, Theorem 1.3.7, if the topological ring A is such that $\widehat{A}$ is strongly admissible, then the analogue of Theorem 2.3.6 above holds for $\mathtt{Ker} f$, with "complete" replacing "C-complete." (But, of course, this condition does not always hold: e.g., see Chapter 1, §3, Remark 8, following Definition 2.) But the assumption that $\widehat{A}$ is strongly admissible is not really a severe restriction; all of the current applications to algebraic geometry and commutative algebra fall into this class — i.e., involve topological rings such that the completion is strongly admissible.

Completion of Proof 2 of Proposition 2.3.4: Given A, M, and $(x_i)_{i \in I}$ as in Proposition 2.3.4, define N as in the first sentence of the beginning of the proof of Proposition 2.3.4. Then we must show that N is a C-complete left A-submodule of N.

In fact, let

$$\sigma_{(x_i)_{i \in I}} : \widehat{A^{(I)}} \to M$$

be defined as in §2, Remark 7. Then the proof of Theorem 2.3.5 shows that $\sigma_{(x_i)_{i \in I}}$ is infinitely linear. Therefore, by Theorem 2.3.6, we have that the set-theoretic image $\mathtt{Im}(\sigma_{(x_i)_{i \in I}})$ of the infinitely linear function $\sigma_{(x_i)_{i \in I}}$ is a C-complete left A-submodule of M. But clearly $\mathtt{Im}(\sigma_{(x_i)_{i \in I}}) = N$. $\qquad\square$

Theorem 2.3.8. *Let M be a C-complete left A-module and let N be a C-compete left A-submodule. Then (by §2, Definition 2) M is also an abstract left A-module and by §2, Remark 4 N is also an abstract left A-submodule of M; so that we have the quotient abstract A-module M/N.*

Then there exists a unique infinite sum structure (§2, Definition 1) on the quotient abstract left A-module M/N, such that the natural mapping:

$$\pi : M \to M/N$$

is an infinitely linear function.

Proof. Let I be a set, let $x_i \in M/N$ for all $i \in I$, and let $(a_i)_{i \in I}$ be a zero-bound family of elements of $\widehat{A}$. Then, if there is an infinite sum structure on M/N such that π is infinitely linear, then if we choose elements $x_i' \in M$ such that $\pi(x_i') = x_i$, we would have, by infinite linearity of π, that

$$(1): \quad \sum_{i \in I} a_i \cdot x_i = \sum_{i \in I} a_i \cdot \pi(x_i') = \pi\left(\sum_{i \in I} a_i \cdot x_i'\right).$$

Therefore, there is clearly at most one infinite sum structure on M/N with the given property.

Next, given $(x_i)_{i \in I}$ and $(a_i)_{i \in I}$ as above, define

$$(2): \quad \sum_{i \in I} a_i x_i = \pi\left(\sum_{i \in I} a_i x_i'\right),$$

where $x_i' \in M$ are elements such that $\pi(x_i') = x_i$, for all $i \in I$. Then I claim that the definition given in equation (2) is well-defined.

In fact, if also $y_i' \in M$ are elements such that

$$\pi(y_i') = x_i, \text{ for all } i \in I,$$

then (since π is a homomorphism of A-modules)

$$\pi(y_i' - x_i') = \pi(y_i') - \pi(x_i') = x_i - x_i = 0 \text{ for all } i \in I$$

so that

$$y_i' - x_i' \in N.$$

But then, since N is a C-complete left A-submodule of M, it follows that

$$\sum_{i \in I} a_i \cdot (y_i' - x_i') \in N.$$

In the C-complete left A-module M, we have that

$$\sum_{i \in I} a_i(y_i' - x_i') = \sum_{i \in I} a_i y_i' - \sum_{i \in I} a_i x_i'$$

(this is because, by §2, Proposition 2.2.1, equation $(IS5)$, we have that the function: $(z_i)_{i \in I} \to \sum_{i \in I} a_i z_i$ from M^I into M is a homomorphism of additive abelian groups. And a homomorphism of additive abelian groups preserves the difference of two elements). Therefore,

$$\sum_{i \in I} a_i y_i' = \sum_{i \in I} a_i x_i' + n$$

where $n \in N$. Throwing through the homomorphism of abstract left A-modules π, since $\pi(n) = 0$, this gives

$$\pi\left(\sum_{i \in I} a_i y_i'\right) = \pi\left(\sum_{i \in I} a_i x_i'\right).$$

Hence the definition in (2) is indeed well-defined. Therefore we have a datum as in (S) of §2, Definition 1, for the abstract left A-module M/N. Using equation (2), and the fact that Axioms $(IS1)$ - $(IS4)$ hold in M, it follows that Axioms $(IS1)$ - $(IS4)$ hold in M/N for the thus defined datum (S) given by equation (2). Therefore, the datum (S) defined by equation (2) endows the abstract left A-module M/N with an infinite sum structure as defined in §2, Definition 1. Finally, equation (2) implies that the function π is infinitely linear. $\square$

Given A, M, and N as in Theorem 2.3.8, the C-complete left A-module M/N thus defined will be called the *quotient C-complete left A-module of the C-complete left A-module M by the C-complete left A-submodule N*.

Remark 4. Theorem 2.3.8 above can be thought of as being a basic property enjoyed by the category of C-complete left A-modules, that is *almost never* enjoyed by the category of complete (in the ususal sense) left A-modules.

Theorem 2.3.8 can also easily be used to construct examples of C-complete left A-modules that are *not* Hausdorff for the A-adic topology:

Example 3. Let $\mathscr{O}$ be a complete discrete valuation ring not a field, and let t be a generator for the maximal ideal of $\mathscr{O}$. Then let M be the completion of the free $\mathscr{O}$-module with basis e_i, $i \geq 0$; and let N be the C-complete left submodule of M generated by the elements: $te_1 - e_0$, $t^2 e_2 - e_0$, ..., $t^i e_i - e_0$, ... Then I claim that $e_0 \notin N$.

In fact, if $e_0 \in N$, then by Proposition 2.3.4 there exist elements $a_i \in \mathscr{O}$, $i \geq 0$, such that $(a_i)_{i \geq 0}$ converges to 0 in $\mathscr{O}$, and such that

$$(1): \quad e_0 = \sum_{i \in I} a_i (t^i e_i - e_0)$$

in M. Let M' be the completion of the free $\mathscr{O}$-module with basis e_i, $i \geq 1$, and let $\rho : M \to M'$ be the $\mathscr{O}$-homomorphism that sends $e_i \in M$ into $e_i \in M'$, for $i \geq 1$, and that sends e_0 into zero. Then ρ (being a continuous homomorphism of completions of free $\mathscr{O}$-modules) is infinitely linear. If equation (1) holds, then throwing through ρ we would have that

$$0 = \sum_{i \in I} a_i t^i e_i$$

in M'. Since M' is the completion of the free $\mathscr{O}$ module with basis e_i, $i \geq 1$, this would imply that $a_i t^i = 0$ in $\mathscr{O}$, $i \geq 1$, i.e., that $a_i = 0$ for $i \geq 1$. Substituting into equation (1), we would have that $e_0 = 0$ in M, an absurdity. Thus,

$$(2): \quad e_0 \notin N,$$

as asserted. Consider the natural map

$$\pi : M \to M/N$$

from M into the quotient C-complete $\mathscr{O}$-module M/N, as defined in Theorem 2.3.8. By equation (2), the image $\pi(e_0)$ of e_0 in M/N is not zero. However, since

$$a_i t^i e_i - e_0 \in N$$

we have that

$$t^i \pi(a_i e_i) = \pi(e_0) \text{ in } M/N.$$

Therefore the element $\pi(e_0)$ of M/N is a non-zero t-divisible element of M/N, i.e., is an element of the $\mathscr{O}$-adic closure of $\{0\}$ in M/N. Therefore the C-complete $\mathscr{O}$-module M/N is not $\mathscr{O}$-adically Hausdorff.

Remark 5. Notice that, the $\mathscr{O}$-module $Q = M/N$ constructed in Example 3, has a natural infinite sum structure; so that we have a natural way of taking infinite sums of the form $\sum_{i \in I} t^i a_i$, for any sequence $a_0, a_1, a_2, \ldots, a_i, \ldots$ of elements of Q; yet these infinite sums *cannot* be completely described using the topology of Q; since that topology is not Hausdorff. (We will see in a later section that, in fact, the $\mathscr{O}$-module Q admits *exactly one* infinite sum structure [5], so that this notion of "infinite sums" in Q is very natural, and in fact is inherent in the $\mathscr{O}$-module structure of Q.)

Remark 6. Let A be a topological ring such that the topology is given by right ideals and let M be a C-complete left A-module. If I is a set, if $(a_i)_{i \in I}$ is a family of elements of A that converges to zero in A and if $(y_i)_{i \in I}$ is any family of elements of M, then by the infinite sum structure on M, we have the infinite linear combination:

$$(1) : \sum_{i \in I} a_i y_i$$

a well-defined element of M. If we let $x_i = a_i y_i$, then $(x_i)_{i \in I}$ is a family of elements of M that converges to zero for the A-adic topology. In general, can one make sense of the infinite sum

$$(2) : \sum_{i \in I} x_i$$

in M — i.e., does the value of the infinite sum (1) depend only on the elements $x_i = a_i y_i \in M$, or does it depend in general on the specific formulation

$$(3) : x_i = a_i \cdot y_i, \quad i \in I,$$

where $(a_i)_{i \in I}$ is zero-bound in A?

If the former, one could make a well-defined definition of infinite sums (2): $\sum_{i \in I} x_i$ in M, whenever $x_i \in M$ are elements such that the sequence $(x_i)_{i \in I}$ converges to 0 for the A-adic topology of M. We will show that in general the expression (1) does depend on the factorizations (3). (Therefore, the term "infinite sum structure" in defining C-complete modules is technically a misnomer. We are really dealing with an "infinite linear combination" structure.) In fact, we even show a counterexample in the simplest possible case (when A is a complete discrete valuation ring not a field).

Notice, however, that if M is A-adically Hausdorff and if the topological ring A is admissible, then the above definition for the infinite sum (2) is, clearly, well-defined. (Namely, (see the proof of the note to Corollary 2.4.2 of §4 below), the infinite linear combination (1) must coincide with the limit in A of this infinite series for the $\widehat{A}$-adic topology. And when M is A-adically Hausdorff, such topological limits are well-defined.) But if M is not A-adically Hausdorff, this definition of infinite sums of the form (2) is not in general well-defined, not even if A is a complete discrete valuation ring, $\mathscr{O}$, that is not a field, see the example immediately following.

---

[5]In fact, we will see later that, given any $\mathscr{O}$-module, it admits *at most one* infinite sum structure in the sense of §2, Definition 1. And in fact that this observation generalizes, from the topological ring $\mathscr{O}$, to all topological rings such that the completion is admissible and has a denumerable neighborhood base at zero.

Example 4. Let $\mathcal{O}$ be a complete discrete valuation ring that is not a field, and let t be a generator for the maximal ideal of $\mathcal{O}$. Let $\widehat{F}$ be the completion of a free $\mathcal{O}$-module F of denumerable rank with basis e_i, $i \geq 0$. Let N be the C-complete submodule of the C-complete module $\widehat{F}$ generated by $\{t^i e_i : i \geq 0\}$. Let M be the quotient module $M = \widehat{F}/N$, a C-complete $\mathcal{O}$-module. Explicitly, $\widehat{F}$ is the set of all expressions

$$(4): \ \{\sum_{i \geq 0} b_i e_i : b_i \in \mathcal{O}, i \geq 0, \text{ and such that } (b_i)_{i \geq 0} \text{ converges to 0 in } \mathcal{O}\}$$

and N is the set of all expressions

$$(5): \ \Big\{\sum_{i \geq 0} t^i b_i e_i : b_i \in \mathcal{O}, i \geq 0, \text{ and such that } (b_i)_{i \geq 0} \text{ converges to zero in } \mathcal{O}\Big\}.$$

Equivalently, N is the set of all expressions

$$(5'): \ \Big\{\sum_{i \geq 0} b_i e_i : b_i \in \mathcal{O}, i \geq 0, \text{ such that } t^i | b_i \text{ in } \mathcal{O}, \text{ and such that } (b_i/t^i)_{i \geq 0}$$

$$\text{converges to zero in } \mathcal{O}\Big\}.$$

(Hence N is the completion of the free $\mathcal{O}$-module with basis $t^i e_i$, $i \geq 0$.)

Then in the C-complete $\mathcal{O}$-module M, consider the elements

$$(6): \ y_i = e_i + N, i \geq 0.$$

We have that

$$(7): \ t^i y_i = 0 \text{ in} M, i \geq 0.$$

Yet the infinite sum, for the C-complete module structure of M,

$$(8): \ \sum_{i \geq 0} t^i y_i$$

is the coset of the element $\sum_{i \geq 0} t^i e_i$ of $\widehat{F}$. And by equation $(5')$ this latter is not an element of the submodule N (since the sequence $(1)_{i \geq 0}$ does not converge to zero in $\mathcal{O}$). Therefore, even though $t^i y_i = 0$, $i \geq 0$, in M, the infinite sum $\sum_{i \geq 0} t^i y_i \neq 0$ in M (for the C-complete module sturcture of M). Therefore infinite sums of the sort (2) of the last Remark (Remark 6) are not well-defined in the C-complete left $\mathcal{O}$-module M (not even if $x_i = 0_M$, all $i \in I$. Since we then have the two factorizations:

$$0_M = 0_A \cdot 0_M, i \geq 0, \text{ and}$$

$$0_M = t^i \cdot y_i, i \geq 0$$

$$\text{yet } \sum_{i \in I} 0_A \cdot 0_M = 0_M \text{ in } M$$

$$\text{and } \sum_{i \in I} t^i \cdot y_i \neq 0_M \text{ in } M,$$

for the infinite sum structure on M).

Remark 7. It follows also that, if M is a C-complete left A-module that is not Hausdorff and if $(x_i)_{i \geq 0}$ is a sequence of elements of M that is a Cauchy sequence for the A-adic topology of M, then there is in general no "tricky" way to define some sort of "limit" of $(x_i)_{i \geq 0}$ using the infinite sum structure. Since the construction of such a well-defined "limit" for such a sequence would be equivalent to posing a well-defined definition for the infinite sum

$$x_1 + (x_2 - x_1) + (x_3 - x_2) + (x_4 - x_3) + \ldots,$$

and by the preceding remark (Remark 6) and Example 4, this is definitely in general impossible when M is not also A-adically Hausdorff (not even if A is a c.d.v.r that is not a field).

Theorem 2.3.9. *Let A be a topological ring such that the topology is given by right ideals. Let I be a set and let M_i, $i \in I$, be C-complete left A-modules. Then there exists a unique infinite sum structure on the Cartesian product module*

$$\prod_{i \in I} M_i,$$

such that the natural projections

$$\pi_k : \prod_{i \in I} M_i \to M_k$$

are infinitely linear, for all $k \in I$.

The abstract left A-module $\prod_{i \in I} M_i$, together with this infinite sum structure, is then a direct product of the family of objects $(M_i)_{i \in I}$ in the category $\mathscr{C}_A$.

Proof. Let J be any set, let $(a_j)_{j \in J}$ be a zero-bound family of elements of $\widehat{A}$, and let $(x_{ij})_{i \in I} \in \prod_{i \in I} M_i$ be elements, for all $j \in J$. If we have an infinite sum structure on $\prod_{i \in I} M_i$ such that the kth projection function $\pi_k : \prod_{i \in I} M_i \to M_k$ is infinitely linear for all $k \in I$, we must have that

$$(1) : \sum_{j \in J} a_j \cdot (x_{ij})_{i \in I} = \left(\sum_{j \in J} a_j \cdot x_{ij} \right)_{i \in I} .$$

Therefore there is at most one infinite sum structure on the abstract left A-module $\prod_{i \in I} M_i$ such that the projection functions π_k are all infinitely linear for each $k \in I$. Conversely, given J, $(a_j)_{j \in J}$ and $(x_{ij})_{i \in I}$ for each $j \in J$, as above, define $\sum_{j \in J} a_j \cdot (x_{ij})_{i \in I}$ by equation (1) above. Then equation (1) (for all such J, $(a_j)_{j \in J}$, $(x_{ij})_{i \in I}$, $j \in J$) defines a datum (S) as in §2, Definition 1, for the abstract left A-module $\prod_{i \in I} M_i$. And, since Axioms $(IS1)$ - $(IS4)$ hold in each of the M_i for all $i \in I$, from equation (1), arguing coordinatewise, we see that Axioms $(IS1)$ - $(IS4)$ hold in $\prod_{i \in I} M_i$.

(E.g., let us verify $(IS1)$ in $\prod_{i \in I} M_i$: If $a_j = \delta_{jj_0}$ for some fixed $j_0 \in J$, and all $j \in J$, then

$$\sum_{j \in J} \delta_{jj_0} \cdot (x_{ij})_{i \in I} = \left(\sum_{j \in J} \delta_{jj_0} \cdot x_{ij} \right)_{i \in I} = (x_{ij_0})_{i \in I}$$

$(IS2)$ - $(IS4)$ are proved similarly.)

Therefore the datum (S) described above does indeed endow the abstract left A-module $\prod_{i \in I} M_i$ with an infinite sum structure. Then, by equation (1), we have that the kth projection π_k is infinitely linear, for all $k \in I$. This proves the first part of the theorem.

Finally, let D be any C-complete left A-module and let $\rho_i : D \to M_i$ be any infinitely linear functions, for all $i \in I$. Then (since $\prod_{i \in I} M_i$ as a set is the direct product of the sets M_i, $i \in I$, in the category of sets) there exists a unique function

$$\rho : D \to \prod_{i \in I} M_i$$

such that $\pi_k \circ \rho = \rho_k$ for all $k \in I$. Explicitly, ρ is the function

$$(2): \quad \rho(d) = (\rho_i(d))_{i \in I}, \text{ for all } d \in D.$$

We must show that ρ is infinitely linear. In fact, if J is any set, $(a_j)_{j \in J}$ is any zero-bound family of elements of $\widehat{A}$ and if $(d_j)_{j \in J}$ is any indexed family of elements of D, then by equations (1) and (2) and the fact that ρ_i is infinitely linear for all $i \in I$, we have

$$\rho \left(\sum_{j \in J} a_j \cdot d_j \right) = \left(\rho_i \left(\sum_{j \in J} a_j \cdot d_j \right) \right)_{i \in I} = \left(\sum_{j \in J} a_j \cdot \rho_i(d_j) \right)_{i \in I}$$
$$= \sum_{j \in J} a_j \cdot (\rho_i(d_j))_{i \in I} = \sum_{j \in J} a_j \cdot \rho(d_j)$$

so that ρ is infinitely linear. $\qquad \square$

Corollary 2.3.10. *Let A be a commutative topological ring such that the topology is given by ideals, and let M and N be C-complete left A-modules. Then the set $\mathrm{Hom}_{\mathscr{C}_A}(M, N)$ of all infinitely linear functions from M into N has a natural induced structure as a C-complete left A-module.*

Proof. The elements of $\mathrm{Hom}_{\mathscr{C}_A}(M, N)$ are the infinitely linear functions from M to N, and in particular are functions from M into N. Therefore

$$\mathrm{Hom}_{\mathscr{C}_A}(M, N) \subset N^M.$$

By Theorem 2.3.9, N^M has a natural structure as a C-complete left A-module (this structure depends only on the structure of N). I claim that the subset $\mathrm{Hom}_{\mathscr{C}_A}(M, N)$ is a C-complete A-submodule of N^M.

To prove this claim, it is necessary and sufficient to show that if I is any set, if $(a_i)_{i \in I}$ is any zero-bound family of elements of $\widehat{A}$, and if f_i, $i \in I$, are infinitely linear functions from M into N, then the element

$$\sum_{i \in I} a_i f_i \in N^M$$

is infinitely linear.

In fact, let J be any set, let $(b_j)_{j \in J}$ be any zero-bound family indexed by J elements of $\widehat{A}$, and let $(x_j)_{j \in J}$ be any family of elements indexed by J of M. Then

$$\left(\sum_{i \in I} a_i \cdot f_i \right) \left(\sum_{j \in J} b_j \cdot x_j \right) = \sum_{i \in I} a_i \cdot \left(f_i \left(\sum_{j \in J} b_j \cdot x_j \right) \right)$$

$$\sum_{i \in I} a_i \left(\sum_{j \in J} b_j \cdot f_i(x_j) \right) = \sum_{j \in J} b_j \left(\sum_{i \in I} a_i \cdot f_i(x_j) \right)$$

$$= \sum_{j \in J} b_j \left(\left(\sum_{i \in I} a_i \cdot f_i \right) (x_j) \right) \quad \text{as required.}$$

Note. The first, resp.: last, equalities in the above displayed formula follow, since the kth projection: $N^M \to N$ is infinitely linear for all $k \in M$; and in particular this is true for $k = \sum_{j \in J} b_j \cdot x_j$, resp.: $k = x_j$. The second equality follows since f_i is infinitely linear for all $i \in I$. The third equality follows from the identity:

$$\sum_{i \in I} a_i \left(\sum_{j \in J} b_j y_{ij} \right) = \sum_{j \in J} b_j \left(\sum_{i \in I} a_i y_{ij} \right),$$

in the C-complete left A-module N, by Lemma 2.3.11 immediately below.

Therefore, the subset $\mathrm{Hom}_{\mathscr{C}_A}(M, N)$ of the C-complete A-module N^M is a C-complete A-submodule; and therefore we have the induced infinite sum structure on $\mathrm{Hom}_{\mathscr{C}_A}(M, N)$ from N^M. $\square$

Lemma 2.3.11. *Let A be a topological ring such that the topology is given by right ideals, let I and J be sets, and let $(a_i)_{i \in I}$ and $(b_j)_{j \in J}$ be zero-bound families of elements of $\widehat{A}$. Suppose also that $a_i \cdot b_j = b_j \cdot a_i$, for all $i \in I$, and all $j \in J$.*

Then for every C-complete left A-module N and every family $(y_{ij})_{(i,j)\in I\times J}$ of elements of N indexed by the product set $I \times J$, we have that

$$(1): \sum_{i\in I}a_i\left(\sum_{j\in J}b_jy_{ij}\right) = \sum_{j\in J}b_j\left(\sum_{i\in I}a_iy_{ij}\right).$$

Proof. Let $\widehat{F}$ be the A-adic completion of a free abstract left A-module, regarded as a C-complete A-module as in §2, Example 1, and let $f : \widehat{F} \to N$ be an infinitely linear function that is surjective. By Theorem 2.3.5, such a pair $\widehat{F}, f$ exists. Then, as in the proof of §2, Proposition 2.2.1; given immediately after Theorem 2.3.5; to prove the above identity for N, it suffices to prove it for $\widehat{F}$. But, the infinite sum structure in $\widehat{F}$ is given by taking limits for the uniform topology of $\widehat{F}$ (that is, for the topology such that $\widehat{F}$ with this topology is the completion of F as a topological abelian group — or, equivalently, as a topological left A-module). Since we have that $a_i \cdot b_j = b_j \cdot a_i$ for all $i \in I, j \in J$,

$$\sum_{i\in G}a_i\left(\sum_{j\in H}b_jy_{ij}\right) = \sum_{j\in H}b_j\left(\sum_{i\in G}a_iy_{ij}\right)$$

in $\widehat{F}$, whenever G, respectively: H, is a finite subset of I, respectively J. Then, passing to the limit in the complete topological left A-module $(\widehat{F}, unif)$, over the directed set consisting of all pairs (G, H), where G is a finite subset of I and H is a finite subset of J, the ordering being: $(G, H) \leq (G', H')$ iff $G \subset G'$ and $H \subset H'$; we obtain the identity (1) in $\widehat{F}$. □

Remark 8. Let A be a commutative topological ring such that the topology is given by ideals. If M, N, and L are C-complete A-modules, then a function

$$f : M \times N \to L$$

will be called *infinitely bilinear* iff, for every $m \in M$, the function $n \to f(m, n)$ from N into L is infinitely linear; and also for every $n \in N$, the function $m \to f(m, n)$ from M into L is infinitely linear. Then, it is not difficult to show that there exists a C-complete A-module, call it $M \overset{C}{\underset{A}{\otimes}} N$, together with an infinitely bilinear function from $M \times N$ into $M \overset{C}{\underset{A}{\otimes}} N$ — denote this function $(m, n) \to m \overset{C}{\underset{A}{\otimes}} n$ — that is universal with these properties (i.e., such that given any infinitely bilinear function $f : M \times N \to L$ into any C-complete A-module L, there exists a unique infinitely linear function $g : M \overset{C}{\underset{A}{\otimes}} N \to L$ such that $f(m, n) = g(m \overset{C}{\underset{A}{\otimes}} n)$, for all $m \in M, n \in N$).

In fact, an explicit construction of such a universal $M \overset{C}{\underset{A}{\otimes}} N$ is: Take the A-adic completion $\widehat{F}$ of the free A-module F on the set $M \times N$, and let $M \overset{C}{\underset{A}{\otimes}} N$

be the quotient module of $\widehat{F}$ by the C-complete A-submodule generated by
$\{(m, \sum_{i \in I} a_i \cdot n_i) - \sum_{i \in I} a_i \cdot (m, n_i) : m \in M, I$ is a set, $n_i \in N$, and $a_i \in \widehat{A}$, all $i \in I$,
and $(a_i)_{i \in I}$ is zero-bound in $\widehat{A}\} \cup \{(\sum_{i \in I} a_i \cdot m_i, n) - \sum_{i \in I} a_i(m_i, n) : n \in N, I$ is a
set, $m_i \in M$ and $a_i \in \widehat{A}$, all $i \in I$, and $(a_i)_{i \in I}$ is zero-bound in $\widehat{A}\}$. Then by
Theorem 2.3.8 $M \overset{C}{\underset{A}{\otimes}} N$ is C-complete; and one immediately verifies the indicated
universal mapping property (actually, to verify this latter, one needs to know
Corollary 2.4.10 of §4).

$M \overset{C}{\underset{A}{\otimes}} N$ will be called the *C-complete tensor product* of M and N. By Corollary 2.3.10, the functor $\mathrm{Hom}_{\mathscr{C}_A}$ can be regarded as being a functor, contravariant
in the first variable and covariant in the second, from the category $\mathscr{C}_A$ into itself. Then it is easy to establish that the (covariant functor of two variables,
$(M, N) \rightsquigarrow M \overset{C}{\underset{A}{\otimes}} N$, from $\mathscr{C}_A$ into itself), is the right adjoint of the functor $\mathrm{Hom}_{\mathscr{C}_A}$,
that is, we have the canonical isomorphism of functors in the three variables M,
N, L:

$$\mathrm{Hom}_{\mathscr{C}_A}(M \overset{C}{\underset{A}{\otimes}} N, L) \approx \mathrm{Hom}_{\mathscr{C}_A}(M, \mathrm{Hom}_{\mathscr{C}_A}(N, L)),$$

contravariant in M and N and covariant in L, from $\mathscr{C}_A^o \times \mathscr{C}_A^o \times \mathscr{C}_A$ into the
category of sets (in fact, even into the category $\mathscr{C}_A$).

We will give an alternative construction of the C-complete tensor product
later by a different method (which will be particularly useful in the case in which
the topological ring A is admissible, and also is such that the neighborhood
system at zero has a denumerable base), at which time we will study the C-complete tensor product a bit more.

Remark 9. As in the case of abstract modules, Corollary 2.3.10 has generalizations to *C-complete bimodules*. Namely, let A and B be topological rings,
such that the topology of A is given by right ideals, and such that the topology
of B is given by left ideals. Then a *C-complete (A, B)-bimodule* is an abstract
(A, B)-bimodule M, together with an A-infinite sum structure such that M
becomes a C-complete left A-module; and also a B^o-infinite sum structure, so
that M becomes a C-complete right B-module; and such that, whenever I and
J are sets, $(a_i)_{i \in I}$ is a zero-bound family in $\widehat{A}$, $(b_j)_{j \in J}$ is a zero-bound family
in $\widehat{B}$, and $(x_{ij})_{I \times J}$ is an arbitrary family, indexed by the product set $I \times J$, in
M, then we have that

$$\sum_{j \in J} \left(\sum_{i \in I} a_i x_{ij} \right) b_j = \sum_{i \in I} a_i \left(\sum_{j \in J} x_{ij} b_j \right).$$

We will let $\mathscr{C}_{A,B}$ stand for the category of all such objects (with maps the
functions that are both A-left infinitely linear, and B-right infinitely linear).

A weaker notion is as follows: If say A is as above, and R is any abstract
ring, then by an *abstract (A, R)-bimodule that is C-complete as a left A-module*,

we will mean simply an abstract (A, R)-bimodule N, together with an A-infinite sum structure on the left A-module N, in the sense of §2, Definition 1. We will let $(\mathscr{C}_A, R)$ stand for the category of all such objects (with maps, the homomorphisms of abstract (A, R)-bimodules that are A-infinitely linear.) Then, the proof of Corollary 2.3.10 can be generalized as follows: Let A, B, and R and M be as above. I.e., let A, resp.: B, be a topological ring such that the topology is given by right, resp.: left, ideals, let R be an abstract ring, let M be a C-complete (A, B)-bimodule and let N be an abstract (A, R)-bimodule that is C-complete as a left A-module. Then the set

$$\mathrm{Hom}_{\mathscr{C}_A}(N, M)$$

of all infinitely A-linear functions from M into N, has a natural structure as an abstract (R, B)-bimodule that is C-complete as a right B-module. (I.e., it is an object in $(R, \mathscr{C}_B)$.)

Similarly, an analogue of Remark 8 above can be stated: Let A be an abstract ring, let B be a topological ring, and let R be a topological ring such that the topology of B, resp: R, is given by right, resp: left, ideals. Let M, resp.: N be an abstract (B, A)-, resp.: (A, R)-bimodule, that is C-complete as a left B-, resp.: as a right R-module. Then if T is any C-complete (B, R)-bimodule, then let us call a function:

$$t : M \times N \to L$$

(B, R)-*infinitely bilinear*, if

(1) For all $m \in M$, the function $n \to t(m, n)$ from N into L is R-infinitely linear.

(2) For all $n \in N$, the function $m \to t(m, n)$ from M into L is B-infinitely linear. The function t will also be called A-*balanced* if also

(3) For all $m \in M$, $n \in N$, and $a \in A$, we have that

$$t(ma, n) = t(m, an).$$

(In the presence of (1) and (2), condition (3) is of course equivalent to the function: $M \times N \to L$ being A-balanced in the usual sense, ignoring all C-complete structures — i.e., to the function t factoring through the natural map: $M \times N \to M \underset{A}{\otimes} N$, where $M \underset{A}{\otimes} N$ is the usual tensor product.) Then, there exists a C-complete (B, R) bimodule, denote it $\mathscr{C}_B M \underset{A}{\otimes} N \mathscr{C}_R$, together with a (B, R)-infinitely bilinear map that is A-balanced:

$$M \times N \to \mathscr{C}_B M \underset{A}{\otimes} N \mathscr{C}_R$$

— let us denote this function $(m, n) \to \mathscr{C}_B m \underset{A}{\otimes} n \mathscr{C}_R$ — such that the pair consisting of the C-complete (B, R)-bimodule $\mathscr{C}_B M \underset{A}{\otimes} N \mathscr{C}_R$ together with this infinitely (B, R)-bilinear, A-balanced function, is universal with these properties. Also, given such a pair M, N, the functor $(M, N) \rightsquigarrow \mathscr{C}_B M \underset{A}{\otimes} N \mathscr{C}_R$, from

$(\mathscr{C}_B, A) \times (\mathscr{C}_{R^o}, A^o)$ into $\mathscr{C}_{B,R}$, is the right adjoint of the functor $(N, L) \rightsquigarrow$ $\mathrm{Hom}_{\mathscr{C}_R}(N, L)$, from $(\mathscr{C}_{R^o}, A^o) \times \mathscr{C}_{B,R}$ into $(\mathscr{C}_B, A)$. That is, we have an isomorphism of functors:

$$\mathrm{Hom}_{\mathscr{C}_{B,R}}(\mathscr{C}_B M \underset{A}{\otimes} N_{\mathscr{C}_R}, L) \approx \mathrm{Hom}_{(\mathscr{C}_B, A)}(M, \mathrm{Hom}_{\mathscr{C}_R}(N, L))$$

from $(\mathscr{C}_B, A)^o \times (\mathscr{C}_{R^o}, A^o)^o \times \mathscr{C}_{B,R}$ into the category of sets (or, in fact, into the category of abelian groups).

Remark 10. The analogue of Theorem 2.3.9, with A-adically complete left A-modules instead of C-complete left A-modules, is false without additional hypotheses. For example, by Chapter 1, §2, Remark 8.1, if A is the commutative complete topological ring described in Chapter 1, §2, Example 2, then the direct product of denumerably many copies of A is not A-adically complete. Most likely, the analogue of Corollary 2.3.10 in this case is also false.

However, if one makes the very mild assumption on the topological ring A, that the topology of A be given by (an arbitrary set) of *right finitely generated* ideals, then by Chapter 1, §3, Theorem 1.3.3, we have that the analogue of Theorem 2.3.9 holds good, with A-adically complete instead of C-complete. Also, if one makes the still-quite-mild assumption, that the topological ring A is such that $\hat{A}$ is strongly admissible, then by Chapter 1, §3, Theorem 1.3.7, one obtains that the analogue of Corollary 2.3.10 above then holds with A-adically complete replacing C-complete. (— since $\mathrm{Hom}_A(N, M)$ is A-adically closed in M^N.)

Concerning the construction of the C-complete tensor product in Remark 6 above, it is perhaps interesting that the "complete tensor product" of two p-adic Banach spaces has been studied with some success in the literature. This is (a special case of) the A-adically complete analogue of the C-complete tensor product, in the case that the topological ring A is the complete discrete valuation ring $\widehat{\mathbf{Z}_p}$ of p-adic integers. As we shall see later, some very special, nice properties hold for the C-complete tensor product, when the topological ring A is, e.g., such that the topology is the t-adic from some element $t \in A$, such that the t-torsion in A is bounded below (i.e., such that there exists an integer n such that $x \in A$, $t^i \cdot x = 0$ for some $i \geq 0$, implies $t^n \cdot x = 0$. This is always true if the ring A is right Noetherian). This is the underlying reason why that "complete tensor prodlulct of p-adic Banach spaces" is at least somewhat well-behaved. (But of course, one should really work with the C-complete tensor product, as constructed above.)

The next Theorem is very important.

Theorem 2.3.12. *Let A be an arbitrary topological ring such that the topology of A is given by right ideals. Let $\mathscr{C}_A$ be the category of all C-complete left A-modules, with infinitely linear functions as maps.*

Then $\mathscr{C}_A$ is an abelian category. [6] *Moreover, the "stripping functor" from $\mathscr{C}_A$ into the category of abstract left A-modules, that to every C-complete left A-module associates the underlying abstract left A-module, is exact.*

---

[6]See, e.g., [COC], Introduction, Chapter 1, for a definition and description of the basic properties of abelian categories.

Notes:

1. It follows that the (underlying abstract left A-module of the) kernel, cokernel, image, or coimage of any infinitely linear function f of C-complete left A-modules, is also a kernel, cokernel, image, or coimage of this function f considered as being an abstract homomorphism of abstract left A-modules.

2. By Theorem 2.3.9, the category $\mathscr{C}_A$ is also closed under taking arbitrary direct products of objects; and the "stripping functor" described above also preserves arbitrary direct products of objects. It follows that $\mathscr{C}_A$ is closed under taking arbitrary inverse limits of arbitrary functors over set-theoretically legitimate categories; and that, after taking any such generalized inverse limit in the category $\mathscr{C}_A$, that the underlying abstract left A-module of this generalized inverse limit is simply the generalized inverse limit of the corresponding abstract left A-modules. (More about this later.)

Proof. We have already observed that $\mathscr{C}_A$ is an additive category. By Theorem 2.3.9, $\mathscr{C}_A$ is closed under arbitrary direct products of objects, and in particular $\mathscr{C}_A$ is closed under taking finite direct products of objects.

Let $f : M \to N$ be an infinitely linear function of C-complete left A-modules. Then by Theorem 2.3.6, the set-theoretic kernel $K = f^{-1}(\{0\})$ is a C-complete left A-submodule of M, and therefore is an object in $\mathscr{C}_A$. It follows readily that K obeys the universal mapping property for being a kernel of f in the additive category $\mathscr{C}_A$. Also by Theorem 2.3.6, the set-theoretic image $I = \{f(x) : x \in M\}$ is a C-complete left A-submodule of N. Then by Theorem 2.3.8, we have the quotient C-complete left A-module N/I. Then it is again immediate that N/I obeys the universal mapping property for being a cokernel of f in $\mathscr{C}_A$. Since K, resp.: N/I, is of course also the kernel, resp.: cokernel, of f in the category of abstract left A-modules, it follows that the additive functor of additive categories, "underlying abstract left A-module,"

$$U : \mathscr{C}_A \rightsquigarrow \{ \text{ category of abstract left } A\text{-modules } \},$$

preserves the taking of both kernels and cokernels of maps; and therefore also the taking of images and coimages. Let

$$\eta : \mathtt{Coim}(f) \to \mathtt{Im}(f)$$

be the factorization map (from the additive category-theoretic coimage of f into the additive category-theoretic image of f) in the category $\mathscr{C}_A$. Then it follows that

$$U(\eta) : \mathtt{Coim}(U(f)) \to \mathtt{Im}(U(f))$$

is the factorization map of $U(f)$ in the category of abstract left A-modules. Since this latter category is abelian, we have that $U(\eta)$ is an isomorphism of left A-modules; i.e., that η is bijective. But a bijective infinitely linear function is always an isomorphism in $\mathscr{C}_A$ (since the set-theoretic two-sided inverse of such a function is also infinitely linear). Therefore, η is an isomorphism $\mathscr{C}_A$.

This proves that $\mathscr{C}_A$ is an abelian category. Since the functor U is additive, and preserves kernels and cokernels, U is exact.

The proofs of Notes 1 and 2 are left as an easy exercise in category theory. $\square$

Remark 11. It is not too difficult to prove right now that the category $\mathscr{C}_A$ of C-complete left A-modules is closed under arbitrary direct sums of objects; and therefore is closed under arbitrary direct limits of functors over arbitrary set-theoretically legitimate categories. However, as we shall see, the direct sum of objects in $\mathscr{C}_A$ almost always has very pathological properties. (In particular, the "stripping functor" U from $\mathscr{C}_A$ into the category of abstract left A-modules is virtually always very far from preserving infinite direct sums.) E.g., we will show that for A a complete discrete valuation ring not a field, $\mathscr{C}_A$ obeys the Eilenberg–Moore Axiom $(P1)^o$ but not $(P2)^o$. See [EM] (or [COC], Introduction, Chapter 1). (This may be the first such example known. In this specific example, direct sum over any set is an exact functor.) And, for A just very slightly more complicated (and, e.g., Noetherian commmutative and such that the topology is the I-adic for some ideal I), we have that, typically, even $(P1)^o$ fails. (In consequence, we will show that, if A is a c.d.v.r (or, more generally, if A is commutative and is such that direct sums over arbitrary sets is an exact functor), then the functor "C-complete tensor product" is balanced. But, in the case that direct sums over arbitrary sets in $\mathscr{C}_A$ is not an exact functor — and that is the typical situation — then we shall see that the functor "C-complete tensor product" is not balanced.)

We will see, also, in the next section, that the category of C-complete left A-modules always has enough projectives. However, if the Eilenberg–Moore Axiom (dual of $(P2)$) fails — and it almost always fails — then, see [EM], it follows as we shall see that the category of C-complete left modules in this (typical) case, does not have enough injectives.

Remark 12. Of course, the overwhelming advantage of the category $\mathscr{C}_A$ of C-complete left A-modules, over the category of A-adically complete abstract left A-modules, is that $\mathscr{C}_A$ is an *abelian* category. The category of A-adically complete abstract left A-modules almost never enjoys this property (although, of course, it will have this property in the trivial case in which the topology of A is the discrete topology). For example, even if A is a complete discrete valuation ring not a field (this is the simplest case in which A does not have the discrete topology), then the category of A-adically complete abstract left A-modules is not abelian. (Of course, the basic property that is lacking for arbitrary completes is Theorem 2.3.8.)

2.4 Generators and Relations for C-complete Left A-modules

Theorem 2.4.1. *Let A be a topological ring such that the topology is given by right ideals. Then if M is any C-complete left A-module, there exist free abstract left A-modules G and F, and an infinitely linear function:*

$$f : \widehat{G} \to \widehat{F}$$

such that $M \approx \mathrm{Cok}(f)$ as C-complete left A-modules.

Proof. By §3, Theorem 2.3.5, there exist a free abstract left A-module F and a surjective function

$$\rho : \widehat{F} \to M$$

such that ρ is infinitely linear. Let $K = \text{Ker}(\rho)$. Then by §3, Theorem 2.3.6, we have that K is a C-complete left A-module. Therefore, again by §3, Theorem 2.3.5, there exists a free abstract left A-module G and an epimorphism $p : \widehat{G} \to K$ that is infinitely linear. Then clearly $M \approx \text{Cok}(\widehat{G} \xrightarrow{p} K \xrightarrow{i} \widehat{F})$ in the category of C-complete left A-modules (where i is the inclusion map). $\quad\square$

Remark 1. Let A be an arbitrary topological ring such that the topology is given by right ideals, and let M be an abstract left A-module. Let us consider conditions on M under which there might exist an infinite sum structure on M (as defined in §2, Definition 1). In fact, by §3, Theorem 2.3.5, if such a structure exists on M, then there exists an epimorphism of abstract left A-modules

$$\pi : \widehat{F} \to M$$

where F is the completion of a free module. (This is already a condition on the left A-module M. Not every abstract left A-module obeys this condition. E.g., if the topology of A is given by denumerably many right ideals — so that a Hausdorff quotient of an A-adically complete left A-module is always A-adically complete — and if the topological ring A is admissible — so that $\widehat{F}$ is A-adically complete — then this condition would imply that $M/(\text{closure of } \{0\})$ is A-adically complete.) Necessary and sufficient conditions that M admit an infinite sum structure, is that there be such an epimorphism of abstract left A-modules π, such that the infinite sum structure of $\widehat{F}$ induces a well-defined datum (S) — in the sense of §2, Definition 1 — on the abstract left A-module M. (For, if this is the case, then Axioms $(IS1)$ - $(IS4)$ for M then follow from the corresponding Axioms for $\widehat{F}$.) On the other hand, if we have $\pi : \widehat{F} \to M$ is an epimorphism of abstract left A-modules, then by §3, Theorem 2.3.6 and Theorem 2.3.8, the infinite sum structure on $\widehat{F}$ will induce an infinite sum structure on M by passing to the quotient iff the kernel, $K = \text{Ker}(\pi)$, of π is a C-complete left A-submodule of $\widehat{F}$. E.g., this will be the case if K is closed in $\widehat{F}$ for the uniform topology. (*Proof.* For all $x \in K$, $A \cdot x \subset K$. Since the multiplication: $(\widehat{A}, unif) \times (\widehat{K}, unif) \to (\widehat{K}, unif)$ is continuous, it follows that K is an $\widehat{A}$-submodule of $\widehat{F}$. And since the infinite sum structure of $\widehat{F}$ is deduced by taking the limit of infinite series for the uniform topology, it follows that K is closed under taking infinite linear combinations.)

Suppose now that the topological ring A is *admissible*. If M is A-adically Hausdorff, and if there exists an epimorphism $\pi : \widehat{F} \to M$ of abstract left A-modules where F is a free abstract left A-module, then the kernel K of π is A-adically closed in $\widehat{F}$, and therefore is also closed in $\widehat{F}$ for the uniform topology of $\widehat{F}$. Therefore, we have proved:

Corollary 2.4.2. *Let A be an admissible topological ring, and let M be an abstract left A-module that is A-adically Hausdorff. Then M admits an infinite* sum structure iff there exists an epimorphism of abstract left A-modules: $\widehat{F} \to M$.

Note: Again, if the topological ring A is admissible, then an A-adically Hausdorff abstract left A-module admits at most one infinite sum structure.

Proof of the Note to Corollary 2.4.2: By Lemma 2.4.3 below, if M admits an infinite sum structure then the structure of M as an $\widehat{A}$-module is uniquely determined by its structure as an A-module. Given an infinite sum structure on M, let $\pi : \widehat{F} \to M$ be as in §3, Theorem 2.3.5. Then, since A is admissible, the A-adic and uniform topologies of $\widehat{F}$ coincide. (See Chapter 1, §3, Definition 1.) But in $\widehat{F}$, the infinite sum structure is deduced by taking limits of infinite series for the uniform topology of $\widehat{F}$ (see §2, Definition 1). Since π is infinitely linear, and is also A-adically continuous, it follows that the infinite sum structure in M is deduced by taking limits of infinite series for the $\widehat{A}$-adic topology of M, (and therefore in particular is uniquely determined). □

Remark For a description of the analogous situation to Corollary 2.4.2, in the case that the topological ring A is not admissible, see Chapter 3, §2, Remark 6.

Lemma 2.4.3. *Let A be a topological ring such that the topology is given by right ideals. Suppose that the A-adic and uniform topologies coincide in the A-adic completion $\widehat{A}$ of A regarded as an abstract left A-module. If M is any A-adically Hausdorff abstract left A-module, then there exists at most one structure of abstract left $\widehat{A}$-module on the abstract abelian group of M such that the induced abstract A-module structure is the given one of M.*

Proof 1. Let M' be an abstract $\widehat{A}$-module such that the underlying A-module structure is M. Then, since the uniform and A-adic topologies coincide in $\widehat{A}$, we have that the $\widehat{A}$-adic topology of M' coincides with the A-adic topology of M. Since the underlying A-module of M' is M, it follows that the $\widehat{A}$-adic completion of M' is A-isomorphic to the A-adic completion $\widehat{M}$ of M.

By Chapter 1, §3, Remark 0.3, since the A-adic and $\widehat{A}$-adic topologies are the same in $\widehat{A}$, we have, for every open ideal I in A that

$$A/I \approx \widehat{A}/(I \cdot \widehat{A}).$$

Therefore

$$(A/I) \underset{A}{\otimes} (\widehat{A} \underset{A}{\otimes} M) = ((A/I) \underset{A}{\otimes} \widehat{A}) \underset{A}{\otimes} M = (\widehat{A}/(I \cdot \widehat{A})) \underset{A}{\otimes} M \approx (A/(IA)) \underset{A}{\otimes} M.$$

Passing to the inverse limit over I, we have that the natural map on the A-adic completions:

$$(0) : \widehat{M} \to \widehat{\widehat{A} \underset{A}{\otimes} M}$$

is an isomorphism. (This map is the image of the natural map: $M \to \widehat{A} \underset{A}{\otimes} M$ under the functor "A-adic completion".) Therefore the A-adic completion of $\widehat{A} \underset{A}{\otimes} M$ is $\widehat{M}$.

We have the natural map of $\widehat{A}$-modules

$$(1): \ \widehat{A} \underset{A}{\otimes} M \to M'$$

Since we have verified that both modules have the same A-adic completion, and since the $\widehat{A}$-adic completions are (as A-modules) the same as the A-adic completions, it follows that the map induced by (1) on the $\widehat{A}$-adic completions is an isomorphism of abelian groups — and therefore also of $\widehat{A}$-modules. Therefore

$$\text{(Completion of } M' \text{ as } \widehat{A}\text{-module)} \approx \text{(Completion of } \widehat{A} \underset{A}{\otimes} M \text{ as } \widehat{A}\text{-module)}$$

as $\widehat{A}$-modules. By equation (0), the right side of this isomorphism is $\widehat{M}$. So

$$\text{(Completion of } M' \text{ as } \widehat{A}\text{-module)} \approx \widehat{M}.$$

Therefore the $\widehat{A}$-module structure of M' is the one deduced from the inclusion: $M \to \widehat{M}$ and the $\widehat{A}$-module structure of $\widehat{M}$ — and therefore is uniquely determined from the A-module of M. $\qquad\square$

Proof 2. Let M be an abstract left $\widehat{A}$-module such that the underlying abstract left A-module of M is A-adically Hausdorff. We must show that the structure of M as an abstract left $\widehat{A}$-module is determined by the structure of M as an abstract left A-module.

If $a \in \widehat{A}$, then let $(a_i)_{i \in D}$ be a net in the topological space A such that $\lim_{i \in D}(\iota(a_i)) = a$ in $\widehat{A}$, where $\iota : A \to \widehat{A}$ is the natural map. Since M is an abstract left $\widehat{A}$-module, the scalar multiplication: $\widehat{A} \times M \to M$ is continuous (see Chapter 1). Therefore, if $x \in M$, then

$$(2): \ \lim_{i \in D}(\iota(a_i \cdot x)) = a \cdot x,$$

the limit being taken for the $\widehat{A}$-adic topology of M. Since by hypothesis the uniform and A-adic topologies coincide in $\widehat{A}$, we have that a complete system of neighborhoods of zero for the uniform topology of $\widehat{A}$ is $\{I \cdot \widehat{A} : I$ is an open right ideal in $A\}$. Therefore the $\widehat{A}$-adic and A-adic topologies of M coincide. Thus equation (2) is also valid in M for the A-adic topology. Since the left side of equation (2) then depends only on the structure of M as an abstract left A-module, it follows from equation (2) that the structure of M as an abstract left $\widehat{A}$-module is completely determined by the structure as an abstract left A-module. $\qquad\square$

Lemma 2.4.4. *The hypotheses being as in Lemma 2.4.3, if M and N are abstract left $\widehat{A}$-modules and if N is A-adically Hausdorff, then*

$$\operatorname{Hom}_{\widehat{A}}(M, N) = \operatorname{Hom}_A(M, N).$$

Proof 1. Replacing M by $M/(\text{closure of }0)$ if necessary, we can assume that M is A-adically Hausdorff also. Let $f : M \to N$ be A-linear. To show that f is also $\widehat{A}$-linear: We have the commutative diagram:

$$
\begin{array}{ccc}
\widehat{M} & \xrightarrow{\;\hat{f}\;} & \widehat{N} \\
{\scriptstyle\iota_M}\big\uparrow & & {\scriptstyle\iota_N}\big\uparrow \\
M & \xrightarrow{\;f\;} & N
\end{array}
$$

where ι_M and ι_N are the natural maps. By the proof of Lemma 2.4.3, the $\widehat{A}$ structure of M, resp. N, is deduced from the $\widehat{A}$-structure of $\widehat{M}$, resp. $\widehat{N}$; and this latter depends only on the A-module structure of M, resp.: N. Therefore $\hat{f}$ is $\widehat{A}$-linear. From the commutative diagram, f is $\widehat{A}$-linear. □

Proof 2. Let $f : M \to N$ be A-linear. To show that f is $\widehat{A}$-linear. Let $a \in \widehat{A}$ and $x \in M$. Then choose $(a_i)_{i \in D}$ a net in A such that $\lim_{i \in D}(\iota(a_i)) = a$ in $\widehat{A}$, where $\iota : A \to \widehat{A}$ is the natural map. Then we have by the proof of Lemma 2.4.3 that $a \cdot x = \text{limit of the net } (\iota(a_i) \cdot x)_{i \in I}$ in M for the A-adic topology. Since f is a homomorphism of abstract left A-modules, f is continuous for the A-adic topology. Therefore $f(a \cdot x) = \text{limit of the net } \iota((a_i \cdot f(x)))_{i \in D}$ for the A-adic topology. By equation (1) of Proof 2 of Lemma 2.4.3, this latter is $a \cdot f(x)$. (Of course we are using the fact that, since M and N are A-adically Hausdorff, A-adic limits are unique in both M and N.) □

Remark 2. Therefore, if A is a topological ring that is admissible, then the Hausdorff C-complete left A-modules are identical to the set of all Hausdorff quotients of A-adically complete abstract left A-modules. If we have in addition that the topology of A is given by *denumerably many* right ideals, then the Hausdorff quotient of an A-adically complete left A-module is complete. Therefore, in this case:

Corollary 2.4.5. *If A is a topological ring such that the topology is given by denumerably many right ideals, and also such that A is admissible, then the full subcategory of the category $\mathscr{C}_A$ of all C-complete left A-modules, consisting of those that are A-adically Hausdorff, is identical to the category of all A-adically complete abstract left A-modules and all homomorphisms of abstract left A-modules.*

Proof. We have already observed that the objects are the same. Let $f : M \to N$ be a homomorphism of abstract left A-modules. To show that f is infinitely linear. By Lemma 2.4.4, f is $\widehat{A}$-linear. In the proof of the Note to Corollary 2.4.2, we have observed that the infinite sum structure in M, resp.: N is determined by the abstract $\widehat{A}$-module structure and $\widehat{A}$-adic topology of M, resp.: N — namely, an infinite sum $\sum_{i \in I} a_i x_i$, where $(a_i)_{i \in I}$ in A^I is zero-bound, and where $x_i \in M$, resp.: $x_i \in N$, all $i \in I$, is the limit of this sum for the $\widehat{A}$-adic topology of M, resp.: N — it follows that f is infinitely linear. $\square$

A related result is:

Corollary 2.4.6. *Let A be a topological ring such that the topology is given by denumerably many right ideals, and such that A is admissible. Let M be a left A-module such that there exists an epimorphism*

$$\pi : \widehat{F} \to M$$

of abstract left A-modules. Then the infinite sum structure of $\widehat{F}$ induces an infinite sum structure on M, iff the kernel $\mathrm{Ker}(\pi)$ of π is A-adically complete as an abstract left A-module.

Proof. We have already observed in Remark 1 that the infinite sum structure of $\widehat{F}$ induces such a structure in M by means of π, iff $\mathrm{Ker}(\pi)$ is a C-complete left A-submodule of $\widehat{F}$. $K = \mathrm{Ker}(\pi)$, being an A-submodule of the A-adically Hausdorff abstract left A-module $\widehat{F}$, is itself A-adically Hausdorff. Therefore, by Corollary 2.4.5, K admits an infinite sum structure iff K is A-adically complete. (Notice, of course, if K admits an infinite sum structure, then the inclusion: $K \to \widehat{F}$ must be infinitely linear by Corollary 2.4.5.) $\square$

Remark: Corollary 2.4.2 fails if M is not Hausdorff. E.g., if F is any free module of infinite rank, then $\widehat{F}/F$ is a quotient of a complete abstract left A-module — but, if the topological ring A admits a denumerable neighborhood base at zero, then we shall show later that $\widehat{F}/F$ does not admit an infinite sum structure.

Remark 3. We will see later, however, that if the topological ring A admits a denumerable neighborhood base, and if also A is admissible, then:

(1) Every abstract left A-module admits at most one infinite sum structure.

(2) Every function of C-complete left A-modules, that is a homomorphism as abstract left A-modules, is infinitely linear; and therefore, in this case,

(3) the category $\mathscr{C}_A$ of all C-complete left A-modules together with all infinitely linear functions, is an exact, *full* subcategory of the category of all abstract left A-modules and all homomorphisms of left A-modules.

(Notice that the hypotheses of Remarks 2 and 3 are very mild. E.g., if I is any two-sided ideal that is finitely generated as right ideal in A, then A together with the I-adic topology obeys these conditions; as we have observed, e.g., in Chapter 1, §3, Example 6 (following Definition 2).)

We return to the general theory.

Lemma 2.4.7. *Let A be a topological ring such that the topology is given by right ideals. Let M be an abstract left A-module, and let $\iota : M \to \widehat{M}$ be the natural map into the A-adic completion. Suppose that* either

(1) M is projective as an abstract left A-module, or
(2) The neighborhood system at zero in the topological ring A admits a de-numerable base.

Then the image $\iota(M)$ in $\widehat{M}$ is a set of generators for the C-complete left A-module $\widehat{M}$.

Proof. First, consider

Case 1. M is the free left A-module $A^{(I)}$, for some set I. Then $\widehat{M} = \widehat{A^{(I)}} = \{(a_i)_{i \in I} \in \widehat{A}^I : (a_i)_{i \in I}$ is zero-bound in $\widehat{A}\}$. Therefore, if e_i, $i \in I$, are the canonical basis for $A^{(I)}$, then given any element $x \in \widehat{M}$, say $x = (a_i)_{i \in I}$, then there exists a zero-bound family of elements of $\widehat{A}$ — namely, the family $(a_i)_{i \in I}$ — such that

$$x = \sum_{i \in I} a_i \iota(e_i).$$

Therefore, $\{\iota(e_i) : i \in I\}$ generates $\widehat{M}$ as a C-complete left A-module; and therefore so does the larger subset $\iota(M)$ of $\widehat{M}$.

Case 2. M is a projective abstract left A-module. Then let N be another abstract left A-module such that $M \oplus N$ is a free left A-module. Then $\widehat{M \oplus N} \approx \widehat{M} \oplus \widehat{N}$ as C-complete left A-modules. If S is any C-complete left A-submodule of $\widehat{M}$ containing $\iota(M)$, then $S \oplus \widehat{N}$ is a C-complete left A-submodule of $\widehat{M \oplus N}$ containing the image of $M \oplus N$. But, by Case 1, the image of $M \oplus N$ in $\widehat{M \oplus N}$ generates $\widehat{M \oplus N}$ as a C-complete left A-module. Therefore $S \oplus \widehat{N} = \widehat{M} \oplus \widehat{N}$. Therefore $S = \widehat{M}$. S being an arbitrary C-complete left A-submodule of $\widehat{M}$ that contains $\iota(M)$, it follows that M generates $\widehat{M}$ as a C-complete left A-module.

Case 3. Suppose that the neighborhood system at zero in the topological ring A admits a denumerable neighborhood base. Then choose an epimorphism of abstract left A-modules:

$$f : F \to M,$$

where F is a free abstract left A-module. Then since the neighborhood system at zero in A admits a denumerable base, by Chapter 1, §2, Corollary 1.2.7, we have that the induced homomorphism on the A-adic completions:

$$\widehat{f} : \widehat{F} \to \widehat{M}$$

is surjective. But then, $\widehat{f}$ is infinitely linear. By Case 1, the image of F in $\widehat{F}$ generates $\widehat{F}$ as a C-complete left A-module. Therefore

$$\iota(M) = \widehat{f}(\text{image of } F \text{ in } \widehat{F})$$

must generate $\widehat{M} = \widehat{f}(\widehat{F})$ as a C-complete left A-module. $\square$

Theorem 2.4.8. *Let A be an arbitrary topological ring such that the topology is given by right ideals. Then*

(1) If P is a projective abstract left A-module, and if M is any C-complete left A-module, then given any homomorphism

$$f : P \to M$$

of abstract left A-modules, there exists a unique extension of f to an infinitely linear function

$$g : \widehat{P} \to M$$

(2) If P is an abstract projective left A-module, then $\widehat{P}$ is a projective object in the category $\mathscr{C}_A$ of C-complete left A-modules.

(3) The category of C-complete left A-modules has enough projectives.

<u>Note</u>. In fact, the objects $\widehat{P}$, where P is a projective abstract left A-module, are a collection of enough projectives in $\mathscr{C}_A$.

Proof. Let P, M, and $f : P \to M$ be as in (1). Then by §3, Theorem 2.3.5, there exists a surjective, infinitely linear function

$$\rho : \widehat{F} \to M$$

where F is a free left A-module. Then by §3, Proposition 2.3.3, we have that ρ is also a homomorphism of abstract left A-modules. Since ρ is surjective and P is projective, there exists a homomorphism $\overline{g} : P \to \widehat{F}$ of abstract left A-modules such that the diagram:

$$
\begin{array}{ccc}
P & \xrightarrow{\overline{g}} & \widehat{F} \\
{\scriptstyle f}\downarrow & \swarrow{\scriptstyle \rho} & \\
M & &
\end{array}
$$

is commutative. Let $(\widehat{F}, unif)$ denote $\widehat{F}$ with its uniform topology; i.e., the topology such that $\widehat{F}$ is the completion of F in the category of topological left A-modules. And let P, resp.: $\widehat{F}$ denote P, resp.: $\widehat{F}$ regarded as topological left A-modules with their A-adic topologies. Then $\overline{g} : P \to \widehat{F}$ and (the identity of $\widehat{F}$): $\widehat{F} \to (\widehat{F}, unif)$ are continuous. Therefore the composite:

$$\overline{g} : P \to (\widehat{F}, unif)$$

is continuous. By the universal mapping property of the completion $(\widehat{P}, unif)$ of P (considered as the completion of P in the category of topological left A-modules), there therefore exists a unique continuous homomorphism of topological left A-modules:

$$h : (\widehat{P}, unif) \to (\widehat{F}, unif)$$

extending $\overline{g}$. Then, since the infinite sum structure in $\widehat{P}$ and $\widehat{F}$ are deduced from limits of infinite series for their uniform topologies, it follows that the function $h : \widehat{P} \to \widehat{F}$ is infinitely linear. But then $g = \rho \circ h$ is an infinitely linear function from $\widehat{P}$ into M extending f. To prove (1), it remains to prove uniqueness of g. In fact, if g_1 and g_2 are both infinitely linear functions from $\widehat{P}$ into M that extend f, then $K = \mathtt{Ker}(g_1 - g_2)$ is a C-complete left A-submodule of $\widehat{P}$ that contains the image of P. But then, by Lemma 2.4.7, it follows that $K = \widehat{P}$. I.e., $g_1 = g_2$.

This proves (1). If now P is any projective abstract left A-module, then for every C-complete left A-module M, if $U(M)$ denotes the underlying abstract left A-module of M, then we have that

$$(4) : \operatorname{Hom}_A(P, U(M)) \approx \operatorname{Hom}_{\mathscr{C}_A}(\widehat{P}, M).$$

Since P is projective, the functor $\operatorname{Hom}_A(P, -)$ is exact on the category of abstract left A-modules. Therefore the composite with the exact functor, the "stripping functor"

$$U : \mathscr{C}_A \rightsquigarrow \{\text{category of abstract left } A\text{-modules}\},$$

is exact. By equation (4), this implies that the functor $\operatorname{Hom}_{\mathscr{C}_A}(\widehat{P}, -)$ on the category $\mathscr{C}_A$ is exact — i.e., that $\widehat{P}$ is projective in $\mathscr{C}_A$.

Finally, if M is any C-complete left A-module, then by §3, Theorem 2.3.5, we have that M is isomorphic to a quotient in the category $\mathscr{C}_A$ of $\widehat{F}$, for some free abstract left A-module F. Therefore part (2) of the Theorem implies the Note to the Theorem, and therefore also part (3) of the Theorem   □

Corollary 2.4.9. *A as in Theorem 2.4.8, an object of the category $\mathscr{C}_A$ is projective iff it is isomorphic to a direct summand in $\mathscr{C}_A$ of $\widehat{F}$, for some free abstract left A-module F.*

Proof. By Theorem 2.4.8, part (2), if F is a free abstract left A-module, then $\widehat{F}$ is projective in $\mathscr{C}_A$. Therefore every direct summand of $\widehat{F}$ is projective in $\mathscr{C}_A$.

Conversely, if P_0 is a projective object in $\mathscr{C}_A$, then by §3, Theorem 2.3.5, there exists an abstract free left A-module F, and an epimorphism in the category $\mathscr{C}_A$

$$\rho : \widehat{F} \to P_0.$$

But then, since P_0 is projective in $\mathscr{C}_A$, the epimorphism ρ splits. And therefore P_0 is isomorphic to a direct summand of $\widehat{F}$.   □

Corollary 2.4.10. *Let A be a topological ring such that the topology is given by right ideals, let S be any set, let F be the free abstract left A-module on the set S, and let ι be the composite*

$$\iota : S \to F \to \widehat{F}.$$

Then

(1) $\widehat{F}$ is a C-complete left A-module.

(2) $\iota : S \to \widehat{F}$ is a function.

(3) The pair $(\widehat{F}, \iota)$ is universal with these properties. (That is, given any C-complete left A-module M, and any function $f : S \to M$, there exists a unique extension of f to an infinitely linear function: $\widehat{F} \to M$.)

Proof. (1) and (2) are clear. Let M be any C-complete left A-module and let $f : S \to M$ be any function. Then, since M is also an abstract left A-module and F is the free abstract left A-module on the set S, there exists a unique extension of f to a homomorphism of abstract left A-modules

$$g : F \to M.$$

But then, by Theorem 2.4.8, part (1), there exists a unique extension of g to an infinitely linear function: $\widehat{F} \to M$. $\qquad\qquad\square$

Remark 4. Of course, in Corollary 2.4.10, properties (1), (2), and (3) of the pair $(\widehat{F}, \iota)$ characterize this pair up to canonical isomorphisms. Given a set S and a topological ring A such that the topology is given by right ideals as in Corollary 2.4.10, we will call the pair $(\widehat{F}, \iota)$ *the free C-complete left A-module on the set S.*

Definition 1. Let A be any topological ring such that the topology of A is given by right ideals. Let I be any set, and let $(e_i)_{i \in I}$ be an indexed family of distinct elements of some set. Let J be any set, and suppose that, for every $j \in J$, we are given an element r_j of the free C-complete left A-module on $\{e_i : i \in I\}$. Thus, explicitly,

$$r_j = \sum_{i \in I} a_{ji} e_i,$$

where $(a_{ji})_{i \in I}$ are uniquely determined zero-bound families indexed by the set I of elements of $\widehat{A}$. Then by *the C-complete left A-module given by the generators $\{e_i : i \in I\}$ and the relations $\{r_j : j \in J\}$,* we will mean
   (1) A C-complete left A-module M, together with
   (2) A function $\iota : \{e_i : i \in I\} \to M$, such that
   (3) The image of r_j in M is zero for all $j \in J$, (it is equivalent to saying, "such that $\sum_{i \in I} a_{ji}\iota(e_i) = 0$ in M for all $j \in J$"), and such that
   (4) The pair (M, ι) is universal with these properties.

By Corollary 2.4.10, such an (M, ι) always exists — in fact, the quotient C-complete left A-module of the free C-complete left A-module on $\{e_i : i \in I\}$ by the C-complete left submodule generated by $\{r_j : j \in J\}$ satisfies the universal mapping property. Another, equivalent description, of M is:
   The cokernel of the infinitely linear function

$$\alpha : \widehat{A^{(J)}} \to \widehat{A^{(I)}},$$

given by the matrix $(a_{ji})_{(j,i) \in J \times I}$. (By saying that α is "given by the matrix $(a_{ji})_{(j,i) \in J \times I}$," we mean that

$$\alpha(j\text{th element of the canonical basis of } A^{(J)})$$

$$= \sum_{i \in I} a_{ji}(\,i\text{th element of the canonical basis of } A^{(I)})$$

in $\widehat{A^{(I)}}$ for all $j \in J$.) Thus, given M as in Theorem 2.4.1, the construction of G, F and f is equivalent to finding generators and relations for M as a C-complete left A-module.

Proposition 2.4.11. *Let A be a complete topological ring such that the topology is given by right ideals. Let M be an abstract left A-module that is finitely generated as an abstract left A-module.*

Then there exists at most one infinite sum structure on M, so that M becomes a C-complete left A-module. When this is the case, then for every (not necessarily finitely generated) C-complete left A-module N, and every homomorphism f of abstract left A-modules from M into N, we have that f is infinitely linear.

Note. If M is a finitely generated abstract left A-module, and if we choose an integer $n \geq 0$ and an epimorphism

$$f : A^n \to M$$

of abstract left A-modules, then M is C-complete iff $\mathtt{Ker}(f)$ is a C-complete left A-submodule of A^n.

Proof. Step 1. Suppose that n is an integer ≥ 0, that N is a C-complete left A-module, and that $g : A^n \to N$ is a homomorphism of abstract left A-modules. Then g is infinitely linear.

Proof of Step 1: Since A^n is projective as an abstract left A-module, by Theorem 2.4.8, part 1, we have that there exists a unique extension of g to an infinitely linear function:

$$h : \widehat{A^n} \to N.$$

Since A is complete, $\widehat{A^n} = A^n$, and therefore $g = h$. Therefore g is infinitely linear.

Step 2. Let M be a C-complete left A-module, and suppose that M is finitely generated as an abstract left A-module. Let N be any C-complete left A-module, and let $f : M \to N$ be a homomorphism of abstract left A-modules. Then f is infinitely linear.

Proof of Step 2: Choose an integer $n \geq 0$ and an epimorphism of abstract left A-modules $k : A^n \to M$. Then by Step 1, the functions k and $f \circ k$ are infinitely linear. Since k is surjective, it folows that f is infinitely linear.

Step 3. Let M be an abstract left A-module finitely generated as an abstract left A-module. Then M admits at most one infinite sum structure.

Proof of Step 3: Suppose that (S_1) and (S_2) are two infinite sum structures in the sense of §2, Definition 1, on the abstract left A-module M, and let $(M, (S_1))$ and $(M, (S_2))$ denote the C-complete left A-modules thus obtained. Then by Step 2, the identity function of M:

$$(M, (S_1)) \to (M, (S_2))$$

is infinitely linear. Therefore the infinite sum structures (S_1) and (S_2) coincide. Steps 2 and 3 imply the Theorem. Finally, let M, n and $f : A^n \to M$ be as in the Note. If M is C-complete, then by §3, Theorem 2.3.6, $\mathrm{Ker}(f)$ is a C-complete left A-submodule of A^n. Conversely, if $\mathrm{Ker}(f)$ is a C-complete left A-submodule of A^n, then by §3, Theorem 2.3.8, we have that M admits an infinite sum structure. $\square$

Corollary 2.4.12. *Let A be an arbitrary topological ring such that the topology is given by right ideals, and let M be any finitely presented abstract left module over the completion $\widehat{A}$ of A. Then there exists a unique structure of C-complete left A-module on M, such that the induced structure of abstract left $\widehat{A}$-module of M, given by §2, Corollary 2.2.2, is the given abstract $\widehat{A}$-module structure on M.*

In this way, the category of finitely presented abstract left $\widehat{A}$ modules (and all homomorphisms of abstract left $\widehat{A}$-modules) is identified with a full subcategory of the category $\mathscr{C}_A$ of all C-complete left A-modules.

Proof. First, replacing A with $\widehat{A}$ if necessary, we can assume that the topological ring A is complete. Then, by Proposition 2.4.11, the Corollary will follow if every finitely presented left A-module M admits an infinite sum structure. But, given such an M, by definition of "finitely presented" we have

$$M = \mathrm{Cok}(f : A^n \to A^m)$$

where f is some homomorphism of abstract left A-modules and $n, m \geq 0$. By Proposition 2.4.11 applied to the finitely generated left A-module A^n, we have that f is infinitely linear. Therefore by the first note to §3, Theorem 2.3.12, we have that M admits an infinite sum structure. $\square$

<u>Remark 5</u>. Let A be as in Proposition 2.4.11. Then by Corollary 2.4.12, the category Abs_A of abstract left A-modules (and all homomorphisms of abstract left A-modules), and also the category $\mathscr{C}_A$ of C-complete left A-modules (and all infinitely linear functions), both contain the same full subcategory, namely the category of all finitely presented abstract left A-modules (and all homomorphisms of abstract left A-modules).

In fact, by Proposition 2.4.11, Abs_A and $\mathscr{C}_A$ also both contain the, in general larger, full subcategory, consisting of all finitely generated abstract left A-modules that obey the condition in the Note to Proposition 2.4.11 (and all homomorphisms of such abstract left A-modules).

Using Corollary 2.4.10, we can now prove that the category of C-complete left A-modules is closed under direct sums of objects (see, e.g., [COC] for the definition of category-theoretic "direct sum" of objects).

Theorem 2.4.13. *Let A be any topological ring such that the topology is given by right ideals. Then the category of C-complete left A-modules is closed under taking direct sums of families of objects, indexed by arbitrary sets.*

Proof. Let $(M_i)_{i \in I}$ be an arbitrary family of C-complete left A-modules. Then let $\widehat{F}$ be the A-adic completion of the free A-module F on the set

$$S = \bigcup_{i \in I} (\{i\} \times M_i).$$

Let R be the subset of $\widehat{F}$, consisting of the elements: $R = \Big\{ \Big(\sum_{j \in J} a_j \cdot (\iota, x_j) \Big) - \Big(i, \sum_{j \in J} a_j \cdot x_j \Big) : J$ is a set, $(a_j)_{j \in J}$ is a zero-bound family indexed by the set J of elements of $\widehat{A}$, $i \in I$ is any element, and $(x_j)_{j \in J}$ is a family indexed by the set J of elements of $M_i \Big\}$. Then let M be the quotient C-complete left A-module of $\widehat{F}$ by the C-complete left A-submodule of $\widehat{F}$ generated by the set R. In the terminology of Definition 1, M is the C-complete left A-module given by the set of generators S and the set of relations R. Then, as in Definition 1, M obeys the properties (1) - (4) of Definition 1, including the universal mapping property (3). It follows readily that M is a direct sum of the M_i, $i \in I$, in the category of C-complete left A-modules. $\quad\square$

Corollary 2.4.14. *Let $\mathscr{D}$ be any category such that the objects of $\mathscr{D}$ form a set, and let $F : \mathscr{D} \rightsquigarrow \mathscr{C}_A$ be any covariant functor from $\mathscr{D}$ into the category $\mathscr{C}_A$ of C-complete left A-modules and all infinitely linear functions. Then the direct limit of the functor F,*

$$\varinjlim_{d \in \mathscr{D}} F(d)$$

exists in the category $\mathscr{C}_A$.

Proof. This trivially follows, as is the case for every abelian category (or even any category closed under finite direct limits) such that infinite direct sums exist. (E.g., see [COC], Introduction, Chapter 1.) $\quad\square$

Remark 6. Although Theorem 2.4.13 guarantees the existence of a category-theoretic direct sum of any family of objects, it should be noted that the infinite direct sum in the category of C-completes is in general usually a pathological construction. E.g., we will give a very simple, Noetherian commutative example of a ring A, such that the topology is the I-adic for an ideal I generated by two elements, yet such that the denumerable direct sum in the category of C-completes is not an exact functor. Also, in the (rare) cases in which the denumerable direct sum is exact (e.g., if the topology is the t-adic for some element t that is not a right divisor of zero, and is not an invertible element in the ring A, and is such that the t-adic topology makes A into a topological ring — an example of this is any commutative ring A and any non-invertible, non-zero divisor t in A) —, then usually the dual of the Eilenberg–Moore Axiom $(P2)$ fails (see, e.g., either [EM] or [COC], Introduction, Chapter 1) — so that, there exists a sequence $M_1 \overset{f_1}{\rightarrow} M_2 \overset{f_2}{\rightarrow} M_3 \overset{f_3}{\rightarrow} \ldots$ of C-completes and injective infinitely linear functions f_i, $i \geq 1$, such that the direct limit *in the category of C-completee left A-modules* is zero.

Remark 7. From Remark 6, it is of course evident that the "stripping functor" U rarely preserves either infinite direct sums, or infinite generalized direct limits. (However, since the functor U is exact, see §3, Theorem 2.3.12, it is of course true that U preserves finite direct limits.) Also, as follows from the second note to §3, Theorem 2.3.12, U preserves arbitrary infinite generalized inverse limits. Thus, the *inverse limit* in the category of C-completes is reasonably well behaved. (And in particular, "arbitrary direct prodect" is always an exact functor, and Axiom $(P2)$ always holds.)

Remark 8. The proof of Theorem 2.4.13 gives an explicit, but not in general very useful, construction of the direct sum in the category of C-completes. We will give an alternative such construction in §5, which is more useful; and which is particularly illuminating in the case in which the topological ring A admits a denumerable neighborhood base at zero, and also is admissible.

2.5 The C-completion of an Abstract Left Module

Definition 1. Let A be a topological ring such that the topology is given by right ideals. Let M be any abstract left A-module. Then *the C-completion of* M is:

(1) A C-complete left A-module, denoted $C(M)$, together with

(2) A homomorphism of abstract left A-modules $\iota_M : M \to C(M)$, such that

(3) The pair $(C(M), \iota_M)$ is universal with these properties. That is, such that:

If N is any C-complete left A-module, and if $f : M \to N$ is any homomorphism of abstract left A-modules, then there exists a unique infinitely linear function

$$g : C(M) \to N$$

such that the diagram

$$C(M) \xrightarrow{\quad g \quad} N$$
$$\iota_M \Big\uparrow \quad \nearrow f$$
$$M$$

is commutative.

Clearly, as with other universal mapping properties, if the C-completion of M exists in the sense of Definition 1, then it is uniquely determined up to canonical isomorphism.

Proposition 2.5.1. *If $h : M \to R$ is any homomorphism of abstract left A-modules such that $C(M)$ and $C(R)$ both exist, then there is a unique infinitely* *linear function, which we denote C(h), such that the diagram:*

$$
\begin{array}{ccc}
C(M) & \xrightarrow{\;C(h)\;} & C(R) \\[2pt]
\;\Big\uparrow{\scriptstyle \iota_M} & & \;\Big\uparrow{\scriptstyle \iota_R} \\[2pt]
M & \xrightarrow{\;\;h\;\;} & R
\end{array}
$$

is commutative. And the assignment: $M \rightsquigarrow C(M)$, $h \rightsquigarrow C(h)$ defines a functor C from the category of abstract left A-modules M such that $C(M)$ exists into the category $\mathscr{C}_A$.

Proof. Follows immediately from the universal mapping property. $\square$

Theorem 2.5.2. *Let A be any topological ring such that the topology of A is given by right ideals.*
  (1) For every abstract left A-module M, the C-completion $C(M)$ exists.
  (2) For every projective abstract left A-module P, we have that

$$
C(P) = \widehat{P},
$$

the A-adic completion of P, together with the infinite sum structure described in §2, Example 1.
  (3) The functor C, from the category of abstract left A-modules into the category of C-complete left A-modules, is a right exact functor.

Proof. Step 1: By §4, Theorem 2.4.8, part 1, and by Definition 1 above, we have that part (2) of Theorem 2.5.2 holds.

Step 2: I claim that: If $f : M \to N$ is any abstract homomorphism of abstract left A-modules, such that $C(M)$ and $C(N)$ exist in the sense of Definition 1, then $C(\mathrm{Cok}(f))$ exists in the sense of Definition 1, and in fact $C(\mathrm{Cok}(f)) = \mathrm{Cok}(C(M) \xrightarrow{C(f)} C(N))$.

Proof of Step 2: Let $H = \mathrm{Cok}(f : M \to N)$ in the category of abstract left A-modules and let $K = \mathrm{Cok}(C(f) : C(M) \to C(N))$, in the category $\mathscr{C}_A$. Then we have the commutative diagram with exact rows:

$$
(1): \quad
\begin{array}{ccccccc}
C(M) & \xrightarrow{\;C(f)\;} & C(N) & \xrightarrow{\;p\;} & K & \longrightarrow & 0 \\[2pt]
\;\Big\uparrow{\scriptstyle \iota_M} & & \;\Big\uparrow{\scriptstyle \iota_N} & & & & \\[2pt]
M & \xrightarrow{\;f\;} & N & \xrightarrow{\;q\;} & H & \longrightarrow & 0
\end{array}
$$

in the category of abstract left A-modules. From which we deduce the existence of a unique homomorphism of abstract left A-modules $j : H \to K$ such that the rightmost square of diagram (1) becomes commutative. If now L is any C-complete left A-module and $g : H \to L$ is any homomorphism of abstract left A-modules, then $g \circ q : N \to L$ is a homomorphism of left A-modules. Then, by the universal mapping property, condition (3) of Definition 1 applied to N, there exists a unique infinitely linear function $g_1 : C(N) \to L$ such that $g_1 \circ \iota_N = g \circ q$. Then

$$g_1 \circ C(f) \circ \iota_M = g_1 \circ \iota_N \circ f = g \circ q \circ f = 0 = 0 \circ \iota_M.$$

Therefore $g_1 \circ C(f)$ and 0 are both infinitely linear functions from $C(M)$ into L that compose with ι_M to make zero. By the uniqueness part of universality, part (3) of Definition 1 applied to M, we therefore have that

$$g_1 \circ C(f) = 0.$$

Since $K = \text{Cok}(C(f))$, there exists a unique infinitely linear function $\overline{g} : K \to L$ such that

$$\overline{g} \circ p = g_1.$$

But then

$$\overline{g} \circ j \circ q = \overline{g} \circ p \circ \iota_N = g_1 \circ \iota_N = g \circ q.$$

Since q is an epimorphism of abstract left A-modules, this implies that

$$\overline{g} \circ j = g.$$

On the other hand, if $\overline{g}'$ is any other infinitely linear function from K into L such that

$$\overline{g}' \circ j = g, \text{ then}$$

$$(\overline{g} - \overline{g}') \circ p \circ \iota_N = (\overline{g} - \overline{g}') \circ j \circ q = (g - g) \circ q = 0 = 0 \circ \iota_N$$

whence by the uniqueness part of universality in Definition 1 (Definition 1, part (3)) for N, we have that

$$(\overline{g} - \overline{g}') \circ p = 0.$$

Since p is surjective, this implies $\overline{g} = \overline{g}'$, completing the proof of Step 2.

Step 3. Proof of part (1) of the Theorem: If M is any abstract left A-module, then let

$$f : F_1 \to F_2$$

be a homomorphism of free left A-modules such that

$$M \approx \text{Cok}(f).$$

Then by Step 2 applied to $f : F_1 \to F_2$, we have that $C(M)$ exists.

Step 4. Proof of Part (3) of the Theorem: Let

$$M \xrightarrow{f} N \xrightarrow{q} H \to 0$$

be exact in the category of abstract left A-modules. Then by Step 2, we have that

$$C(H) = \text{Cok}(C(f) : C(M) \to C(N))$$

in $\mathscr{C}_A$ — and therefore, that the sequence:

$$C(M) \xrightarrow{C(f)} C(N) \xrightarrow{C(q)} C(H) \to 0$$

is exact in $\mathscr{C}_A$. Since also from Definition 1 it follows readily that the functor C is additive, we have therefore that C is right exact. $\qquad\square$

Remark 1. An alternative construction of $C(M)$, for M an abstract left A-module, is as follows. Let M be given, as an abstract left A-module, by the generators $(e_i)_{i \in I}$ and the relations $(r_j)_{j \in J}$, where r_j is an element of the free abstract left A-module on the set $\{e_i : i \in I\}$. Then by §4, Definition 1, it follows readily that the C-complete left A-module N given by the same family $(e_i)_{i \in I}$ of generators, and the same family $(r_j)_{j \in J}$ of relations, obeys the universal mapping property for a C-completion of M as given in Definition 1 above; and that therefore $C(M)$ exists, and is in fact canonically isomorphic to the N so constructed.

Remark 2. By §4, Theorem 2.4.1, and Definition 1, every C-complete left A-module can be determined by a matrix $(a_{ij})_{(i,j) \in I \times J}$, where I and J are sets, and the matrix (a_{ij}) is such that, for every $i \in I$, the ith column $(a_{ij})_{j \in J}$ of the matrix is a zero-bound family of elments of $\widehat{A}$. (Of course, such a matrix (a_{ij}) is not uniquely determined by the C-complete left A-module M; but rather every such M, up to isomorphism in $\mathscr{C}_A$, comes from a (not uniquely determined) such matrix.) By Remark 1 above, it is clear that: A C-complete left A-module M is isomorphic (as a C-complete left A-module) to $C(N)$, for some abstract left A-module N, iff there is a matrix $(a_{ij})_{(i,j) \in I \times J}$ that gives M, such that, for every $i \in I$, the ith column: $(a_{ij})_{j \in J}$ is (not merely zero-bound in $\widehat{A}$, but) such that (all but finitely many entries are zero, and also such that) all entires are elements of the image of A in $\widehat{A}$. (We will see later that if the topological ring A obeys the extremely mild hypotheses that $\widehat{A}$ is *admissible*, and is such that $\widehat{A}$ has a *denumerable neighborhood base at zero*, then this is always the case for every C-complete left A-module M — i.e., that every such M is isomorphic to $C(N)$ for some abstract left A-module N. In fact, in that case, $M \approx C(M)$. Where, in this latter formula, $C(M)$ means "C(abstract left A-module of M)", i.e., $C(U(M))$.

Corollary 2.5.3. *The hypotheses being as in Theorem 2.5.2, we have that the functor C, from the category of abstract left A-modules into the category $\mathscr{C}_A$, preserves direct limits of covariant functors over arbitrary set-theoretically legitimate categories.*

Proof. The proof is entirely similar to that of the proof of Step 2 of Theorem 2.5.2.   □

Remark 3. A special case of Corollary 2.5.3 is that the functor C preserves arbitrary direct sums of objects. (In fact, this observation, plus right exactness of C, is equivalent to proving Corollary 2.5.3.)

Remark 4. In Corollary 2.5.3 and Remark 3 above, it is important to remember that, in both of these statements, C is regarded as a functor from the category of abstract left A-modules, *into the category $\mathscr{C}_A$ of C-complete left A-modules*. If one thinks of C as being a functor *from the category of abstract left A-modules into itself*, then both Corollary 2.5.3 and Remark 3 fail miserably. (For, e.g., if the topological ring A is complete, then by Theorem 2.5.2, part (2), for every free left A-module of finite rank F, (e.g., for $F = A$), we have

$C(F) = F$. If C preserved infinite direct sums from the category of abstract left A-modules into itself, then $C(F) = F$ for all free left A-modules F. But, by conclusion (2) of Theorem 2.5.2, $C(F) = \widehat{F}$. Therefore $\widehat{F} = F$, for all free left A-modules F – and, e.g., if A admits a denumerable neighborhood base at zero, this is so iff the topology on A is the discrete topology. Therefore, it follows, except in such degenerate cases as that in which the closure of $\{0\}$ in A is open in the topological ring A, that the functor C does *not* preserve arbitrary direct sums, considered as a functor from the category of abstract left A-modules into itself.)

Corollary 2.5.4. *Let M be an arbitrary abstract left A-module, and let $\iota_M :$ $M \to C(M)$ be the natural homomorphism (of abstract left A-modules) from M into its C-completion $C(M)$. Then the (abstract left A-submodule) $\iota_M(M)$ is a set of generators for $C(M)$ as a C-complete left A-module.*

Proof. By Theorem 2.5.2, part (2), we have that this is true if M is a free abstract left A-module. Otherwise, choose an epimorphsim:

$$f : F \to M$$

of abstract left A-modules with F free. Then since $\iota_F(F)$ generates $C(F)$ as a C-complete left A-module and since $C(f) : C(F) \to C(M)$ is onto, it follows that the image $\iota_M(M)$ of this set in $C(M)$ likewise generates $C(M)$ as a C-complete left A-module.   □

The next two Corollaries of Theorem 2.5.2 are important.

Corollary 2.5.5. *The functor C, from the category of abstract left A-modules into the category of C-complete left A-modules, is the zeroth left derived functor of the functor: $M \rightsquigarrow \widehat{M}$ (constructed in §2, Example 1).*

Corollary 2.5.6. *The functor C, considered as a functor from the category of abstract left A-modules into itself, is the zeroth left derived functor of the functor: $M \rightsquigarrow \widehat{M}$ considered as a functor from the category of abstract left A-modules into itself.*

Proof of Corollary 2.5.5 and Corollary 2.5.6. By Theorem 2.5.2, part (2), the functors "C" and "$\widehat{}$" (whether considered as functors into $\mathscr{C}_A$ or into the category of abstract left A-modules), agree on projective objects. Conclusion (3) of Theorem 2.5.2 completes the proof.   □

Another way of phrasing Corollary 2.5.5 (respectively, Corollary 2.5.6), is that (1), (2) and (3) of Theorem 1, is an axiomatic description of the functor C, from the category of abstract left A-modules into $\mathscr{C}_A$ (respectively: into itself).

Remark 5. We will see later that under the extremely mild hypotheses; e.g., that the topological ring A is admissible and admits a denumerable neighborhood base at zero; then for every abstract left A-module M, we have that $\widehat{M}$ can be recovered from $C(M)$ by the formula:

$$C(M)/(A\text{-divisible elements}) \approx \widehat{M}.$$

Therefore, $C(M)$ is a stronger, more subtle, invariant than $\widehat{M}$. (For in these very general cases, $\widehat{M}$ can be interpreted as being the "cruder" object, "$C(M)$ made Hausdorff".)

Remark 6. A great advantage of $C(M)$ over $\widehat{M}$, is that the functor $C(M)$ is right exact. (This is almost never true for the functor $\widehat{M}$.) For, using right exactness of C — (and also, occasionally, that the category $\mathscr{C}_A$ is abelian) — one can apply the well-established machinery of homological algebra to study and prove theorems about C; that are all but unapproachable for "$\widehat{}$". For example, a further corollary of Corollary 2.5.5 is,

Corollary 2.5.7. *Let M be an arbitrary abstract left A-module. Then we have a natural infinitely linear function:*

$$\rho_M : C(M) \to \widehat{M}.$$

The functions, ρ_M, for all abstract left A-modules M, constitute a natural transformation of functors from the functor "C" into the functor "$\widehat{}$", where "C" and "$\widehat{}$" are regarded as functors from the category of abstract left A-modules into the category $\mathscr{C}_A$ of C-complete left A-modules.

Note. Also, for every abstract left A-module M, the diagram:

is commutative (in the category of abstract left A-modules), where $M \to \widehat{M}$ is the natural map from M into its A-adic completion.

Proof. Follows immediately from Corollary 2.5.5 and the general theory of derived functors. (E.g., see [CEHA].)   □

Proposition 2.5.8. *Let A be an arbitrary topological ring such that the topology is given by right ideals and let $\widehat{A}$ be the completion of A. For every abstract left A-module (resp.: left $\widehat{A}$-module), let $C^A(M)$ (resp.: $C^{\widehat{A}}(M)$) denote the C-completion of M regarded as an abstract left A-(resp.: $\widehat{A}$-) module. Then for every abstract left A-module M, we have the canonical isomorphism of C-complete left A-modules (in fact, of C-complete left $\widehat{A}$-modules)*

$$C^A(M) \approx C^{\widehat{A}}(\widehat{A} \underset{A}{\otimes} M).$$

Proof. We have observed in §2 that the categories of C-complete left A-modules, and of C-complete left $\widehat{A}$-modules, are canonically isomorphic. The functor $M \rightsquigarrow C^{\widehat{A}}(\widehat{A} \underset{A}{\otimes} M)$ is right exact, and maps every free abstract left A-module F into $\widehat{F}$. The proposition therefore follows from Corollary 2.5.5.   □

Remark 7. Proposition 2.5.8 shows, quite clearly, why the study of "C-completion over A of abstract left A-modules" reduces to that of "C-completion over $\widehat{A}$ of abstract left $\widehat{A}$-modules"; and why, in many cases, to prove certain theorems it suffices to consider the case in which the topological ring A is complete.

Chapter 3

The Case of an Admissible Topological Ring

3.1 Topological Rings that are C-O.K.

<u>Construction 1</u>. Let A be a topological ring such that the topology is given by right ideals, and let M be an abstract left A-module. Then, see Chapter 2, §5, we have the C-completion $C(M)$ of M and the homomorphism of abstract left A-modules, $\iota_M : M \to C(M)$. Therefore in particular we have the homomorphism of abstract left A-modules

$$(1): \quad \iota_{C(M)} : C(M) \to C(C(M)).$$

On the other hand, the image of ι_M under the functor C is also a homomorphism of abstract left A-modules

$$(2): \quad C(\iota_M) : C(M) \to C(C(M)).$$

This latter, $C(\iota_M)$, is infinitely linear.

$C(M)$ is C-complete, and the identity mapping from $C(M)$ into itself is a homomorphsim of abstract left A-modules. Therefore, by the universal mapping property definition of the C-completion $((C(C(M))), \iota_{C(M)})$ of $C(M)$ regarded as an abstract left A-module, there exists a unique infinitely linear function

$$(3): \quad k_M : C(C(M)) \to C(M)$$

such that

$$(4): \quad k_M \circ \iota_{C(M)} = (\text{the identity of } C(M)).$$

Therefore, the homomorphism of abstract left A-modules $\iota_{C(M)}$ is always a split monomorphism, and k_M is a splitting.

The definition of $C(\iota_M)$, by universal mapping properties, is that it is the unique infinitely linear function such that the diagram

$$(5): \quad \begin{array}{ccc} C(M) & \xrightarrow{\; C(\iota_M) \;} & C(C(M)) \\[4pt] {\scriptstyle \iota_M}\Big\uparrow & & \Big\uparrow{\scriptstyle \iota_{C(M)}} \\[4pt] M & \xrightarrow{\; \iota_M \;} & C(M) \end{array}$$

is commutative. Therefore the functions $C(\iota_M)$ and $\iota_{C(M)}$ agree on the subset $\iota_M(M)$ of $C(M)$. But then, using equations (4) and (5),

$$k_M \circ C(\iota_M) \circ \iota_M = k_M \circ \iota_{C(M)} \circ \iota_M = \iota_M.$$

Therefore the infinitely linear function,

$$k_M \circ C(\iota_M) : C(M) \to C(M),$$

agrees with the identity of $C(M)$ on the subset $\iota_M(M)$. Since $\iota_M(M)$ is a set of generators for $C(M)$ as a C-complete left A-module, it follows that

$$(6): \quad k_M \circ C(\iota_M) = (\text{the identity of } C(M)).$$

Since k_M and $C(\iota_M)$ are both infinitely linear, this implies that $C(\iota_M)$ is a split monomorphism in the category of C-complete left A-modules, with the canonical splitting k_M.

Combining (4) and (6), it follows that:

Lemma 3.1.1. *There exists a unique automorphism*

$$(7): \quad \alpha_M : C(C(M)) \xrightarrow{\approx} C(C(M))$$

of $C(C(M))$ as an abstract left A-module such that

$$(8): \quad \alpha_M \circ \iota_{C(M)} = C(\iota_M)$$

and $(9): \quad k_M \circ \alpha_M = k_M.$

Proof. Equation (4) implies that the homomorphism of abstract left A-modules:

$$(10): \quad C(M) \oplus \mathrm{Ker}(k_M) \to C(C(M))$$

that induces $\iota_{C(M)}$ on $C(M)$, and the inclusion on $\mathrm{Ker}(k_M)$, is an isomorphism of abstract left A-modules. Likewise, equation (6) implies that the homomorphism of C-complete left A-modules

$$(11): \quad C(M) \oplus \mathrm{Ker}(k_M) \to C(C(M))$$

that induces $C(\iota_M)$ on $C(M)$, and the inclusion on $\mathrm{Ker}(k_M)$, is an isomorphism of C-complete left A-modules. Define α_M to be the inverse of (10) followed by (11). Then (7), (8), and (9) hold, and α_M is clearly uniquely determined by (8) and (9). □

Next, consider the exact sequence of abstract left A-modules

$$M \overset{\iota_M}{\to} C(M) \to C(M)/(\iota_M(M)) \to 0.$$

The image under the right-exact functor C is the exact sequence of C-complete left A-modules

$$C(M) \overset{C(\iota_M)}{\to} C(C(M)) \to C(C(M)/\iota_M(M)) \to 0.$$

Since $C(\iota_M)$ is a split monomorphism, this sequence implies that

$$(12): \ \mathtt{Ker}(k_M) \approx C(C(M)/\iota_M(M))$$

canonically as C-complete left A-modules. Combining with the isomorphism (11), it follows that

$$(13): \ C(C(M)) \approx C(M) \oplus C(C(M)/\iota_M(M))$$

canonically in the category of C-complete left A-modules (in this isomorphism, the projection: $C(C(M)) \to C(C(M)/\iota_M(M))$ is the image under C of the natural map: $C(M) \to C(M)/\iota_M(M)$; and the inclusion $C(M) \to C(C(M))$ is $C(\iota_M)$).

Remark 1. Consider $C(\iota_M) - \iota_{C(M)}$, a map from $C(M)$ into $C(C(M))$. By the commutative diagram (5), the map $C(\iota_M) - \iota_{C(M)}$ induces a homomorphism of abstract left A-modules

$$C(M)/\iota_M(M) \to C(C(M)).$$

Then, by the universal mapping property definition of the C-completion of $C(M)/\iota_M(M)$, there is therefore an induced infinitely linear function

$$C(C(M)/\iota_M(M)) \to C(C(M)).$$

Most likely, this latter infinitely linear function corresponds under the canonical isomorphism (13), to the inclusion

$$C(C(M)/\iota_M(M)) \to C(C(M)).$$

Proposition 3.1.2. *Let A be a topological ring such that the topology is given by right ideals, and let M be an abstract left A-module. Then the following eight conditions are equivalent.*

[1] The natural map $\iota_{C(M)} : C(M) \to C(C(M))$ from $C(M)$ into its C-completion $C(C(M))$ is an isomorphism of abstract left A-modules; equivalently, of C-complete left A-modules.

[2] The image under the functor C of the natural mapping $\iota_M : M \to C(M)$ is an isomorphism, $C(\iota_M) : C(M) \overset{\approx}{\to} C(C(M))$.

[3] $\iota_{C(M)}$ is infinitely linear.

[4] The automorphism α_M of the abstract left A-module $C(C(M))$ is infinitely linear.

[5] The automorphism α_M of $C(C(M))$ is the identity.

[6] $\iota_{C(M)} = C(\iota_M)$.

[7] The map k_M is an isomorphism $k_M : C(C(M)) \xrightarrow{\approx} C(M)$.

[8] $C(C(M)/\iota_M(M)) = 0$.

Proof. By equation (4), condition **[1]** iff condition **[7]**. By equation (6), condition **[2]** iff condition **[7]**. Since by equation (4), k_M is an epimorphism, by equation (12), condition **[8]** iff condition **[7]**.

[7] implies **[6]**: If k_M is an isomorphism, then by equations (4) and (6), $C(\iota_M) = k_M^{-1} = \iota_{C(M)}$.

[6] implies **[3]**: Since $C(\iota_M)$ is infinitely linear, if $\iota_{C(M)} = C(\iota_M)$, then $\iota_{C(M)}$ is infinitely linear.

[3] implies **[1]**: $\iota_{C(M)}$ is the natural map from the abstract left A-module $C(M)$ into its C-completion $C(C(M))$. Therefore the set-theoretic image $\iota_{C(M)}(C(M))$ of $\iota_{C(M)}$ generates $C(C(M))$ as a C-complete left A-module. If $\iota_{C(M)}$ is also infinitely linear, then, as for every infinitely linear function, the set-theoretic image of $\iota_{C(M)}$ is a C-complete left A-submodule of $C(C(M))$. But since this set-theoretic image generates $C(C(M))$ as a C-complete left A-module, it follows that the set-theoretic image is all of $C(C(M))$. I.e., that $\iota_{C(M)}$ is surjective. Then by equation (4), $\iota_{C(M)}$ is an isomorphism.

By equation (8), we have that **[5]** iff **[6]**.

Also by equation (8), **[4]** implies **[3]**. And **[5]** implies **[4]** is obvious. $\qquad\square$

Definition 1. Under the hypotheses of Proposition 3.1.2, when the eight equivalent conditions of Proposition 3.1.2 hold, we say that

$$C(M) = C(C(M)).$$

Construction 2. Let A be a topological ring such that the topology is given by right ideals. Let M be a C-complete left A-module. Then, if we regard M as being an abstract left A-module, we have the natural mapping

$$(1): \quad \iota_M : M \to C(M)$$

from M into its C-completion. This is a homomorphism of abstract left A-modules. Since M is C-complete, by the universal mapping property definition of the C-completion, applied to the identity mapping of M, $M \to M$, we deduce that there exists a unique infinitely linear function

$$(2): \quad j_M : C(M) \to M$$

such that

$$(3): \quad j_m \circ \iota_M = \text{(identity of } M\text{)}.$$

Equation (3) implies that we have a canonical isomorphism of abstract left A-modules

$$(4): \quad \beta_M : C(M) \xrightarrow{\approx} M \oplus \text{Ker}(j_M)$$

(β_M is such that, the injection $M \to C(M)$ is ι_M; the inclusion: $\text{Ker}(j_M) \to C(M)$ is the set-theoretic inclusion; and the projection: $C(M) \to M$ is j_M).

Remark 2. Notice that, although both sides of equation (4) are C-complete left A-modules, that the isomorphism β_M is not, in general, infinitely linear (see Proposition 3.1.3 below).

Remark 3. In the special case in which $M = C(N)$ for some abstract left A-module N, then the notations of Constructions 1 and 2 above correspond as follows: $j_M = k_N$, and β_M is the isomorphism (10) of Construction 1 (where in Construction 1, "M" is replaced by "N").

Proposition 3.1.3. *Let A be a topological ring such that the topology is given by right ideals, and let M be an arbitrary C-complete left A-module. Then the following conditions are equivalent.*

[1] The natural map, from the underlying abstract left A-module of M into its C-completion, is an isomorphism (of abstract left A-modules; or, equivalently, of C-complete left A-modules)

$$\iota_M : M \overset{\approx}{\to} C(M).$$

[2] ι_M is infinitely linear.
[3] The infinitely linear function $j_M : C(M) \to M$ is an isomorphism.
[4] The isomorphism β_M is infinitely linear.

Proof. The proof is similar to that of Proposition 3.1.2, but a little bit easier. □

Definition 2. Let A be a topological ring such that the topology is given by right ideals and let M be an abstract left A-module. Then we say that $M = C(M)$ iff the natural homomorphism of abstract left A-modules

$$\iota_M : M \to C(M)$$

from M into its C-completion is an isomorphism of abstract left A-modules.

Remark 4. Let the hypotheses and notations be as in Definition 2. Then, in the special case that $M = C(N)$ for some abstract left A-module N, then $M = C(M)$ in the sense of Definition 2 above iff $C(N) = C(C(N))$ in the sense of Definition 1. This is so, by condition [1] of Proposition 1.

Therefore, Definition 2 generalizes Definition 1. (Definition 1 makes sense only for C-complete left A-modules of the form $C(M)$; while Definition 2 makes sense for arbitrary abstract left A-modules, whether of the form $C(M)$ or not.)

Remark 5. Let the hypotheses and notations be as in Proposition 3.1.3. Then the C-complete left A-module M obeys the four equivalent conditions of Proposition 2 iff $M = C(M)$ in the sense of Definition 2 (or, more precisely, iff the underlying abstract left A-module of M obeys the condition of Definition 2). This follows from Proposition 3.1.3, condition [1].

Proposition 3.1.4. *Let A and M be as in Definition 2, and suppose that $M = C(M)$ in the sense of Definition 2. Then there exists one and only one infinite sum structure on the abstract left A-module M.*

Proof. Since ι_M : $M \to C(M)$ is an isomorphism, and since $C(M)$ is C-complete, there exists an infinte sum structure on M. Suppose that there is another infinite sum structure on the abstract left A-module M, and let M' denote the C-complete left A-module thus obtained. Regard M as being a C-complete left A-module by requiring that ι_M be infinitely linear. Then, the set-theoretic identity: $M \to M'$ is a homomorphism of abstract left A-modules into a C-complete. By the universal mapping property, there exists a unique infinitely linear function

$$\gamma : C(M) \to M'$$

such that $\gamma \circ \iota_M$ = (identity of M). Therefore $\gamma \circ \iota_M$ is infinitely linear; or, otherwise stated, the identity of M: $M \to M'$ is infinitely linear. Therefore the infinite sum structure of M' must be the one just defined on M.   □

Example 1. Let A be a complete topological ring such that the topology is given by right ideals. Let M be a finitely generated left A-module. Then M admits an infinite sum structure iff $M = C(M)$.

Proof. Clearly, if $M = C(M)$, then M admits an infinite sum structure. Conversely, suppose that M admits an infinite sum structure. Then, by Chapter 2, §4, Proposition 2.4.11, since $C(M)$ is C-complete, we have that the natural map

$$\iota_M : M \to C(M)$$

is infinitely linear. But then, condition [2] of Proposition 3.1.3 holds; so that by Remark 5 above, we have that $M = C(M)$.   □

Example 2. Let A be a complete topological ring such that the topology is given by right ideals and let M be a finitely presented abstract left A-module. Then $M = C(M)$.

Proof. By Chapter 2, §4, Corollary 2.4.12, we have that M admits a unique infinite sum structure. Therefore by Example 1 above, $M = C(M)$.   □

Theorem 3.1.5. *Let A be a topological ring such that the topology is given by right ideals. Then the following eight conditions are equivalent.*

[1] For all abstract left A-modules M, we have that $C(M) = C(C(M))$.

[2] For all free left A-modules F, we have that $\widehat{F} = C(\widehat{F})$.

[3] For all C-complete left A-modlues M, we have that $M = C(M)$.

[4] For every abstract left A-module M, there exists at most one infinite sum structure on M.

[5] If M and N are C-complete left A-modules and if f : $M \to N$ is a function, then f is infinitely linear iff f is a homomorphism of abstract left A-modules.

[6] The (abelian) category $\mathscr{C}_A$ of all C-complete left A-modules is a full subcategory of the category of abstract left A-modules.

[7] For all abstract free left A-modules F, we have that

$$C(\widehat{F}/F) = 0.$$

[8] For all abstract left A-modules M, we have that

$$C(C(M)/\iota_M(M)) = 0.$$

Proof. [5] iff [6] is immediate.

[3] implies [1]: $C(M)$ is C-complete. Therefore by [3], with $C(M)$ replacing M, we obtain [1].

[2] implies [3]: By Chapter 2, §4, Theorem 2.4.1, if M is a C-complete left A-module, then M is a cokernel of an infinitely linear map of completions of free abstract left A-modules:

$$M \approx \mathtt{Cok}(\widehat{F} \to \widehat{G})$$

where F and G are free abstract left A-modules. The functor C is right-exact. Therefore it suffices to prove [3] in the case $M = \widehat{F}$, where F is a free left A-module. And this is condition [2].

[3] implies [5]: Given f a homomorphism of abstract left A-modules as in [5], from the commutative diagram

$$
\begin{array}{ccc}
C(M) & \xrightarrow{\;C(f)\;} & C(N) \\
{\scriptstyle \iota_M}\big\uparrow & & \big\uparrow{\scriptstyle \iota_N} \\
M & \xrightarrow{\;\;f\;\;} & N
\end{array}
$$

and the fact that ι_M and ι_N are isomorphisms (by condition [3]), we deduce that f is infinitely linear.

[5] implies [4]: Let (S) and (S') be two infinite sum structures on M. Then the identity function of M is a homomorphism of abstract left A-modules; therefore by condition [5] the identity of M is infinitely linear, from $(M, (S))$ into $(M, (S'))$. Otherwise stated, the infinite sum structures (S) and (S') coincide.

[4] implies [1]: Given M as in [1], we have the automorphism α_M of the abstract left A-module $C(C(M))$, as in Definition 1. Using α_M, we can define a new infinite sum structure on $C(C(M))$. By condition [4], this must be the given infinite sum structure of $C(C(M))$. Otherwise stated, α_M is infinitely linear. Then by Proposition 3.1.2 we have $C(M) = C(C(M))$.

[1] iff [8]: This is a portion of Proposition 3.1.2.

[2] iff [7]: Proposition 3.1.2 applied to $M = \widehat{F}$.

[1] implies [2]: condition [1] in the special case $M = $ a free A-module F, is condition [2]. □

Remark 6. When the eight equivalent conditions of Theorem 3.1.5 hold, then, e.g., by condition [6], the property of "being a C-complete left A-module" can be thought of as being a *condition* on an abstract left A-module, rather than an additional structure.

Definition 3. Let A be a topological ring such that the topology is given by right ideals. Then we say that the topological ring A is *C-O.K.* iff A obeys the eight equivalent conditions of Theorem 3.1.5.

The following theorem shows that virtually every topological ring that one might expect to encounter in commutative algebra or in algebraic geometry is C-O.K.

Theorem 3.1.6. *Let A be a topological ring such that the topology is given by denumerably many right ideals. Then the following three conditions are equivalent.*

[1] A is C-O.K.
[2] A is admissible.
[3] A is strongly admissible.

Proof. [1] implies [3]: Let M be an abstract left A-module. Then, since by hypothesis the topology of A is such that A admits a denumerable neighborhood base at zero, it follows that every element of $\widehat{M}$ is the limit of a denumerable Cauchy series in M. Therefore the natural infinitely linear function

$$(1): \quad \rho_M : C(M) \to \widehat{M}$$

is surjective. Similarly,

$$(2): \quad \rho_{\widehat{M}} : C(\widehat{M}) \to \widehat{\widehat{M}}$$

is surjective. But, since the ring A is C-O.K. and $\widehat{M}$ is C-complete, we have that $\widehat{M} = C(\widehat{M})$, i.e., that the natural map is an isomorphism

$$(3): \quad \iota_{\widehat{M}} : \widehat{M} \to C(\widehat{M}).$$

Since (2) is surjective and (3) is an isomorphism, the composite

$$(4): \quad \widehat{M} \overset{\iota_{\widehat{M}}}{\to} C(\widehat{M}) \overset{\rho_{\widehat{M}}}{\to} \widehat{\widehat{M}}$$

is surjective. But this latter composite is the natural map from $\widehat{M}$ into its A-adic completion. (See Chapter 2, §5, Note to Corollary 2.5.7, with $\widehat{M}$ replacing M.) Since also $\widehat{M}$ is A-adically Hausdorff (since in fact $\widehat{M}$ is even Hausdorff (since complete) for its uniform topology, and the A-adic topology is finer than the uniform topology), it follows that the natural map from the abstract left A-module $\widehat{M}$ into its A-adic completion $\widehat{\widehat{M}}$ is an isomorphism. Otherwise stated, $\widehat{M}$ is A-adically complete. M being an arbitrary abstract left A-module, by Chapter 1, §3, Definition 2 this implies that A is strongly admissible.
[3] implies [2]: This is immediate.
[2] implies [1]: This proof is delayed until the next section. □

Corollary 3.1.7. *Let A be a topological ring such that the topology is given by right ideals. If A is C-O.K., then A is admissible.*

Proof. Let F be a free abstract left A-module. Then, by Chapter 1, §2, Corollary 1.2.9, we have the homomorphisms of abstract left A-modules ι and $\widehat{k}$, from $\widehat{F}$ into $\widehat{\widehat{F}}$. Since $\widehat{F}$ and $\widehat{\widehat{F}}$ are A-adic completions of abstract left A-modules, they are objects in $\mathscr{C}_A$; and since the ring A is C-O.K., by Theorem 3.1.5 we have that any homomorphism of abstract left A-modules, from one C-complete to another is infinitely linear. Therefore ι and $\hat{k}$ are infinitely linear. (Of course, $\hat{k}$ is always infinitely linear, whether or not the topological ring A is C-O.K.) If $(e_i)_{i \in I}$ is a basis of F as an abstract left A-module, then every $x \in \hat{F}$ can be written uniquely in the form $\sum_{i \in I} a_i e_i$, where $(a_i)_{i \in I}$ is a family indexed by I of elements in $\hat{A}$ that is zero-bound; the sum being the one deduced from the infinite sum structure derived from the uniform topology of the $\hat{A}$-module $\hat{F}$. In particular, it follows that the image $\iota_F(F)$ of F in $\hat{F}$ generates $\hat{F}$ as a C-complete left A-module. By Chapter 1, §2, Corollary 1.2.9 we have that $\hat{k}$ and ι agree on the subset $\iota_F(F)$. Since $\hat{k}$ and ι are infinitely linear, and $\iota_F(F)$ generates $\hat{F}$ as a C-complete left A-module, it follows that $\iota = \hat{k}$. But then, by Chapter 1, §3, Definition 1, we have that the topological ring A is admissible.   □

Remark 7. Theorem 3.1.6 above shows the true significance of a topological ring, such that the topology is given by right ideals, being admissible.

Remark 8. The proof of the implication, "[1] implies [3]," in Theorem 3.1.6 above, used that the topology of the ring A is such that there is a denumerable neighborhood base at zero, in only one point: Namely, to know that every element of $\hat{M}$ is the limit of some denumerable Cauchy series of elements of M. Perhaps some weaker condition than the hypothesis of a denumerable neighborhood base might also imply this. One might then hope to generalize Theorem 3.1.6 to rings obeying such a milder condition.

Remark 9. Let G be a topological abelian group. Then let us call G *series complete* iff for every set I and every indexed family $(a_i)_{i \in I}$ of elements of G such that the infinite series

$$(1): \quad \sum_{i \in I} a_i$$

in G is a Cauchy series, we have that the Cauchy series (1) has a unique limit in G. Define the *series completion* of a topological abelian group G, to be a universal series complete topological abelian group into which G maps. Then it is easy to see that:

(1) A complete topological abelian group is series complete.

(2) Every topological abelian group G admits a series completion unique up to isomorphism. The series completion is canonically isomorphic to a subgroup of the completion $\hat{G}$, together with the induced topology.

(3) I do not know whether or not *in general* the conditions of "series complete" and "complete" are identical for topological abelian groups. It seems very unlikely (see, however, the next Remark).

(4) Let A be a topological ring such that the topology is given by right ideals. Then the series completion A' of A is a topological subring (with the induced topology) of $\hat{A}$.

(5) One could attempt to generalize the notion of "infinite sum structure on an A-module M" by requiring only that the datum (S) of Chapter 2, §2, Definition 1, be only defined for those $(a_i)_{i \in I}$ that are elements of $\underline{A'}$ (rather than

$\widehat{A}$) and that are zero-bound in $\widehat{A'}$. It appears that almost all of the formalism of the last chapter, and also of this section, should go through to such a generalized notion of "series C-complete left A-modules," and "series C-completion." (The corresponding notion of "series strongly admissible" is always equivalent to "series admissible").

(6) However, I do not know if such a definition would actually be a generalization (since, in (3) above, I have noted that I do not know whether or not "series complete" is a strictly more general condition than "complete" — although it probably is).

(7) In the denumerable neighborhood base case, of course, "series complete" is the same as "complete", "series C-completion" is the same as "C-completion", etc.

Remark 10. A topological abelian group is *sequentially complete* iff every Cauchy sequence (equivalently, every denumerable Cauchy series) has a unique limit. If A is a topological abelian group, then the *sequential completion* of A is a universal sequentially complete topological abelian group into which A maps. The sequential completion of A always exists, and is canonically isomorphic to the set of all limits of convergent sequences (equivalently: convergent denumerable series) of elements of A in the completion $\widehat{A}$ of A. If A is a topological ring, then the sequential completion of A has a natural structure as a topological ring. If A is a topological ring such that the topology is given by right ideals, and if M is an abstract left A-module, then one can define a *sequential infinite sum structure* on M to be, a datum as in Chapter 2, §2, Definition 1, but in which all of the indexing sets I are required to be of cardinality $\leq \aleph_0$. This gives rise to a notion, likewise, of the *sequential C-completion of an* abstract left A-module. Most, if not all, of the formalism that we have developed until now goes through to this "sequential C-completion". (There is no question that this construction would be in general different from the C-completion.) And one then has the mild advantage, that in Theorem 3.1.6 above, one can eliminate the hypothesis that "the neighborhood system at zero in the topological ring A admits a denumerable base".

Remark 11. However, in the case that the topological ring A is such that there exists a denumerable neighborhood base at zero, then the notions of "C-completion," "series C-completion" and "sequential C-completion" all coincide. And to my knowledge the case in which there exists a denumerable neighborhood base at zero includes all cases and examples studied seriously in commutative algebra and algebraic geometry to date.

Example 3. Let A be a commutative ring, and let I be a finitely generated ideal in A. Regard A as a topological ring with the I-adic topology. Then we have seen (see Chapter 1, §3, Example 6 following Definition 2) that A is strongly admissible. Therefore by Theorem 3.1.6 A is C-O.K.

Example 4. Let A be a ring, and let $I_1, I_2, \ldots$, be a sequence of two-sided ideals in the ring A, each of which is finitely generated as a right ideal. Let τ be the topology on the ring A, such that a neighborhood base at zero is given by the ideals, the set of all finite products of the ideals $I_1, I_2, \ldots$. Then A is a topological ring, and A is strongly admissible. (See Chapter 1, §3, Example 2 following Definition 2.) Therefore, by Theorem 3.1.6, A is C-O.K.

Of course, Example 3 above includes all examples seriously studied to date in commutative algebra and algebraic geometry.

(An example more general than Example 4 above is Example 3 in Chapter 1, §3, immediately after Definition 2.)

Proposition 3.1.8. *Let A be a topological ring such that the topology is given by right ideals. Let C^A, resp. $C^{\widehat{A}}$, denote the functor "C-completion" from the category of abstract left A-modules (resp. the category of abstract left $\widehat{A}$-modules) into the category $\mathscr{C}_A$ of C-complete left A-modules.[1] Then the following two conditions are equivalent.*

[1] $C^A(\widehat{A}) = \widehat{A}$.

[2] For every abstract left $\widehat{A}$-module M, we have that the natural map:

$$C^A(M) \to C^{\widehat{A}}(M)$$

is an isomorphism of C-complete left A-modules (equivalently: of abstract left A-modules).

Note. Let $K = \text{Ker}(\mu : \widehat{A} \underset{A}{\otimes} \widehat{A} \to \widehat{A})$, where $\mu(a \underset{A}{\otimes} b) = ab$, for all $a, b \in \widehat{A}$. Regard K as an abstract left $\widehat{A}$-module. Then another condition equivalent to [1] and [2] is:

[3] $C^{\widehat{A}}(K) = 0$.

Proof of the Proposition. First, note that by Chapter 2, §5, Proposition 2.5.8, if M is any abstract left $\widehat{A}$ module $C^A(M) = C^{\widehat{A}}(\widehat{A} \underset{A}{\otimes} M)$. Therefore the natural map: $\widehat{A} \underset{A}{\otimes} M \to M$ induces a map:

$$C^A(M) = C^{\widehat{A}}(\widehat{A} \underset{A}{\otimes} M) \to C^{\widehat{A}}(M),$$

which is the "natural map" referred to in condition [2]. (The natural map: $C^A(M) \to C^{\widehat{A}}(M)$ can also be constructed by using universal mapping properties.)

[2] implies [1]: Take $M = \widehat{A}$. Then $C^A(\widehat{A}) = C^{\widehat{A}}(\widehat{A}) = \widehat{A}$.

[1] implies [2]: By Chapter 2, §5, Corollary 2.5.3 and Remark 4, we have that the functor C^A, from the category of abstract left A-modules into the category $\mathscr{C}_A$ (but *not* into the category of abstract left A-modules — see Chapter 2, §5, end of Remark 4), preserves infinite direct sums of objects. Therefore C^A, restricted to the category $Abs_{\widehat{A}}$ of abstract left $\widehat{A}$-modules, likewise preserves

---

[1] As noted in Chapter 2, §5, Proposition 2.5.8, for every topological ring A such that the topology is given by right ideals, we have that $\mathscr{C}_A = \mathscr{C}_{\widehat{A}}$ — i.e., the category of C-complete left A-modules is canonically isomorphic to the categroy of C-complete left $\widehat{A}$-modules. (Therefore, the functors C^A and $C^{\widehat{A}}$ are both functors mapping into the same category, $\mathscr{C}_A = \mathscr{C}_{\widehat{A}}$.)

direct sums of objects. Similarly the functor $C^{\widehat{A}}$ from $Abs_{\widehat{A}}$ inot $\mathscr{C}_A$ preserves direct sums of objects. Condition [1] says the functors, $C^A | Abs_{\widehat{A}}$ and $C^{\widehat{A}}$, agree on the abstract $\widehat{A}$-module $\widehat{A}$. Since these two functors preserve infinite direct sums, they therefore agree on all free abstract left $\widehat{A}$-modules. If M is any abstract left $\widehat{A}$-module, then there exists an exact sequence in $Abs_{\widehat{A}}$

$$F \to G \to M \to 0$$

with F and G free. Since C^A and $C^{\widehat{A}}$ are both right exact (Chapter 2, §5, Theorem 2.5.2, part (3)), it therefore follows that $C^A(M) = C^{\widehat{A}}(M)$.

□

Proof of the Note. [3] iff [1]: We have the short exact sequence of abstract left $\widehat{A}$-modules

$$(1): \ 0 \to K \to \widehat{A} \underset{A}{\otimes} \widehat{A} \to \widehat{A} \to 0.$$

The function: $b \to b \underset{A}{\otimes} 1$ from $\widehat{A}$ into $\widehat{A} \underset{A}{\otimes} \widehat{A}$ is a map in $Abs_{\widehat{A}}$ that splits the epimorphism in the short exact sequence (1); therefore the sequence (1) splits in $Abs_{\widehat{A}}$.

By Chapter 2, §5, Proposition 2.5.8, condition [1] of this Proposition is equivalent to: The map induced by the epimorphism in the short exact sequence (1),

$$C^{\widehat{A}}(\widehat{A} \underset{A}{\otimes} \widehat{A}) \to C^{\widehat{A}}(\widehat{A})$$

is an isomorphism. Since the short exact sequence (1) splits, and since $C^{\widehat{A}}$ is an additive functor, it is equivalent to saying:

$$C^{\widehat{A}}(K) = 0.$$

□

A strengthening of Proposition 3.1.8 is possible, in the case that the topological ring A admits a denumerable neighborhood base at zero. The proof uses a result that will be proved in the next section (§2, Corollary 3.2.4).

Corollary 3.1.9. *Let A be a topological ring such that A admits a denumerable neighborhood base at zero consisting of right ideals. Then the three conditions of Proposition 3.1.8 are also each equivalent, to each of the following conditions*
[4] The uniform and A-adic topologies coincide in $\widehat{A}$.
[5] K is divisible as an abstract left $\widehat{A}$-module.
[6] K is divisible as an abstract left A-module.

Note: The proof shows that [4], [5], and [6] are always equivalent, whether or not A admits a denumerable neighborhood base at zero. And that the three conditions [1], [2], and [3] of Proposition 3.1.8 always imply [4], [5], and [6] whether or not A admits a denumerable neighborhood base at zero.

Proof. [4] iff [5]. By Chapter 1, §3, Remark 0.3, we have that condition [4] holds iff for every open right ideal I in A, we have that

$$(1): \quad I \cdot \widehat{A} = \overline{I},$$

where $\overline{I}$ is the closure of the image of I in $\widehat{A}$ for the uniform topology. The short exact sequence of abstract left $\widehat{A}$-modules

$$(2): \quad 0 \to K \to \widehat{A} \underset{A}{\otimes} \widehat{A} \to \widehat{A} \to 0$$

splits; namely, the function $x \to x \underset{A}{\otimes} 1$ from $\widehat{A}$ into $\widehat{A} \underset{A}{\otimes} \widehat{A}$ is such a splitting. By definition of "divisible as an abstract left $\widehat{A}$-module," we have that condition [5] holds iff for every open right ideal J in $\widehat{A}$, we have that

$$\widehat{A}/J \underset{A}{\otimes} K = 0.$$

Since (2) is split exact, it is equivalent to saying:

$$(\widehat{A}/J) \underset{\widehat{A}}{\otimes} (\widehat{A} \underset{A}{\otimes} \widehat{A}) \to (\widehat{A}/J) \underset{\widehat{A}}{\otimes} \widehat{A}$$

is an isomorphism. Every open right ideal in $\widehat{A}$ $(= (\widehat{A}, unif))$ is of the form $J = \overline{I}$ for an open right ideal I in A. Therefore, condition [5] holds iff for every open right ideal I in A, we have that the natural map

$$(3): \quad (\widehat{A}/\overline{I}) \underset{\widehat{A}}{\otimes} (\widehat{A} \underset{A}{\otimes} \widehat{A}) \to (\widehat{A}/\overline{I}) \underset{\widehat{A}}{\otimes} \widehat{A}$$

is an isomorphism. But

$$(\widehat{A}/\overline{I}) \underset{\widehat{A}}{\otimes} (\widehat{A} \underset{A}{\otimes} \widehat{A}) = (\widehat{A}/\overline{I}) \underset{A}{\otimes} \widehat{A} = (A/I) \underset{A}{\otimes} \widehat{A} = \widehat{A}/(I \cdot \widehat{A}), \text{ and}$$

$$(\widehat{A}/\overline{I}) \underset{\widehat{A}}{\otimes} \widehat{A} = \widehat{A}/\overline{I}.$$

Therefore the map (3) is an isomorphism iff $\overline{I} = I \cdot \widehat{A}$, for all open right ideals I in A. Considering equation (1), therefore, [4] iff [5].

[6] implies [5]: For every open right ideal I in A, we clearly have that $I \cdot \widehat{A} \subset \overline{I}$. If K is divisible as an abstract left A-module, then $I \cdot K = K$, for all open right ideals I in A. Then

$$(I \cdot \widehat{A}) \cdot K = I \cdot (\widehat{A} \cdot K) = I \cdot K = K,$$

and therefore

$$\overline{I} \cdot K \supset (I \cdot \widehat{A}) \cdot K = K,$$

$$\overline{I} \cdot K = K,$$

for all open right ideals I in A. And therefore K is divisible as an abstract left $\widehat{A}$-module.

[4] implies [6]: The uniform and A-adic topologies coincide in $\widehat{A}$ iff for every open right ideal I in A we have that

$$I \cdot \widehat{A} = \overline{I}.$$

(See Chapter 1, §3, Remark 0.3.) Since we have shown that [4] implies [5], we know that K is divisible as an abstract left $\widehat{A}$-module. That is,

$$\overline{I} \cdot K = K.$$

But then

$$I \cdot K = I \cdot (\widehat{A} \cdot K) = (I\widehat{A}) \cdot K = \overline{I} \cdot K = K,$$

for all open right ideals I in A. Therefore K is divisible as an abstract left A-module.

[3] implies [5]: If $C^{\widehat{A}}(K) = 0$, then since the natural map: $K \to (\widehat{A}$-adic completion of K) factors through $C^{\widehat{A}}(K)$ (Chapter 2, §5, Corollary 2.5.7, with K replacing M and $\widehat{A}$ replacing A), it follows that the natural map: $K \to (\widehat{A}$-adic completion of K) is the zero map. Therefore the kernel of this map — which is the divisible part of the abstract left $\widehat{A}$-module K — is all of K. That is, K is a divisible module as an abstract left $\widehat{A}$-module.

If A admits a denumerable base at zero, then [5] implies [3]: This follows from §2, Corollary 3.2.4, below.   □

Proposition 3.1.8 enables us to give the relationship between the conditions "A is C-O.K." and "$\widehat{A}$ is C-O.K." in general:

Theorem 3.1.10. *Let A be a topological ring such that the topology is given by right ideals. Then the following are equivalent.*

[1] A is C-O.K.

[2] Both conditions [2a] and [2b] hold:

[2a] $\widehat{A}$ is C-O.K. and

[2b] The three equivalent conditions of Proposition 3.1.8 hold.

Proof. First, let us show that [1] implies [2b]: If A is C-O.K., then by Definition 3 and Theorem 3.1.5, condition [2], we have that

$$C^A(\widehat{F}) = \widehat{F},$$

for all free abstract left A-modules F. In particular, taking $F = A$, we obtain condition [1] of Proposition 3.1.8; therefore, [2b] holds. Hence, to complete the proof, we can, and do, assume that condition [2b] holds. Therefore condition [2] of Proposition 3.1.8 holds; i.e., $C^A(M) = C^{\widehat{A}}(M)$, for all abstract left $\widehat{A}$-modules M. Let F be a free abstract left A-module. Then $\widehat{F}$ is an abstract

left $\widehat{A}$-module (and is both the A-adic completion of the free abstract left A-module F; and the $\widehat{A}$-adic completion of the free abstract left $\widehat{A}$-module $\widehat{A} \underset{A}{\otimes} F$).
Therefore by condition [2] of Proposition 3.1.8 with $M = \widehat{F}$, we have that

$$C^A(\widehat{F}) = \widehat{F} \text{ iff } C^{\widehat{A}}(\widehat{F}) = \widehat{F}.$$

As F runs through the set of all free abstract left A-modules, $\widehat{F}$ runs through representatives for the set of all isomorphism classes of A-adic completions (respectively: $\widehat{A}$-adic completions) of free A- (respectively: $\widehat{A}$-) modules. Therefore A obeys condition [2] of Theorem 3.1.5 iff $\widehat{A}$ obeys condition [2] of Theorem 3.1.5. That is, A is C-O.K. iff $\widehat{A}$ is C-O.K. □

Corollary 3.1.11. *Let A be a topological ring such that the topology is given by denumerably many right ideals. Then the two equivalent conditions of Theorem 3.1.10 are also equivalent to*
[3] Both conditions [3a] and [3b] hold.
[3a] $\widehat{A}$ is C-O.K.
[3b] The uniform and A-adic topologies in $\widehat{A}$ coincide.

Proof. By Theorem 3.1.6, A, resp. $\widehat{A}$, is C-O.K. iff A, resp. $\widehat{A}$, is admissible. Therefore the equivalence of [1] and [3] follows from Chapter 1, §3, Proposition 1.3.1. □

Another way of writing Corollary 3.1.11 is:

Corollary 3.1.12. *Let A be a topological ring such that the topology is given by denumerably many right ideals. Suppose also that $\widehat{A}$ is C-O.K. Then the following conditions are equivalent:*
[1] A is C-O.K.
[2] $C^A(\widehat{A}) = \widehat{A}$.
[3] $C^{\widehat{A}}(K) = 0$.
[4] The uniform and A-adic topologies coincide in $\widehat{A}$.
[5] For every abstract left $\widehat{A}$-module M, the natural map is an isomorphism: $C^A(M) \overset{\approx}{\to} C^{\widehat{A}}(M)$.
[6] K is divisible as an abstract left $\widehat{A}$-module.

Proof. By Proposition 3.1.8, [2] iff [3] iff [5]. By Theorem 3.1.10, these three equivalent conditions are equivalent to [1]. And by Corollary 3.1.11, [1] is equivalent to [4]. And by Corollary 3.1.9, [6] iff [4] □

(Notice, amusingly, that the proofs of Theorem 3.1.10, and of Corollary 3.1.11 and Corollary 3.1.12, do not make use of §2, Corollary 3.2.4, below.)

Proposition 3.1.13. *Let A be a topological ring such that the topology is given by right ideals. Suppose that A is C-O.K. Then, for every abstract left A-module M, the following conditions are equivalent.*

[1] $M \approx C(N)$ *as abstract left A-modules for some abstract left A-module* N.

[2] The natural homomorphism $\iota_M : M \to C(M)$ *is an isomorphism of abstract left A-modules.*

[3] There exists an infinite sum structure (necessarily unique) on M.

[4] M *is C-complete.*

[5] There exist free abstract left A-modules F *and* G, *and a homomorphism of abstract left A-modules* $f : \widehat{F} \to \widehat{G}$ *of the A-adic completions, such that* $M \approx \mathrm{Cok}(f)$ *as abstract left A-modules.*

[5'] (Same as [5], but delete the condition that the abstract left A-modules F *and* G *are free).*

[6] There exist free abstract left A-modules F *and* G, *and a homomorphism* $f : F \to G$ *of abstract left A-modules, such that if* $\widehat{f} : \widehat{F} \to \widehat{G}$ *is the induced homomorphism on the A-adic completions, then* $M \approx \mathrm{Cok}(\widehat{f})$ *as abstract left A-modules.*

[6'] (Same as [6], but delete the condition that the abstract left A-modules F *and* G *be free).*

[7] Both conditions [7a] and [7b] below hold.

[7a] There exists a free abstract left A-module F *and an epimorphism of abstract left A-modules*

$$f : \widehat{F} \to M$$

from the A-adic completion $\widehat{F}$ *of* F *onto* M *and*

[7b] For every (or, equivalently, for at least one) pair F, f *as in [7a], we have that the kernel* $\mathrm{Ker}(f)$ *of* f *is such that, there exists a free abstract left A-module* G, *and an epimorphism of abstract left A-modules*

$$g : \widehat{G} \to \mathrm{Ker}(f)$$

(I.e., $\mathrm{Ker}(f)$ *obeys condition [7a]).*

[7'] Both conditions [7'a] and [7'b] below hold.

[7'a] (Same as [7a], but delete the condition that the abstract left A-module F *be free).*

[7'b] (Same as [7b], but delete the condition that F *and* G *be free).*

[8] There exist A-adically complete abstract left A-modules F *and* G, *and a homomorphism* $f : F \to G$ *of abstract left A-modules, such that* $M \approx \mathrm{Cok}(f)$ *as abstract left A-modules.*

[8'] (Same as [8], but assume in addition that F *and* G *are isomorphic as abstract left A-modules to the A-adic completions of free left A-modules.)*

Note: If the topological ring A is such that there exists a denumerable neighborhood base at zero, then:

(1) Condition [7b] simplifies to: For every (or, equivalently, for at least one) pair (F, f) as in [7a], we have that $\mathrm{Ker}(f)$ is A-adically complete.

(2) Condition [7'b] simplifies similarly.

Proof. Clearly, [2] implies [1] implies [3] implies [4]. (We have observed, in Theorem 3.1.5, that under the hypotheses of condition [3], if A is C-O.K., then the infinite sum structure is unique.) By Theorem 3.1.5, [4] implies [2]. By Chapter 2, §4, Theorem 2.4.1, [4] implies [5].

[5] implies [5′]: Obvious.

[5′] implies [4]: Since by Theorem 3.1.5, $\mathscr{C}_A$ is an abelian category and is an exact full subcategory of the category of all abstract left A-modules, and $\widehat{F}$ and $\widehat{G}$ are in $\mathscr{C}_A$, and by Theorem 3.1.5 f is in $\mathscr{C}_A$, therefore $M \approx \mathrm{Cok}(f)$, is in $\mathscr{C}_A$. Clearly, [6] implies [6′] implies [5].

[2] implies [6]: Let

$$F \xrightarrow{f} G \to M \to 0$$

be an exact sequence of abstract left A-modules with F and G free. Then throwing through the right exact functor C, and using that $M = C(M)$ by condition [2], we obtain that $M \approx \mathrm{Cok}(\widehat{f} : \widehat{F} \to \widehat{G})$.

[6] implies [6′]: Obvious.

[6′] implies [5′]: Obvious.

Let us use the notation "[7∀b]" to stand for condition [7b] with "For every ..."; let "[7∃b]" stand for condition [7b] with "For at least one...". Similarly for "[7′∀b]" and "[7′∃b]." Then clearly,

[7a] implies [7′a], and
[7′∀b] implies [7∀b] implies [7∃b] imply [7′∃b].

([7′a] and [7′∃b] imply [5′]): Let h be the composite

$$\widehat{G} \xrightarrow{g} (\mathrm{Ker}(f)) \hookrightarrow \widehat{F} \xrightarrow{f} M.$$

Then since f and g are surjective,

$$M \approx \mathrm{Cok}(h : \widehat{G} \to \widehat{F})$$

so that M obeys condition [5′].

[4] implies [7a] and [7′∀b]: By Chapter 2, §4, Theorem 2.4.1, we have that [7a] holds. Given any epimorphism $f : \widehat{F} \to M$ where F is an abstract left A-module, then since $\widehat{F}$ is C-complete, and since by Theorem 3.1.5 the category $\mathscr{C}_A$ is an exact, full subcategory of the category of abstract left A-modules, we therefore have the $\mathrm{Ker}(f)$ is C-complete. But then, by Chapter 2, §4, Theorem 2.4.1, there exists a free abstract left A-module G and an epimorphism $g : \widehat{G} \to \mathrm{Ker}(f)$. Therefore [7′∀b] and [7∀b] both hold.

[5] implies [8′]: Since the topological ring A is C-O.K., by Corollary 3.1.7 we have that A is admissible. Therefore (by Chapter 1, §3, Definition 1, Condition [1]) we have, for every abstract free left A-module F, that $\widehat{F}$ is C-complete.

[8′] implies [8]: Obvious.

[8] implies [5′]: Since F and G are A-adically complete, we have that $F = \widehat{F}$ and $G = \widehat{G}$.

This proves the theorem; it is left to prove the Note. In fact, if A has a denumerable neighborhood base at zero, then as we have seen in Chapter 1, §2, Lemma 1.2.6, in this case a Hausdorff quotient module of an A-adically complete abstract left A-module is A-adically complete. Therefore condition [7'b] is in this case equivalent to asserting that $\mathrm{Ker}(f)$ is A-adically complete. On the other hand, if $\mathrm{Ker}(f)$ is A-adically complete, then $\mathrm{Ker}(f)$ is C-complete, and therefore by Chapter 2, §3, Theorem 2.3.5, we have that [7b] holds. Therefore, likewise, [7b] is equivalent, in this case, to the assertion that $\mathrm{Ker}(f)$ is A-adically complete. $\qquad\qquad\square$

3.2 The Divisible Part of $C(M)$

Proposition 3.2.1. *Let A be a topological ring such that the topology of A is given by* denumerably many *right ideals. Let M be any C-complete left A-module. Then M has no non-zero infinitely divisible elements. (See Chapter 1, §2, Definition 2, for the definition of the "infinitely divisible elements of an abstract left A-module".)*

Proof. By Chapter 2, §4, Theorem 2.4.1, there exists an exact sequence

$$(1): \ \widehat{G} \xrightarrow{f} \widehat{F} \xrightarrow{\pi} M \to 0$$

in the category $\mathscr{C}_A$, where G and F are free left A-modules. Let $(\widehat{F}, unif)$, resp. $(\widehat{G}, unif)$, denote $\widehat{F}$, resp. $\widehat{G}$, with the topology such that $(\widehat{F}, unif)$, resp. $(\widehat{G}, unif)$ is the completion of F, resp. G (together with its A-adic topology) in the category of topological left A-modules. Then I claim that

$$(2): \ f: (\widehat{G}, unif) \to (\widehat{F}, unif)$$

is continuous.
(*Proof.* Let f_0 denote the composite: $G \to \widehat{G} \xrightarrow{f} \widehat{F}$. Then

$$f_0: (G, A\text{-adic topology}) \to (\widehat{F}, A\text{-adic topology})$$

is continuous (as are all homomorphisms of abstract left A-modules for the A-adic topologies). Therefore so is the composite

$$(G, A\text{-adic topology}) \xrightarrow{f_0} (\widehat{F}, A\text{-adic topology}) \to (\widehat{F}, unif).$$

Since $(\widehat{F}, unif)$ is a complete topological left A-module, by universality of $(\widehat{G}, unif)$ there exists a unique extension f_0' of f_0 to $(\widehat{G}, unif)$ such that

$$f_0': (\widehat{G}, unif) \to (\widehat{F}, unif)$$

is a continuous homomorphism of complete topological left A-modules. But then, by definition of the infinite sum structures in $\widehat{G}$ and $\widehat{F}$ (namely, infinite sums being defined as the limits of the corresponding convergent Cauchy series for the uniform topology — see Chapter 2, §2, Example 1), f_0' is infinitely linear. But then, f and f_0' are both infinitely linear, and agree on the subset

$(\text{Im}(G \to \widehat{G}))$ of $\widehat{G}$, a set of generators for $\widehat{G}$ as a C-complete left A-module (Chapter 2, §4, Lemma 2.4.7, hypothesis 2). Therefore $f = f_0'$; and in particular

$$f : (\widehat{G}, unif) \to (\widehat{F}, unif)$$

is continuous.)

Let D be the infinitely divisible part of the abstract left A-module M. (See Chapter 1, §2, Definition 2.) Then from the exact sequence (1) we deduce the exact sequence

$$(3) : \quad \widehat{G} \xrightarrow{\overline{f}} E \xrightarrow{\rho} D \to 0$$

where $E = \pi^{-1}(D)$ and $\rho = \pi|E$, and $\overline{f} = f$ regarded as a function into E. Let $(E, unif)$ denote E together with the induced topology from $(\widehat{F}, unif)$. Then the function

$$(4) : \quad (\widehat{G}, unif) \xrightarrow{\overline{f}} (E, unif)$$

is a continuous homomorphism of topological left A-modules. Also $(\widehat{G}, unif)$ is complete, and $(E, unif)$ is Hausdorff. The A-adic topology of $D = \text{Cok}(\overline{f})$ (as is true for any epimorphism of abstract left A-modules) is deduced, by passing to the quotient, from the A-adic topology of E. Since D is infinitely A-divisible, we have that D is A-adically indiscrete. Therefore, from (3), $\text{Im}(\overline{f})$ is dense in E for the A-adic topology.

(Proof.

$$\dfrac{E}{\text{the } A\text{-adic closure of } \text{Im}(\overline{f})} \text{ is a quotient } A\text{-module of } \dfrac{E}{\text{Im}(\overline{f})} \approx D.$$

Since D is A-divisible, so is every quotient module. But the leftmost A-module in the displayed equation is Hausdorff. Therefore it is both Hausdorff and divisible, therefore zero. I.e., $\text{Im}(\overline{f})$ is dense in E.) $\qquad\square$

Since also there exists a denumerable base for the neighborhood system at zero in A, we have that all of the hypotheses of Lemma 1.2.13 hold for the continuous homomorphism (4). Therefore, by that Lemma, we have that $\overline{f}$ is surjective. But then by equation (3) $D = \{0\}$. $\qquad\square$

Remark: In the course of proving Proposition 3.2.1, we have proved some other results worth recording: namely, Corollaries 3.2.2 and 3.2.3 below.

Corollary 3.2.2. *Let A be a topological ring such that the topology is given by denumerably many right ideals and let G and F be abstract left A-modules. Let $(\widehat{G}, unif)$ and $(\widehat{F}, unif)$ be the completions of G and F as topological left A-modules. Let $f : \widehat{G} \to \widehat{F}$ be a homomorphism of abstract left A-modules. Then f is continuous (for the uniform topologies of $\widehat{G}$ and $\widehat{F}$) iff f is infinitely linear (for the structures of $\widehat{G}$ and $\widehat{F}$ as C-complete left A-modules, as defined in Chapter 2, §2, Example 1).*

Proof. If f is continuous, by the definition of the infinite sum structures in $\widehat{F}$ and $\widehat{G}$ (Chapter 2, §2, Example 1), using limits of Cauchy series for the uniform topologies, then clearly f is infinitely linear. Conversely, if f is infinitely linear, then by the parenthetical argument within the proof of Proposition 3.2.1, we have that f is continuous. □

Corollary 3.2.3. *Let A be a topological ring such that the topology is given by denumerably many right ideals. Let $\mathscr{H}_A$ denote the category of all topological left A-modules each of which is isomorphic (as topological left A-module) to the completion of an abstract left A-module, together with all continuous homomorphisms of such topological left A-modules. Then $\mathscr{H}_A$ is a full subcategory of the category $\mathscr{C}_A$ of all C-complete left A-modules and infinitely linear functions.*
Note: In the statement of this corollary, every topological left A-module in $\mathscr{H}_A$ is regarded as a C-complete left A-module by means of the construction of Chapter 2, §2, Example 1.

Proof. This is a restatement of Corollary 3.2.2. □

Corollary 3.2.4. *Let A be a topological ring such that the topology is given by denumerably many right ideals. Let M be any abstract left A-module. Then the following three conditions are equivalent:*
[1] M is A-divisible.
[2] $\widehat{M} = 0$.
[3] $C(M) = 0$.

Proof. Condition [1] says $I \cdot M = M$, for all right ideals I. Therefore [1] iff [2].
[3] implies [1]: By Chapter 2, §5, Corollary 2.5.7, the natural map $M \to \widehat{M}$ factors: $M \overset{\iota_M}{\to} C(M) \overset{\rho_M}{\to} \widehat{M}$. So if $C(M) = 0$, then the natural map: $M \to \widehat{M}$ is zero. Therefore every element of M is A-divisible — i.e., M is an A-divisible module.
[1] implies [3]: The set-theoretic image of an A-divisible module by a homomorphism of abstract left A-modules is A-divisible. Therefore, $\iota_M(M)$ is an A-divisible, abstract A-submodule of $C(M)$. But then every element of $\iota_M(M)$ is infinitely divisible in $C(M)$. By Proposition 3.2.1, it follows that $\iota_M(M) = 0$. Equivalently, $\iota_M = 0$. But then by the universal mapping property definition of $C(M)$ (Chapter 2, §5, Definition 1), it follows that every abstract A-homomorphism from M into a C-complete left A-module is zero. Therefore $\{0\}$ obeys the universal mapping property for $C(M)$, and $C(M) = \{0\}$. □

Corollary 3.2.5. *Let A be a topological ring such that the topology is given by denumerably many right ideals. Let M be a C-complete left A-module such that M is A-divisible. Then $M = \{0\}$.*

Proof. By Proposition 3.2.1, M has no non-zero infinitely divisible elements. Since M is A-divisible, M consists entirely of A-divisible elements. Therefore $M = \{0\}$. □

Lemma 3.2.6. *Let A be a topological ring that is admissible, let M be any C-complete left A-module, and let N be any complete topological left A-module such that the topology of N is coarser than the A-adic topology. Then every homomorphism of abstract left A-modules*

$$f : M \to N$$

is infinitely linear (for the infinite sum structure on N defined in Chapter 2, §2, Example 1.1).

Proof. By Chapter 2, §3, Theorem 2.3.5, there exists F a free abstract left A-module and a surjective infinitely linear function

$$g : \widehat{F} \to M.$$

Then, since g is surjective and infinitely linear, to show that f is infinitely linear it suffices to show that the composite

$$\widehat{F} \xrightarrow{g} M \xrightarrow{f} N$$

is infinitely linear. Therefore, replacing M by $\widehat{F}$, and f by $f \circ g$ if necessary, we can, and do, assume that $M = \widehat{F}$ for some free abstract left A-module F. Then let $(\widehat{F}, unif)$ denote $\widehat{F}$ with its uniform topology (i.e., with the topology making $\widehat{F}$ into the completion of F, as a topological left A-module (with the A-adic topology)). Let $\widehat{F}$, resp. $(N, A\text{-adic})$ denote $\widehat{F}$, resp. N, together with its A-adic topology. Then the composite: $\widehat{F} \xrightarrow{f} (N, A\text{-adic}) \xrightarrow{id_N} N$ is continuous. Since A is by hypothesis admissible, by Chapter 1, §3, Definition 1, we have that the A-aidc and uniform topologies on $\widehat{F}$ coincide. I.e., that $\widehat{F} = (\widehat{F}, unif)$. Therefore the composite:

$$(\widehat{F}, unif) = \widehat{F} \xrightarrow{f} (N, A\text{-adic}) \xrightarrow{id_N} N$$

is continuous — i.e., f is a continuous homomorphism of topological A-modules from $(\widehat{F}, unif)$ into N. For every fixed $x \in \widehat{F}$, consider the diagram

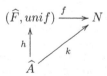

where $h(a) = a \cdot x$ and $k(a) = a \cdot f(x)$, for all $a \in \widehat{A}$. Since $(\widehat{F}, unif)$ is a topological left $\widehat{A}$-module, h is continuous. By Chapter 2, §2, Example 1.1, we have that k is continuous. Also, since f is a homomorphism of A-modules, $f \circ h|A = k|A$. Thus, the continuous functions $f \circ h$ and k agree on the dense subset $\text{Im}(A \to \widehat{A})$ of $\widehat{A}$. Since N is Hausdorff (being complete), it follows that $f \circ h = k$. I.e., f is a homomorphism of $\widehat{A}$-modules. Since the infinite sum structure of $\widehat{F}$, resp. of N, is deduced from the $\widehat{A}$-module structure of $\widehat{F}$, resp of N, and from the topology of $(\widehat{F}, unif)$, resp. of N, it follows that

$$f : \widehat{F} \to N$$

is infinitely linear.   □

Corollary 3.2.7. *Let A be a topological ring that is admissible, and let N be an abstract left A-module. Suppose that there exists some topology τ on N, coarser than the A-adic topology, such that (N, τ) becomes a complete topological left A-module. Then there exists a* unique *infinite sum structure, in the sense of Chapter 2, §2, Definition 1, on the abstract left A-module N.*

Proof. Let (S) be any infinite sum structure on N as an A-module. Let (C) denote the infinite sum structure on N deduced from Chapter 2, §2, Definition 1.1. Then the identity of N

$$id_N : (N, (S)) \to (N, (C))$$

is a homomorphism of abstract left A-modules. But then, by Lemma 3.2.6, id_N is infinitely linear. Otherwise stated, the infinite sum structures (S) and (C) on the abstract left A-module N coincide.   □

Corollary 3.2.8. *Let A be a topological ring that is admissible. Then for every abstract left A-module N, there is a* unique *infinite sum structure on the abstract left A-module $\widehat{N}$.*

Proof. Let τ be the topology on $\widehat{N}$ such that $(\widehat{N}, \tau)$ is the completion of N in the category of topological left A-modules. Then the conclusion of Corollary 3.2.8 follows from Corollary 3.2.7 applied to $\widehat{N}$.   □

Remark 1. Corollary 3.2.7 or Corollary 3.2.8 is *equivalent* to A being admissible. More precisely: Let A be a topological ring such that the topology is given by right ideals. Then the following four conditions are equivalent:
[1] A is admissible.
[2] For every abstract left A-module N, if there exists a topology τ coarser than the A-adic, such that (N, τ) is complete as a topological left A-module, then there exists a unique infinite sum structure on the abstract left A-module N.
[3] For every abstract left A-module N, we have that there is only one structure of C-complete left A-module on the abstract left A-module $\widehat{N}$.
[4] For every abstract free left A-module F, there is only one structure of C-complete left A-module on $\widehat{\widehat{F}}$.

Proof. Corollary 3.2.7 states that [1] implies [2]. [2] implies [3] and [3] implies [4] are clear. On the other hand, suppose [4]. If A were not admissible, then by Chapter 1, §3, Proposition 1.3.1, either $\widehat{A}$ is not admissible, or else the uniform and A-adic topologies on $\widehat{A}$ are different. In the latter case, by

Chapter 2, §3, Example 2, there are *two different* infinite sum structures on $\widehat{A}$, contradicting [4]. And in the former case, by Chapter 2, §3, Example 1, there exists a free left A-module F, such that there are *two different* infinite sum structures on $\widehat{F}$, which contradicts [4]. Therefore A is admissible, proving [1]. □

Lemma 3.2.9. *Let A be an arbitrary topological ring such that the topology is given by right ideals and let M be an abstract left A-module. Then there are induced natural homomorphisms:*

$$(1):\ \widehat{C(M)} \to \widehat{M},\ \text{and}\ (2):\ \widehat{M} \to \widehat{C(M)}$$

of C-complete left A-modules. The composite: $\widehat{M} \to \widehat{C(M)} \to \widehat{M}$ is the identity.

Note: The Proof of Lemma 3.2.9 also shows that the homomorphism (1) (resp. (2)) can be characterized as being the unique homomorphism of abstract left A-modules that is continuous for the uniform topologies of $\widehat{M}$ and of $\widehat{C(M)}$, and also is such that the triangle:

is commutative (resp. and also is such that the square:

$$
\begin{array}{ccc}
\widehat{M} & \longrightarrow & \widehat{C(M)} \\
\uparrow & & \uparrow \\
M & \xrightarrow{\ \iota_M\ } & C(M)
\end{array}
$$

is commutative).

Proof. Consider the natural map

$$(0):\ \rho_M : C(M) \to \widehat{M}.$$

Since this is a homomorphism of abstract left A-modules, it is continuous for the A-adic topology. Since the uniform topology of $\widehat{M}$ is coarser than the A-adic, it follows that the natural homomorphism: $C(M) \to (\widehat{M}, unif)$ is continuous, where $C(M)$ is regarded as a topological left A-module with the A-adic topology, and $(\widehat{M}, unif)$ is $\widehat{M}$ together with its uniform topology. Since $(\widehat{M}, unif)$ is a complete topological left A-module, we therefore deduce the existence of a unique homomorphism of topological left A-modules

$$(1):\ (\widehat{C(M)}, unif) \to (\widehat{M}, unif)$$

that is part of a commutative triangle together with the map: $C(M) \to \widehat{C(M)}$ and the natural map ρ_M from $C(M)$ into $\widehat{M}$. Similarly, if we regard M as a topological left A-module with its A-adic topology, then the composite:

$$M \to C(M) \to (\widehat{C(M)}, unif)$$

is continuous. Since $(\widehat{C(M)}, unif)$ is complete, it follows that there exists a unique continuous homomorphism of topological left A-modules

$$(2): \ (\widehat{M}, unif) \to (\widehat{C(M)}, unif)$$

that is compatible with the natural map $\iota_M : M \to C(M)$. It follows that the composite:

$$(\widehat{M}, unif) \to (\widehat{C(M)}, unif) \to (\widehat{M}, unif)$$

is a continuous endomorphism of the topological left A-module $(\widehat{M}, unif)$ compatible with the natural map: $M \to (\widehat{M}, unif)$ from M into its completion. But then, by definition of the completion of a uniform space, it follows that this composite must be the identity.

Finally, since the homomorphisms (1) and (2) are continuous for the uniform topologies of $\widehat{C(M)}$ and of $\widehat{M}$, it follows that they are also homomorphisms of C-complete left A-modules. (Since, recall in Chapter 2, §2, Remark following Example 1.1, observation (2), that if N is any abstract left A-module, the structure of $\widehat{N}$ as an $\widehat{A}$-module is determined by continuity of the scalar product $a \to a \cdot x$, $\widehat{A} \to \widehat{N}$, for each $x \in \widehat{N}$; and see Chapter 2, §2, Remark following Example 1.1, observation (3), that the definition of the infinite sum structure on $\widehat{N}$, is such that, infinite sums are defined to be the limits of the corresponding Cauchy series for the uniform topology of $\widehat{N}$.) □

Corollary 3.2.10. *The hypotheses being as in Lemma 3.2.9, we have that* $\widehat{M}$ *is canonically isomorphic to a direct summand of* $\widehat{C(M)}$, *in the abelian category of C-complete left A-modules.*

Proof. Immediate from Lemma 3.2.9. □

Theorem 3.2.11. *Let A be a topological ring that is admissible. Let M be any abstract left A-module. Then the canonical infinitely linear function*

$$\rho_M : C(M) \to \widehat{M}$$

extends naturally to an isomorphism of abstract left A-modules

$$\widehat{C(M)} \overset{\approx}{\to} \widehat{M}.$$

Notes 1. The proof shows, more precisely, that $(\widehat{M}, unif)$ (i.e., $\widehat{M}$ together with the topology making $\widehat{M}$ into the completion of M in the category of topological left A-modules), together with ρ_M, is the completion of $C(M)$ (regarded as a topological left A-module with the A-adic topology).

2. Another way of stating the conclusion of Note 1, is that the natural homomorphism of abstract left A-modules

$$\iota_M : M \to C(M)$$

induces an isomorphism on the A-adic completions

$$\widehat{\iota_M} : \widehat{M} \xrightarrow{\approx} \widehat{C(M)}$$

as complete topological left A-modules.

Proof 1. In general, let $\iota : M \to N$ be a map of abstract left A-modules. Then ι is such that, the induced map $\widehat{\iota} : (\widehat{M}, unif) \to (\widehat{N}, unif)$ is an isomorphism iff for every open right ideal I, we have that the induced homomorphism:

$$M/(I \cdot M) \to N/(I \cdot N)$$

of abelian groups is an isomorphism.
(*Proof.* Let ι/I denote the map induced by ι,

$$\iota/I = \iota \otimes_A A/I : M/(I \cdot M) \to N/(I \cdot N).$$

If $\overline{I \cdot M}$ denotes the closure of the image of $I \cdot M$ in $\widehat{M}$, then $\overline{I \cdot M}$ is also the closure of $I \cdot \widehat{M}$ in $\widehat{M}$, and $M/(I \cdot M) \approx \widehat{M}/(\overline{I \cdot M})$. Therefore if $\widehat{\iota} : (\widehat{M}, unif) \to (\widehat{N}, unif)$ is an isomorphism, the maps: $M/(IM) \to N/(IN)$ must be isomorphisms of abelian groups. Conversely, we have that

$$(\widehat{M}, unif) \approx \varprojlim_{I \in \mathscr{R}} M/(I \cdot M)$$

as topological groups (where $\mathscr{R}$ is the collection of open right ideals of A). Therefore, if $\iota/I : M/(IM) \to N/(IN)$ is an isomorphism for all open right ideals I, then $\widehat{\iota}$ is an isomorphism of topological A-modules).

We prove the theorem by proving Note 2. By the above, it is necessary and sufficient to show, for every open right ideal I in A, and for every abstract left A-module M, that the homomorphism of abelian groups:

$$\iota_M/I : M/(IM) \to C(M)/(I \cdot C(M))$$

is an isomorphism.

Fix I and let M vary. Then the functors: $M \rightsquigarrow M/(IM)(= (A/I) \underset{A}{\otimes} M)$ and $M \rightsquigarrow C(M)/(I \cdot C(M)) = ((A/I) \underset{A}{\otimes} C(M))$ are both right exact functors from the category of abstract left A-modules into the category of abelian groups. Given an abstract left A-module M, choose free abstract left A-modules, F and G and an exact sequence:

$$F \to G \to M \to 0.$$

Then since the functors: $M \rightsquigarrow M/(IM)$, $M \rightsquigarrow C(M)/(I \cdot C(M))$ preserve cokernels, to prove that ι_M/I is an isomorphism, it suffices to prove that ι_F/I

and ι_G/I are isomorphisms. Therefore, it is necessary and sufficient to prove the assertion in the case that M is a free abstract left A-module. But then (Chapter 2, §5, Theorem 2.5.2, assertion (2)) $C(F) = \widehat{F}$. Therefore, the assertion of Note 2 (for all abstract left A-modules M), is equivalent to the assertion that:

For every free abstract left A-module F, and for every open right ideal I in A, we have that the natural map:

$$F/(I \cdot F) \to \widehat{F}/(I \cdot \widehat{F})$$

is an isomorphism of abelian groups. But this latter condition is condition [10] in Chapter 1, §3, Definition 1 of "admissible." $\qquad\square$

Proof 2. In general, if N is any abstract left A-module, then one characterization of the A-adic completion of N as a complete topological left A-module is:

(1) A complete topological left A-module N', such that the topology of N' is coarser than the A-adic, together with

(2) A homomorphism

$$f : N \to N'$$

of abstract left A-modules, such that

(3) The pair (N', f) is universal with these properties.

(To prove this, note first that given (1), any homomorphism of abstract left A-modules as in (2) is continuous, where N is given its A-adic topology — since f is the composite, $(N, A\text{-adic}) \xrightarrow{f} (N', A\text{-adic}) \to N'$. Of course, the A-adic completion $\widehat{N}$ of N with its uniform topology obeys (1) and (2), so that by (3) there exists a unique continuous homomorphism: $N' \to \widehat{N}$ such that the composite: $N \to N' \to \widehat{N}$ is the natural map: $N \to \widehat{N}$; and given any (N', f) as in (1) and (2), since as we have observed f is necessarily continuous, by the universal mapping property definition of the completion $\widehat{N}$ of N, there exists a unique continuous homomorphism: $\widehat{N} \to N'$ such that the composite: $N \to \widehat{N} \to N'$ is f. And, by the universality property (3) the composite: $N' \to \widehat{N} \to N'$ is the identity map of N'; and by the universlity property of $\widehat{N}$, the composite $\widehat{N} \to N' \to \widehat{N}$ is the identity of $\widehat{N}$.)

Next, we prove Note 2 by verifying that $(\widehat{M}, unif)$ together with ρ_M : $C(M) \to \widehat{M}$ obeys conditions (1), (2), and (3) for the abstract left A-module $C(M)$. In fact, (1) and (2) are immediate; it remains to verify (3). In fact, let D be any complete topological left A-module such that the topology of D is coarser than the A-adic; and let

$$f : C(M) \to D$$

by any homomorphism of abstract left A-modules. Then, by Lemma 3.2.6, we have that f is infinitely linear, where D is given its structure as a C-complete left A-module by means of Chapter 2, §2, Example 1.1. Consider the commutative diagram:

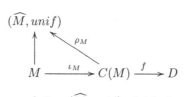

By universality of the completion $(\widehat{M}, unif)$ of M, there exists a unique continuous homomorphism of topological left A-modules $\alpha : (\widehat{M}, unif) \to D$ such that

$$(1): \quad \alpha \circ \rho_M \circ \iota_M = f \circ \iota_M.$$

By Lemma 3.2.6, we have that α is infinitely linear. By Corollary 2.5.7 we have that ρ_M is infinitely linear. But then, from equation (1), we see that the two infinitely linear functions, $\alpha \circ \rho_M$ and f, from $C(M)$ into D, agree on the set of generators $\iota_M(M)$ for the C-complete left A-module M (Chapter 2, §5, Corollary 2.5.4). Therefore (Chapter 2, §3, Corollary 2.3.7),

$$(2): \quad \alpha \circ \rho_M = f.$$

Finally, suppose that α and β are two continuous homomorphisms of topological left A-modules

$$\alpha, \beta : (\widehat{M}, unif) \to D$$

such that

$$(3): \quad \alpha \circ \rho_M = \beta \circ \rho_M = f.$$

Then a fortiori

$$(4): \quad \alpha \circ \rho_M \circ \iota_M = \beta \circ \rho_M \circ \iota_M = f \circ \iota_M$$

Since $(\widehat{M}, unif)$ together with $\rho_M \circ \iota_M$ is the completion of M as a topological left A-module, there is one and only one continuous homomorphism α from $(\widehat{M}, unif)$ into D obeying

$$\alpha \circ \rho_M \circ \iota_M = f \circ \iota_M.$$

Therefore $\alpha = \beta$.

This verifies property (3) for the pair $(\widehat{M}, unif)$ and the homomorphism $\rho_M : C(M) \to \widehat{M}$. Therefore $(\widehat{M}, unif)$ together with ρ_M is the completion of $C(M)$, (together with the A-adic topology of $C(M)$), as a topological left A-module. $\qquad \square$

Remark 2. The property expressed in Theorem 3.2.11 also characterizes admissible topological rings. More precisely:

Let A be a topological ring such that the topology is given by right ideals. Then the following three conditions are equivalent:

[1] A is admissible

[2] For every abstract left A-module M, the natural homomorphism of abstract left A-modules,

$$\iota_M : M \to C(M)$$

is such that the induced homomorphism on the A-adic completion is an isomorphism of abstract left A-modules,

$$\widehat{\iota_M} : \widehat{M} \to \widehat{C(M)}$$

[3] Condition [2] holds for all free abstract left A-modules.

Proof. [1] implies [2] is Theorem 3.2.11. [2] implies [3] is clear.
[3] implies [1]: Condition [3] is condition [4] in Chapter 1, §3, Definition 1, of what it means for the topological ring A to be "admissible". □

Corollary 3.2.12. *Let A be an admissible topological ring, and let M be an abstract left A-module. Then for every open right ideal I in A, we have that the map induced by ι_M is an isomorphism of abelian groups:*

$$M/(I \cdot M) \overset{\approx}{\to} C(M)/(I \cdot C(M)).$$

Proof. We have verified this in the course of Proof 1 of Theorem 3.2.11. □

Corollary 3.2.13. *Let A be an admissible topological ring. If M is an A-adically complete abstract left A-module, and if $div(C(M))$ denotes the set of A-divisible elements of $C(M)$, then*

(1) $C(M)/div(C(M)) \approx M$ canonically, as C-complete left A-modules.
(2) $div(C(M))$ is a C-complete left A-submodule of $C(M)$.
(3) The short exact sequence of C-complete left A-modules

$$0 \to div(C(M)) \to C(M) \to M \to 0$$

splits canonically (ι_M is the splitting) in the category of abstract left A-modules.

Note 1. Also, the infinitely A-divisible part, and the A-divisible part, of the abstract left A-module $C(M)$ coincide.
Note 2. An equivalent formulation of Note 1 is: The abstract left A-module $div(C(M))$ is an A-divisible abstract left A-module.

Proof. By Theorem 3.2.11,

$$(4): \ \widehat{C(M)} \approx \widehat{M}.$$

The composite:

$$(5): \ M \overset{\iota_M}{\to} C(M) \to \widehat{C(M)} \overset{\approx}{\to} \widehat{M}$$

is the natural map from M into $\widehat{M}$. By hypothesis, this is an isomorphism. Therefore, by equation (5), the natural map: $C(M) \to \widehat{C(M)}$ is an epimorphism. I.e.,

$$(6): \ \widehat{C(M)} \approx C(M)/div(C(M)).$$

Since also $\widehat{M} = M$, equations (4) and (6) imply conclusion (1) of the Corollary.

Since $\mathrm{div}(C(M))$ is the kernel of the infinitely linear function $\rho_M : C(M) \to \widehat{M} = M$, we have conclusion (2) of the Corollary.

Under the identification of (the A-adically complete abstract left A-module) M with $\widehat{M}$, the composite of the sequence (5) is the identity of M. Therefore ι_M splits the short exact sequence in conclusion (3) of the Corollary, proving conclusion (3) of the corollary.

Since (3) splits, we have that

$$C(M) \approx M \oplus \mathrm{div}(C(M))$$

canonically *in the category of abstract left A-modules*. Throwing through the additive functor "A-adic completion," this gives that

$$(7): \widehat{C(M)} \approx [M \oplus div(C(M)]\widehat{\quad} .$$

Equations (4) and (7) imply that

$$[div(C(M)]\widehat{\quad} = 0.$$

— i.e., that the abstract left A-module $\mathrm{div}(C(M))$ is A-divisible, proving Note 2. Note 1 is a restatement of Note 2. □

Remark 3. The proof above, that the sequence in (3) splits in the category of abstract left A-modules, does *not* show that the sequence splits *in the category of C-complete left A-modules*.

Corollary 3.2.14. *Let A be an admissible topological ring and let M be a C-complete left A-module. Then in §1, Construction 2, we have built the unique infinitely linear function $j_M : C(M) \to M$ such that $j_M \circ \iota_M = $ identity of M. (Since j_M is infinitely linear, $\mathrm{Ker}(j_M)$ is C-complete.)*

Therefore we have constructed the canonical isomorphism of abstract left A-modules

$$\beta_M : C(M) \xrightarrow{\approx} M \oplus (\mathrm{Ker}(j_M)).$$

Then

(1) $\mathrm{Ker}(j_M)$ *as an abstract left A-module is A-divisible.*

Proof. Throwing the isomorphism (of abstract left A-modules) β_M through the additive functor "A-adic completion," we obtain the isomorphism

$$\widehat{\beta_M} : \widehat{C(M)} \approx \widehat{M} \oplus \widehat{\mathrm{Ker}(j_M)}.$$

Therefore by Theorem 3.2.11, $\widehat{\mathrm{Ker}(j_M)} = 0.$ □

Remark 4. If we delete the hypothesis, in Corollary 3.2.14, that the topological ring A (such that the topology is given by right ideals) is admissible, then it remains true that $\text{Ker}(j_M)$ is C-complete.

Completion of the proof of §1, Theorem 3.1.6. We must show that condition [2] of §1, Theorem 3.1.6, implies condition [1] of §1, Theorem 3.1.6. That is, let A be an admissible topological ring such that there is a denumerable neighborhood base at zero. We must show that A is C-O.K. We verify condition [3] of §1, Theorem 3.1.5.

Let M be a C-complete left A-module. Then since A is admissible by Corollary 3.2.14 we have that the kernel $\text{Ker}(j_M)$ of the natural epimorphism

$$j_M : C(M) \to M$$

is A-divisible and C-complete. But then, since A has a denumerable neighborhood base at zero, by Corollary 3.2.5 we have that $\text{Ker}(j_M) = \{0\}$. Therefore j_M is an isomorphism – that is, condition [3] of §1, Proposition 3.1.3 holds. Therefore by §1, Definition 2, we have that $M = C(M)$. □

Proposition 3.2.15. *Let A be an admissible topological ring. Suppose also that A is commutative. If M is any abstract left A-module, such that there exists an open ideal I in A such that $I \cdot M = 0$, then the natural map*

$$\iota_M : M \to C(M)$$

is an isomorphism.

Proof. Since $I \cdot M = 0$, M is A-adically discrete and therefore A-adically complete. For every $f \in A$, since A is commutative, we have that multiplication by $f : M \to M$ is an endomorphism μ_f of M as abstract left A-module. For $f \in I$, by hypotheses $\mu_f = 0$. Therefore $C(\mu_f) = 0$. But $C(\mu_f)$ is the endomorphism, "multiplication by f", of $C(M)$. Therefore $f \cdot C(M) = 0$. This being true for all $f \in I$, we have that $I \cdot C(M) = 0$. Therefore $C(M)$ is discrete as an abstract left A-module. But then, since M and $C(M)$ are both A-adically discrete, $M = \widehat{M}$, $C(M) = \widehat{C(M)}$. Therefore by Theorem 3.2.11, $C(M) = M$. □

Remark 5. An alternate method of proof exists for Proposition 3.2.15, that goes through to some non-commutative cases: In general, if A is a topological ring such that the topology is given by right ideals, and if M is an abstract left A-module such that the A-adic topology of M is discrete, then the annihilator ideal of M is an open two-sided ideal J. One constructs an exact sequence:

$$(A/J)^{(S)} \to (A/J)^{(T)} \to M \to 0$$

— then the theorem (that $C(M) = M$ for all M that are A-adically discrete) is true iff it is true for abstract left A-modules of the form $(A/J)^{(S)}$ for all open two-sided ideals J in A, and for all sets S. Considering the short exact sequence

$$0 \to J^{(S)} \to A^{(S)} \to (A/J)^{(S)} \to 0$$

and throwing through the right exact functor C, we obtain that

$$C((A/J)^{(S)}) \approx \widehat{A^{(S)}}/\mathcal{J}$$

where $\mathcal{J} = $ the C-complete left submodule of $\widehat{A^{(S)}}$ generated by $J^{(S)}$. Therefore,

$$C((A/J)^{(S)}) = (A/J)^{(S)}$$

iff $\mathcal{J} = J^S \cap \widehat{A^{(S)}}$ in A^S. But this latter holds, iff

(1) Whenever $(a_i)_{i \in S} \in J^S$, and $(a_i)_{i \in S}$ converges to zero in A, then there exists $j_i \in J$ and $b_i \in \hat{A}$, such that $(b_i)_{i \in I}$ converges to zero in $\hat{A}$, and such that $a_i = b_i j_i$, for all $i \in S$. By the Proposition, this condition holds (for all open two-sided ideals J in A and all sets S) if A is commutative and admissible. (This can also be easily verified directly from (1) above.) Another set of conditions under which it is easy to verify (1) is: That A admits a denumerable neighborhood base, and that for every open right ideal I in A, we have that the right ideal $I \cdot J$ of A is open. Therefore:

(2) Let A be a topological ring such that the topology of A is given by denumerably many right ideals, and such that the square of an open right ideal is open. Then, for every abstract left A-module M such that M is A-adically discrete, we have that $C(M) = M$.

Proposition 3.2.16. *Let A be a topological ring such that the topology is given by right ideals. Then*

(1) If A is admissible, then an A-adically closed abstract left A-submodule of a C-complete left A-module, is a C-complete left A-submodule.

(2) Conversely, if the topology of A admits a denumerable neighborhood base at zero consisting of two-sided ideals, then A is admissible iff an A-adically closed abstract left A-submodule of a C-complete left A-module is always a C-complete left A-submodule.

Proof. (1) Let M be a C-complete left A-module and let N be an A-adically closed abstract left A-submodule of M. Then by Chapter 2, §3, Theorem 2.3.5, there exists an epimorphism

$$f : \widehat{F} \to M$$

in the category $\mathscr{C}_A$, where F is a free abstract left A-module. If we can show that $f^{-1}(N)$ is a C-complete left A-submodule of $\widehat{F}$, then it will follow that N is a C-complete left A-submodule of M. (Since then N is the image in the category $\mathscr{C}_A$ of the composite map: $f^{-1}(N) \hookrightarrow \widehat{F} \xrightarrow{f} M$.) Also, $f^{-1}(N)$ is A-adically closed in $\widehat{F}$ (since an infinitely linear function is also a homomorphism of abstract left A-modules, and a homomorphism of abstract left A-modules is continuous for the A-adic topologies). Therefore, replacing M with $\widehat{F}$ and N with $f^{-1}(N)$ if necessary, it suffices to prove the Proposition in the case $M = \widehat{F}$, the completion of a free abstract left A-module. But then, since A is admissible, the uniform and A-adic topologies of $\widehat{F}$ coincide (see Chapter 1, §3, Definition 1, condition [2]). Since N is by hypothesis closed in $\widehat{F}$ for the A-adic topology, N

is therefore closed in $\widehat{F}$ for the uniform topology. Since $(\widehat{F}, unif)$ is a topological $(\widehat{A}, unif)$-module, the scalar multiplication: $(\widehat{A}, unif) \times (\widehat{F}, unif) \to (\widehat{F}, unif)$ is continuous. It follows that any A-submodule of $\widehat{F}$ that is closed for the uniform topology is an $\widehat{A}$-submodule. Therefore N is an $\widehat{A}$-submodule of $\widehat{F}$ (e.g., see Chapter 2, §2, Remark immediately following Example 1.1, observation number (2)). The definition of the infinite sum structure on $\widehat{F}$ is the one deduced by taking limits for the uniform topology of $\widehat{F}$. (E.g., see Chapter 2, Remark immediately following Example 1.1, observation no. (3).) Since N is closed in $\widehat{F}$ for the uniform topology, therefore N is closed under taking infinite linear combinations; i.e., N is a C-complete left A-submodule of $\widehat{F}$.

(2) Conversely, suppose that the topology of A is given by denumerably many two-sided ideals; and that the conclusion of the condition in (1) of the statement of the Proposition holds. Then let F be any free left A-module, and let I be any open two-sided ideal in A. Then $N = I \cdot \widehat{F}$ is an A-adically open left A-submodule of $\widehat{F}$. Therefore $I \cdot \widehat{F}$ is also A-adically closed in $\widehat{F}$. By the property in (1) applied to the C-complete left A-module $\widehat{F}$, therefore we have that $I \cdot \widehat{F}$ is a C-complete left A-submodule of $\widehat{F}$. Since the topology of A admits a denumerable neighborhood base at zero, the same is true for the uniform topology in $\widehat{F}$. Therefore the closure, for the uniform topology of $\widehat{F}$, of $N = I \cdot \widehat{F}$, consists of the set of all limits in $(\widehat{F}, unif)$ of Cauchy sequences, or equivalently of denumerable Cauchy series, of elements of N. Since N is a C-complete left A-submodule of $\widehat{F}$, and since the infinite sum structure of $\widehat{F}$ is by definition deduced from the uniform topology by taking limits of Cauchy series, it follows that the limit in $(\widehat{F}, unif)$ of every denumerable Cauchy series in N, also lies in N. Therefore $N = I \cdot \widehat{F}$ is closed in $\widehat{F}$ for the uniform topology. This being true for every open two-sided ideal I in A, by Chapter 1, §3, Definition 1, condition [9], we have that the topological ring A is admissible.   □

Remark 6. The proof of Proposition 3.2.16 also shows that:

(1) Let A be a topological ring such that the topology is given by right ideals. Let M be a C-complete left A-module, and let N be an abstract left A-submodule of M. If also $\widehat{F}$ is the completion of an arbitrary (not necessarily free) abstract left A-module F, and if $\phi : \widehat{F} \to M$ is any surjective infinitely linear function; then N is a C-complete left A-submodule of M if $\phi^{-1}(N)$ is closed in $\widehat{F}$ for the uniform topology.

(2) And conversely, suppose that the topology of A is given by denumerably many right ideals, and suppose M, N, $\widehat{F}$, and ρ are as above; then N is a C-complete left A-submodule of M iff $\phi^{-1}(N)$ is closed in $\widehat{F}$ for the uniform topology.

Corollary 3.2.17. *Let A be a topological ring such that the topology is given by right ideals; and such that A is admissible. Then, for every C-complete left A-module M, we have that*

(1) The divisible part of M, $div(M)$, is a C-complete left A-submodule of M.

(2) $M/div(M)$ is C-complete and is A-adically Hausdorff.
(3) We have the exact sequence of C-complete left A-modules

$$0 \to div(M) \to M \to \widehat{M}.$$

Proof. $div(M)$ is the closure of $\{0\}$ in M for the A-adic topology. Therefore $div(M)$ is A-adically closed in M. Therefore, by Proposition 3.2.16, part (1), we have that $div(M)$ is a C-complete left A-submodule of M:

The assertion (2) follows immediately.

The sequence (3) is exact for any abstract left A-module M. It remains to show that it is an exact sequence in the category of C-complete left A-modules — i.e., that the natural homomorphism of abstract left A-modules: $M \to \widehat{M}$ is infinitely linear. But this follows from Lemma 3.2.6 (with $N = (\widehat{M}, unif)$). □

Remark 7. Let A be an arbitrary topological ring such that the toplogy is given by right ideals, and let M be any abstract left A-module. Then recall that we have the natural map

$$(1): \quad \rho_M : C(M) \to \widehat{M},$$

a homomorphism of C-complete left A-modules (and, in fact, the only homomorphism of C-complete left A-modules that is compatible with the natural maps: $M \to C(M)$, $M \to \widehat{M}$). $\widehat{M}$ is complete, and therefore Hausdorff, for the uniform topology. The A-adic topology on $\widehat{M}$ is finer than the uniform topology; therefore, $\widehat{M}$ is likewise Hausdorff for the A-adic topology. In general, a homomorphism of abstract left A-modules maps A-divisible elements into A-divisible elements. Therefore, from the homomorphism (1), we deduce a natural homomorphism of abstract left A-modules

$$(2): \quad C(M)/div(C(M)) \to \widehat{M}.$$

If A is an admissible topological ring, then by Corollary 3.2.17 applied to the C-complete left A-module $C(M)$, $C(M)/div(C(M))$ is C-complete. And, in this case, by conclusion (3) of Corollary 3.2.17, the homomorphism (2) is a homomorphism of C-complete left A-modules.

However, in the general case (if A is an arbitrary topological ring such that the topology is given by right ideals, and if M is an arbitrary abstract left A-module), then there is no reason to believe that $C(M)/div(C(M))$ is a quotient C-complete left A-module of $C(M)$. Necessary and sufficient conditions for this to be so is that $div(C(M))$ be a C-complete left submodule of $C(M)$ — i.e., that the infinite sum of A-divisible elements in $C(M)$ be A-divisible. (And there may very well be counterexamples at such a level of generality.) Therefore, in this case, the homomorphism (2) is then obviously only a homomorphism of *abstract* left A-modules.

Corollary 3.2.18. *Let A be an admissible topological ring such that the topology is given by right ideals, and such that the topological ring A admits a denumerable neighborhood base at zero. Then, for every C-complete left A-module M, we have the short exact sequence of C-complete left A-modules:*

$$(1):\ 0 \to (div(M)) \to M \to \widehat{M} \to 0.$$

Also, if N is any abstract left A-module, then we have the short exact sequence of C-complete left A-modules:

$$(2):\ 0 \to (div(C(N))) \to C(N) \to \widehat{N} \to 0.$$

Note: If we delete the hypothesis that "the topological ring A admits a denumerable neighborhood base at zero", then the proof of (2) shows that the sequence:

$$(2'):\ 0 \to (div(C(N))) \to C(N) \to \widehat{N}$$

is exact.

Proof. (1) By Corollary 3.2.20 below, we have that the map $M \to \widehat{M}$ is surjective. The rest of the assertion follows from conclusion (3) of Corollary 3.2.17.

(2) Given any abstract left A-module N, let $M = C(N)$. Then the short exact sequence (1) applies to $M = C(N)$, and yields a short exact sequence

$$(3):\ 0 \to div(C(N)) \to C(N) \to \widehat{C(N)} \to 0.$$

But, since the topological ring A is admissible, by Theorem 3.2.11, we have that $\widehat{C(N)} \approx \widehat{N}$. Substituting into equation (3) gives conclusion (2) of the Corollary.   □

Remark 8. Corollary 3.2.18 tells us that, for a complete topological ring that is admissible, and that admits a denumerable neighborhood base at zero:

The (usual) completion, $\widehat{N}$, of every abstract left A-module N, can be recovered from the C-completion, $C(N)$, by the formula

$$\widehat{N} = C(N)/(\text{divisible elements}).$$

Therefore, in this important case (which includes all serious current applications to algebraic geometry and commutative algebra), we see that the C-completion is perhaps a "better" invariant that the usual completion — it is, in general, a non-Hausdorff, completion $C(N)$; such that $C(N)$ "made Hausdorff" (i.e., after passing to the quotient by the closure of zero) is the usual completion $\widehat{N}$.

Remark 9. In the next Chapter, we will see that, if N is left flat as an A-module, and if the topology of A is given by denumerably many right ideals, then $C(N) = \widehat{N}$.

Lemma 3.2.19. *Let A be a topological ring such that the topology is given by right ideals, and let M be an abstract left A-module. Choose P a projective abstract left A-module and $g : P \to M$ an epimorphism of abstract left A-modules. Then the images in $\widehat{M}$ of the homomorphisms*

$$(1): \ \rho_M : C(M) \to \widehat{M}, \ and$$

$$(2): \widehat{g} : \widehat{P} \to \widehat{M}$$

coincide.

Proof. We have the commutative diagram:

$$
\begin{array}{ccc}
\widehat{P} & \xrightarrow{\ \widehat{g}\ } & \widehat{M} \\
\uparrow & & \uparrow{\scriptstyle \rho_M} \\
C(P) & \xrightarrow{\ C(g)\ } & C(M)
\end{array}
$$

in the category of C-complete left A-modules. Since g is an epimorphism and C is right exact, $C(g)$ is an epimorphism. Since P is projective, $C(P) = \widehat{P}$ — i.e., the leftmost vertical map $C(P) \to \widehat{P}$ is an isomorphism. Therefore from commutativity of the square, $\mathrm{Im}(\widehat{g}) = \mathrm{Im}(\rho_M)$. $\qquad\square$

Corollary 3.2.20. *Let A be a topological ring such that the topology is given by right ideals and such that there exists a denumerable neighborhood base at zero. Then, for every abstract left A-module M, we have that the natural mapping*

$$(1): \rho_M : C(M) \to \widehat{M}$$

of C-complete left A-modules is surjective.

Proof. Let P be an abstract free left A-module and let $g : P \to M$ be an epimorphism of abstract left A-modules. Then by Lemma 3.2.19, it is necessary and sufficient to prove that $\widehat{g} : \widehat{P} \to \widehat{M}$ is surjective. And this follows from the next Lemma. $\qquad\square$

Lemma 3.2.21. *Let A be a topological ring such that the topology is given by denumerably many right ideals. Let $f : M \to N$ be a surjective homomorphism of abstract left A-modules. Then the induced mapping*

$$\widehat{f} : \widehat{M} \to \widehat{N}$$

is surjective.

Proof. (We have actually already proved this several times.)

Since A admits a denumerable neighborhood base at zero, therefore so do M and N. Therefore every element x of $\widehat{N}$ is the limit of the image in $\widehat{N}$ of a Cauchy sequence $(x_i)_{i \geq 1}$ in N for the A-adic topology. Letting $y_i = x_{i+1} - x_i$,

$i \geq 0$ (where $x_0 = 0$), then x is the limit in $(\widehat{N}, unif)$ of the image of the denumerable Cauchy series in $(N, A\text{-adic})$

$$(1): \sum_{i \geq 0} y_i.$$

Let $I_1 \supset I_2 \supset I_3 \supset \ldots$ be a sequence of open right ideals in A that give the topology of A and let $I_0 = A$. For each $i \geq 0$, let

$$n(i) = \begin{cases} \text{the largest integer such that } y_i \in I_{n(i)} \cdot N, \text{ if there is a largest such integer.} \\ i, \text{ if } y_i \in I_n \cdot N \text{ for all } n \geq 0. \end{cases}$$

Since (1) is a Cauchy series in $(N, A\text{-adic})$, we have that

$$(2): \lim_{i \to \infty} n(i) = +\infty.$$

We have that $y_i \in I_{n(i)} \cdot N$. Since f is surjective, there exists $z_i \in I_{n(i)} \cdot M$ such that

$$(3): f(z_i) = y_i.$$

Then by (2), the series

$$(4): \sum_{i \geq 0} z_i$$

is a Cauchy series in $(M, A\text{-adic})$. Let z be the limit of the image of the series (4) in the completion $(\widehat{M}, unif)$ of $(M, A\text{-adic})$. Then by continuity of $\widehat{f}$ and by equation (3), we have that

$$(5): \widehat{f}(z) = x.$$

$\square$

Corollary 3.2.22. *Let A be a topological ring such that the topology is given by denumerably many right ideals. Let N be an abstract left A-module such that N is isomorphic as an abstract left A-module to a quotient module of an A-adically complete abstract left A-module. Then N is A-adically complete iff N is A-adically Hausdorff (i.e., iff $div(N) = \{0\}$).*

Proof. Let K be an A-adically complete left A-module and let $g : K \to N$ be an epimorphism. Then $K = \widehat{K}$, so we have the commutative diagram:

$$
\begin{array}{ccc}
\widehat{K} & \xrightarrow{\widehat{g}} & \widehat{N} \\
\uparrow{\scriptstyle id} & & \uparrow \\
K & \xrightarrow{g} & N
\end{array}
$$

By Lemma 3.2.21, $\widehat{g}$ is an epimorphism. Therefore the natural map: $N \to \widehat{N}$ is an epimorphism. Since the kernel of $N \to \widehat{N}$ is always $div(N)$, it follows that $N \to \widehat{N}$ is an isomorphism iff $div(N) = \{0\}$. $\square$

An immediate corollary of Corollary 3.2.18 is

Corollary 3.2.23. *Let A be a topological ring that is admissible, and such that there exists a denumerable neighborhood base at zero. Then for every C-complete left A-module M, we have that*

(1) $M/\text{div}(M)$ is A-adically complete. (i.e., M is "complete but not Hausdorff"). Also

(2) M is A-adically complete iff M is A-adically Hausdorff.

Proof. Since (§1, Theorem 3.1.6) the ring A is C-O.K., we have that M is C-complete iff the natural map: $M \to C(M)$ is an isomorphism. If M is C-complete, equation (1) of Corollary 3.2.18 becomes

$$(3): \quad M/\text{div}(M) \approx \widehat{M}.$$

By §1, Theorem 3.1.6, the topological ring A is strongly admissible. Therefore (Chapter 1, §3, Definition 2, condition [1]) $\widehat{M}$ is A-adically complete. This and equation (3) implies conclusion (1) of the corollary.

M is Hausdorff iff $\text{div}(M) = \{0\}$. Therefore by equation (1), if M is Hausdorff, then M is complete. $\qquad\square$

Corollary 3.2.24. *Let A be a topological ring such that the topology is given by right ideals. Then*

(1) The kernel of the natural homomorphism of abstract left A-modules

$$\iota_M : M \to C(M)$$

is contained in the divisible elements of M.

(2) If also A admits a denumerable neighborhood base at zero, then the kernel of ι_M contains the infinitely divisible elements of M.

Proof. (1) We have seen in the Note to Corollary 2.5.7, that the natural map: $M \to \widehat{M}$ factors through ι_M: $M \overset{\iota_M}{\to} C(M) \overset{\rho_M}{\to} \widehat{M}$. Therefore $\text{Ker}(\iota_M) \subset \text{Ker}(M \to \widehat{M}) = \{\text{divisible elements of } M\}$.

(2) Since A has a denumerable neighborhhod base at zero, by Proposition 3.2.1 we have seen that the C-complete left A-module $C(M)$ has no non-zero infinitely A-divisible elements. Therefore every infinitely A-divisible element of M must map into zero in $C(M)$. $\qquad\square$

Corollary 3.2.25. *Let A be a topological ring that is admissible. Let M be an abstract left A-module. Then for every element $x \in M$, we have that x is A-divisible in M iff the image $\iota_M(x)$ in $C(M)$ is A-divisible in $C(M)$.*

Note: Another way of stating the conclusion of this Corollary is: $\iota_M^{-1}(\text{div}(C(M))) = \text{div}(M)$.

Proof. By Theorem 3.2.11, we have that

$$\widehat{C(M)} \approx \widehat{M}.$$

Therefore, an element $x \in M$ is A-divisible iff the image of x in $\widehat{M}$ is zero iff the image of x in $\widehat{C(M)}$ is zero iff the image of x in $C(M)$ is A-divisible.   □

Remark 10. As we shall see later, at least in some special cases, e.g., in the case that the topology of A is the right t-adic, for some element t in A, such that, say, either t is not a right zero-divisor of A, or A is right Noetherian — then the kernel of $\iota_M : M \to C(M)$ is precisely the set of all infinitely divisible elements of M. Hence, for example, in this case (which of course includes the case of a complete discrete valuation ring), $C(M)$ is a "more subtle" completion than the usual $\widehat{M}$ in yet another way. By Corollary 3.2.18, equation (2), we have that

$$\widehat{M} = C(M)/(\text{divisible elements}).$$

Also, of course,

$$M/(\text{divisible elements}) \hookrightarrow \widehat{M}.$$

But in this case

$$M/(\text{infinitely divisible elements}) \hookrightarrow C(M),$$

since in this case the kernel of ι_M is the infinitely divisible part of M. Thus, $C(M)$ "keeps alive" those divisible elements of M that are not infinitely divisible, i.e.,

$$\text{div}(M)/(\text{infinitely divisible part of } M) \hookrightarrow C(M).$$

Remark 11. We will give many examples, later, (even if A is any complete valuation ring that is not a field) in which $C(M)$ is *not* Hausdorff, even in cases in which M is Hausdorff.

For the rest of this book, to save space, we will often use the phrase *right standard topological ring* — more briefly, *standard topological ring* — for a topological ring A, such that the topology is given by right ideals.

Chapter 4

The Higher C-completions

4.1 The Spectral Sequence of the C-completion

Let A be a standard topological ring. Then we have constructed the functor C, a right-exact functor from the category of abstract left A-modules (an abelian category with enough projectives) into the category of C-complete left A-modules (an abelian category). C can be characterized as being the zeroth left derived functor of the functor "A-adic completion," $M \to \widehat{M}$. This suggests considering the higher left derived functors of C (or equivalently, of "A-adic completion").

Definition 1. For i an integer ≥ 0, the ith higher C-completion functor, C_i, is the ith left derived functor [CEHA] of the functor "A-adic completion" (or, equivalently, of the functor C), from the abelian category of abstract left A-modules into the abelian category of all C-complete left A-modules.

Thus explicitly for every abstract left A-module M, and for every integer $i \geq 0$, we have $C_i(M)$, the ith higher C-completion of M. This is a C-complete left A-module.

Hence, for $i = 0$, $C_0(M) = C(M)$, the C-completion as constructed in Chapter 2, §5, Definition 1. More generally, for $i \geq 0$, the higher C-completions $C_i(M)$ can be constructed as follows: Let P_* be any acyclic projective resolution of the abstract left A-module M. Then

$$(1): \ C_i(M) = H_i(\widehat{P_*}), i \geq 0.$$

(I.e., take the A-adic completion of the left A-modules P_i, $i \geq 0$, and then take the homology groups of the resulting chain complex of C-complete left A-modules.)

Remark 1. As we have seen in Chapter 2, §3, Theorem 2.3.12, the functor "underlying abstract left A-module," from the category of C-complete left A-modules into the category of abstract left A-modules is exact (and in fact is even also faithful). It follows that, if for every abstract left A-module M, we interpret the ith higher C-completion of M as being an *abstract left A-module* (rather

than as a C-complete left A-module — by taking the underlying abstract left
A-module of $C_i(M)$—), for $i \geq 0$, then once again the functors C_i — this time
regarded as functors from the category of *abstract left A-modules* into itself —
are once again the left derived functors of the functor C — or equivalently of
the functor "A-adic completion" (likewise regarded as being an additive functor
from the category of abstract left A-modules into itself).

In particular, if M is any abstract left A-module, then $C_i(M)$ (regarded as
an abstract left A-module) can again be computed by equation (1) — where the
chain complex $\widehat{P_*}$ is then regarded simply as being a chain complex of abstract
left A-modules (ignoring the C-complete structures).

Remark 2. Of course, in the special case that the standard topological ring
A is C-O.K. (actually a very mild condition — see Chapter 3, §1, Theorem 3.1.6
and Chapter 1, §3, Examples 1, 2, and 3); then the category of C-complete left
A-modules becomes naturally a *full subcategory* of the category of all abstract
left A-modules (in fact, even a full, exact, abelian such subcategory). Therefore,
in this case, the higher C-completions $C_i(M)$, $i \geq 0$, as defined in Definition 1,
are simply abstract left A-modules (that are C-complete), for all integers $i \geq 0$.
(In other words, in this case, — *and in this case only* — there is no difference
between the constructions of Definition 1 and Remark 1 above).

Theorem 4.1.1. *(The spectral sequence of the C-completion)*
Let A be a standard topological ring, and let B_* be a non-negatively indexed
chain complex of abstract left A-modules. Suppose that

$$(2) : C_i(B_q) = 0, \text{ for } i \geq 1, q \geq 0.$$

*Then there is induced a first quadrant, homological spectral sequence (in the
category of C-complete left A-modules), starting with $E^2_{p,q}$, such that*

$$(3) : \quad E^2_{p,q} = C_p(H_q(B_*)),$$

and such that the abutment is the group

$$(4) : \quad H_n(C(B_*)), n \geq 0.$$

Corollary 4.1.2. Let A be a standard topological ring, let B_* and D_* be non-
negative chain complexes of abstract left A-modules, and let

$$(1) : \quad f_* : B_* \to D_*$$

be a homomorphism of chain complexes of abstract left A-modules. Suppose that

$$(2) : \quad C_i(B_q) = C_i(D_q) = 0 \text{ for } i \geq 1, q \geq 0,$$

*and that the maps induced by f_* on homology are isomorphisms of abstract left
A-modules*

$$(3) : \quad H_n(f_*) : H_n(B_*) \overset{\approx}{\to} H_n(D_*),$$

for all integers n. Then the maps induced by $C(f_*)$ *on homology*

$$(4): \quad H_n(C(f_*)) : H_n(C(B_*)) \to H_n(C(D_*))$$

are isomorphisms of C-complete left A-modules, for all integers n.

Proof. The map f_* induces a map from the spectral sequence of C_* into the spectral sequence of D_*. The hypothesis (3) of the Corollary implies that f_* induces an isomorphism

$$C_p(H_q(f_*)) : C_p(H_q(B_*)) \stackrel{\approx}{\to} C_p(H_q(D_*)).$$

By conclusion (3) of the Theorem, this is equivalent to asserting that f_* induces an isomorphism along $E^2_{p,q}$. Therefore f_* induces an isomorphsim on the abutments. □

Remark 3. The above spectral sequence is extremely important in many computations involving cohomology of completions, and p-adic cohomology. In practice, the hypothesis (2) of Theorem 4.1.1 is often obeyed (we will see later, for example, that this is the case if the B_q obey a mild flatness condition). Then the spectral sequence abuts at, essentially, the "homology of the C-completions"; and $E^2_{p,q}$ is the "higher C-completions of the homology". One can imagine the use of such a spectral sequence. E.g., as we shall see, the exact sequence I.8 of [PPWC] (which also dominated Chapter 2 of [COC]), is closely related to a very special case of the spectral sequence of the C-completion. In fact, one might say that it is this spectral sequence which really underlies that short exact sequence I.8.

Remark 4. The reader might suspect that the spectral sequence of the C-completion is really just a minor variant of the well-known short exact sequences:

$$0 \to \left(\varprojlim_{i \geq 0}{}^1 H_{n-1}(B_*/I_i B_*) \right) \to H_n(\widehat{B_*}) \to \left(\varprojlim_{i \geq 0} H_n(B_*/I_i B_*) \right) \to 0, n \geq 0$$

(which are valid under the very mild condition that the topology of A be given by denumerably many right ideals $I_1 \supset I_2 \supset \dots$. See [COC], Chapter 3, Corollary 1.1). However, this is *very definitely not* the case (even in the simplest case in which A is a complete discrete valuation ring not a field). In fact, in the above (familiar) short exact sequences, we have maps:

$$H_n(\widehat{B_*}) \to \left(\varprojlim_{i \geq 0} H_n(B_*/I_i B_*) \right), n \geq 0.$$

However, in the spectral sequence of the C-completion the edge homomorphisms are maps:

$$C(H_n(B_*)) \to H_n(C(B_*)),$$

$n \geq 0$. A moment's reflection will show that essentially these maps are running in the *opposite direction* to the maps,

$$H_n(\widehat{B_*}) \to \left(\varprojlim_{i \geq 0} H_n(B_*/I_i B_*) \right), n \geq 0.$$

In fact, it was in using the essentially *opposite* direction of the short exact sequence I.8 ([COC], Chapter 2) (which short exact sequence is related to a special case of the spectral sequence of Theorem 4.1.1), and of the usual short exact sequences of the inverse limit just described above ([COC], Chapter 3), that the author was able to prove the "finite generation" theorems in Chapter 5 of [COC], such as [COC], Theorem 4, p. 650.

The "short exact sequence I.8" was also used extensively in [PPWC], where it first appeared.

A more general result than Theorem 4.1.1 is

Proposition 4.1.3. *Let A be a standard topological ring and let B_* be an arbitrary non-negative chain complex of abstract left A-modules. Then there are induced two first quadrant homological spectral sequences in the category of C-complete left A-modules. The first such spectral sequence, $^I E_{p,q}^r$, starts with $^I E_{p,q}^1$, where*

$$(5): \quad ^I E_{p,q}^1 = C_q(B_p).$$

The second such spectral sequence, $^{II} E_{p,q}^r$, starts with $^{II} E_{p,q}^2$, where

$$(6): \quad ^{II} E_{p,q}^2 = C_p(H_q(B_*)).$$

Both of these spectral sequences abut at the same sequence of C-complete left A-modules, K_n, $n \geq 0$. (But with, in general, different filtrations on the C-complete left A-modules K_n, $n \geq 0$.)

Note. For every non-negatively indexed chain complex B_* of abstract left A-modules, we therefore have $K_n(B_*)$, $n \geq 0$, a sequence of C-complete left A-modules. In the notation of [PPWC], Chapter I, §2, these functors K_n, $n \geq 0$, are the *left hyperderived functors of the right exact functor C* — (or, equivalently, *of the additive functor "A-adic completion"*) — from the category of abstract left A-modules into the category of C-complete left A-modules.

Corollary 4.1.4. *Let A be a standard topological ring, let M be an abstract left A-module and let B_* be an acyclic homological resolution of M (that is, B_* is a non-negative chain complex, $H_0(B_*) \approx M$ and $H_i(B_*) = 0$, $i \geq 1$). Then there is induced a homological, first quadrant spectral sequence, starting with*

$$(7): \quad E_{p,q}^1 = C_q(B_p),$$

and with abutment

$$(8): \quad C_n(M), n \geq 0.$$

Proof of Theorem 4.1.1, Proposition 4.1.3, and Corollary 4.1.4. Let $\mathscr{A}$ be the dual of the category of abstract left A-modules. Let β be the dual of the category of C-complete left A-modules and let $F : \mathscr{A} \rightsquigarrow \beta$ be the functor "C-completion." Then Proposition 4.1.3 follows from [PPWC], Chapter I, §2, Theorem 1, p. 118; Theorem 4.1.1 follows from [PPWC], p. 119, Example 2; and Corollary 4.1.4 follows from [PPWC], p. 119, Example 1. □

Remark 5. Of course, Theorem 4.1.1 and Corollary 4.1.4 are immediate consequences of Proposition 4.1.3.

Remark 6. Although we did not state it as part of Theorem 4.1.1, Proposition 4.1.3 and Corollary 4.1.4, of course in each of the spectral sequences discussed in these theorems, the filtration on the abutment is finite, and the associated graded of the abutment is $E_{p,q}^{\infty}$. (This is implied in the adjective "first quadrant" — see [COC], Introduction, Chapter 2.)

Also, of course, each of these spectral sequences is "functorial" — e.g., under the hypotheses of Proposition 4.1.3, if $'B_*$ is any other non-negative chain complex of abstract left A-modules, and if $f_* : B_* \to {'B_*}$ is any map of chain complexes, then f_* induces a map of spectral sequences (see [COC] for the definition) from the first (resp.: second) spectral sequence of B_* into the first (resp.: second) spectral sequence of $'B_*$.

Proposition 4.1.5. *Let A be a standard topological ring, let M be an abstract left A-module and let B_* be an acyclic homological resolution of M (i.e., B_* is a non-negative chain complex such that $H_i(B_*) = \{0\}$, $i \geq 1$, and $H_0(B_*) \approx M$). Suppose also that*

$$(9): \ C_i(B_p) = 0, \ for \ i \geq 1, \ and \ p \geq 0.$$

Then there are induced canonical isomorphisms of C-complete left A-modules

$$(10): \ C_n(M) \approx H_n(C(B_*)), n \geq 0.$$

Proof. By Corollary 4.1.4, we have the spectral sequence (7) with abutment (8). By equation (9),

$$E_{p,q}^1 = 0 \text{ for } q \neq 0.$$

Therefore, the spectral sequence degenerates past $E_{p,q}^2$ (since $d_{p,q}^r$ for $r \geq 2$ raises q-degree by $r - 1 \neq 0$, and therefore $d_{p,q}^r = 0$ for $r \geq 2$), and therefore $E_{n,0}^2 = E_{n,0}^{\infty} = (n$th group of the abutment), $n \geq 0$. Equations (7) and (8) complete the proof. □

4.2 Inverse Limits and Higher Inverse Limits of C-complete Left A-modules

Theorem 4.2.1. *Let A be a standard topological ring, let I be a set and let $(M_i)_{i \in I}$ be an indexed family of C-complete left A-modules. Then the Cartesian* *product of abstract left A-modules:*

$$\prod_{i \in I} M_i$$

has a natural infinite sum structure, with respect to which it becomes the direct product of the M_i, $i \in I$, in the category of C-complete left A-modules.

Proof. In fact, we have already proved this theorem (Chapter 2, §3, Theorem 2.3.9). We recall the construction briefly.

If J is a set, and $(a_j)_{j \in J}$ is a zero-bound family of elements of $\widehat{A}$, and if x_j, $j \in J$ are elements of $\prod_{i \in I} M_i$ — say, $x_j = (x_{ij})_{i \in I}$, all $j \in J$ — then define

$$\sum_{j \in J} a_j x_j = \left(\sum_{j \in J} a_j x_{ij} \right)_{i \in I} .$$

It is immediate that this definition of "infinite sums" endows the abstract left A-module $\prod_{i \in I} M_i$ with an infinite sum structure as defined in Chapter 2, so that $\prod_{i \in I} M_i$ becomes a C-complete left A-module. And the C-complete left A-module $\prod_{i \in I} M_i$ thus defined, together with the usual projections

$$\pi_k : \prod_{i \in I} M_i \to M_k, \text{ all } k \in I,$$

obeys the universal mapping property to be a direct product of the M_k, $k \in I$, in the category of C-complete left A-modules and infinitely linear functions. □

Proposition 4.2.2. *Let D be a directed set (respectively: Let $\mathscr{D}$ be a category that is a set), let A be a standard topological ring, and let $(M_i, \alpha_{ij})_{i,j \in D}$ be an inverse system indexed by the directed set D in (respectively: and let F be a contravariant functor from the category $\mathscr{D}$ into) the category of C-complete left A-modules and infinitely linear maps. Let M be the inverse limit of $(M_i, \alpha_{ij})_{i,j \in D}$ (respectively: of F) in the category of abstract left A-modules, and let $\pi_i : M \to M_i$ (respectively: $\pi_i : M \to F(i)$) be the projections, for all $i \in D$ (respectively: all objects i of $\mathscr{D}$). Then there exists a unique infinite sum structure on M such that the functions π_i are infinitely linear, for all $i \in D$ (respectively: all objects i of $\mathscr{D}$). And M together with this infinite sum structure is an inverse limit of the inverse system $(M_i, \alpha_{ij})_{i,j \in D}$ (respectively: of the contravariant functor F) in the category of C-complete left A-modules.*

Proof. Since the parenthetical statement is more general, we prove that.

In fact, an infinite sum structure on M is such that π_i is infinitely linear, for all objects $i \in I$, iff the inclusion:

$$M \to \prod_{i \in I} F(i)$$

is infinitely linear (for the infinite sum structure defined on $\prod_{i \in I} F(i)$ in the proof of Theorem 4.2.1). Therefore such an infinite sum structure, if it exists, is unique — necessarily being the one induced from $\prod_{i \in I} F(i)$; and such an infinite sum structure exists iff the abstract left A-submodule M of $\prod_{i \in I} F(i)$ is a C-complete left A-submodule (— i.e., iff M is "closed under taking infinite sums"). But, if S is the set of all triples (i, j, f), such that i, j are objects in $\mathscr{D}$, and $f : i \to j$ is a map, then the abstract left A-module M is the kernel, in the category of abstract left A-modules, of the obvious A-homomorphism:

$$g : \prod_{i \in I} F(i) \to \prod_{(i,j,f) \in S} F(i)$$

(— the one such that, the (i, j, f)th coordinate is the function that maps $(x_k)_{k \in I} \in \prod_{k \in I} F(k)$ into $x_i - (F(f)(x_j)) \in F(i)$). Since g is infinitely linear, it follows that $M = \text{Ker}(g)$ is a C-complete left A-submodule of $\prod_{i \in I} F(i)$, proving the first assertion.

The second assertion follows from the fact that, in any additive category $\mathscr{A}$ (and a fortiori in any abelian category $\mathscr{A}$) (e.g., in the category of C-complete left A-modules), if $\mathscr{D}$ is any category that is a set and if F is any contravariant functor from $\mathscr{D}$ into $\mathscr{A}$, and if $\prod_{i \in \mathscr{D}} F(i)$ and $\prod_{(i,j,f) \in S} F(i)$ exist, and if $\text{Ker}(g : \prod_{i \in \mathscr{D}} F(i) \to \prod_{(i,j,f) \in S} F(i))$ exist, then $\text{Ker}(g)$, together with the restrictions of the projections: $\prod_{i \in \mathscr{D}} F(i) \to F(k)$, $k \in \mathscr{D}$, is an inverse limit of the contravariant functor F. (This latter assertion is proved, e.g., by an immediate verification of the necessary universal mapping property.) □

Corollary 4.2.3. *Let A be a standard topological ring. Then the abelian category $\mathscr{C}_A$ of all C-complete left A-modules is closed under taking arbitrary infinite direct products of objects (and therefore is also closed under taking arbitrary inverse limits of contravariant functors with domain a category that is a set).*

The "stripping functor," that to every C-complete left A-module associates its underlying abstract left A-module is a functor from $\mathscr{C}_A$ into the category of abstract left A-modules, that (in addition to being exact and faithful), preserves

(1) arbitrary direct products of objects;

(2) arbitrary inverse limits over directed sets; and

(3) arbitrary inverse limits of contravariant functors over categories that are sets.

*(4) For every set I, the functor "I-fold direct product" (from the category
of I-tuples of C-complete left A-modules into the category of C-complete left
A-modules), is exact. In particular, the category of C-complete left A-modules
obeys the Eilenberg–Moore axiom (P1). (For the definition, see [COC], Introduction, Chapter 1, §7, Appendix, p. 88.) And in fact, the category of C-complete
left A-modules obyes the Eilenberg–Moore Axiom (P2). (For the definition, see
[COC], Introduction, Chapter 1, §7, Appendix, p. 92.)*

Proof. Follows immediately from Proposition 4.2.2, and the fact that the stripping functor from the category of C-complete left A-modules into the category
of abstract left A-modules, is exact and faithful. (And therefore, a map of C-
completes is an epimorphism of C-completes iff it is an epimorphism of abstract
left A-modules.)   □

Remark 1. We shall see later, however, that the stripping functor rarely
preserves infinite *direct sums*, or infinite *direct limits* (whether over directed
sets or over categories that are sets).

Proposition 4.2.4. *Let A be a standard topological ring, let M_i be C-complete
left A-modules, for all integers $i \geq 0$, and let $f_{i+1} : M_{i+1} \to M_i$ be infinitely
linear maps, for all integers $i \geq 0$. Then there is induced a natural structure of
C-complete left A-module, on the abstract left A-module:*

$$\varprojlim_{i \geq 0}{}^1 M_i.$$

*The assignments: $(M_i, f_{i+1})_{i \geq 0} \rightsquigarrow \varprojlim_{i \geq 0} M_i, \varprojlim_{i \geq 0}{}^1 M_i$, from the category of inverse
systems indexed by the non-negative integers of C-complete left A-modules into
the category of C-complete left A-modules, is a system of right derived functors
(with the functors in dimensions 2, 3, ..., etc., being taken as zero).*

Proof. Define $\underline{C}^0 = \underline{C}^1 = \prod_{i \geq 0} M_i$, and as in [COC], bottom of p. 81, define

$$d^0 : \underline{C}^0 \to \underline{C}^1$$

such that

$$\pi_n \circ d^0 = f_{n+1} \circ \pi_{n+1} - \pi_n, \quad n \geq 0.$$

Then, by definition (e.g., see [COC], p. 82, line 6),

$$\varprojlim_{i \geq 0}{}^1 M_i = \operatorname{Cok}(d^0).$$

Since, by Theorem 4.2.1, the domain and range of d^0 are C-complete left A-
modules; and since d^0 is infinitely linear (since the composites with the π_n are
infinitely linear, for $n \geq 0$), it follows that $\varprojlim_{i \geq 0}{}^1 M_i$ has a unique structure as

a C-complete left A-module such that the natural map: $\underline{C}^1 \to \left(\varprojlim_{i \geq 0}{}^1 M_i \right)$ is infinitely linear.

Then, by [COC], Introduction, Chapter 1, §7, Theorem 1, p. 82, we have that the functors:

$$(1): \varprojlim_{i \geq 0}, \varprojlim_{i \geq 0}{}^1, 0, 0, \ldots, 0, \ldots$$

are an exact connected sequence of functors from the category of inverse systems indexed by the non-negative integers of C-complete left A-modules into the category of C-complete left A-modules. Also, by [COC], Introduction, Chapter 1, §7, Theorem 2, p. 83, for every such inverse system $(M_i, f_{i+1})_{i \geq 0}$ in which the f_i are surjective, we have that

$$\varprojlim_{i \geq 0}{}^1 M_i = 0.$$

Since every inverse system is isomorphic to a subsystem of such an inverse system (Proof easy; or see [EM]), it follows that the sequence (1) is a system of derived functors. □

Remark 2. Since, as we shall later see, the category of C-complete left A-modules rarely has *enough injectives*, one must make precise what one means by the statement, "The functors: $\varprojlim_{i \geq 0}, \varprojlim_{i \geq 0}{}^1, 0, \ldots, 0, \ldots$ from the abelian category $\mathscr{A}$ of all inverse systems of C-complete left A-modules indexed by the non-negative integers into the abelian category of C-complete left A-modules forms a *system of right derived functors*."

More generally, if $\mathscr{A}$ and β are abelian categories, $\mathscr{A}$ not necessarily having enough injectives, by a *system of cohomological derived functors from $\mathscr{A}$ into $\mathscr{B}$*, we mean: a non-negative cohomological exact connected sequence of functors F^n, $n \geq 0$, from $\mathscr{A}$ into $\mathscr{B}$, such that, for every object A in $\mathscr{A}$, and for every integer $n \geq 1$, there exists an object A' and a monomorphism $f' : A \to A'$ in $\mathscr{A}$ such that $F^n(f') = 0$. (Then it is easy to see that, given a left exact functor $F^0 : \mathscr{A} \rightsquigarrow \beta$, there exists, up to canonical isomorphism, at most one system of derived functors from $\mathscr{A}$ into β that is F^0 in dimension zero.) (Notice also that this definition is equivalent to that given in [COC], Introduction, Chapter 1, §10, Exercise 1, p. 453 — where in Exercise 1, one insists that $F = F^0$, $\eta =$ identity automorphim of F. Notice that Axiom (4) of that Exercise is automatic in the case $\eta =$ identity automorphism of F.)

The following Proposition will be proved in a later paper; we make no use of it in this book; however, it is, in my opinion, useful and interesting:

Proposition 4.2.5. *Let A be a standard topological ring, and let $\mathscr{D}$ be a category that is a set. Let $\mathscr{C}_A$ denote the abelian category of all C-complete left A-modules and infinitely linear functions, and let $\mathscr{A} = (\mathscr{C}_A)^{\mathscr{D}^\circ}$ denote the abelian category consisting of all contravariant functors from $\mathscr{D}$ into $\mathscr{C}_A$ and all natural transformations of functors.*

For every contravariant functor F from $\mathscr{D}$ into the category $\mathscr{C}_A$ of C-complete left A-modules, if we regard F as being a contravariant functor into the category of abstract left A-modules, then we have the usual higher inverse limits:

$$(2): \varprojlim_{d \in \mathscr{D}}^{i} F(d), \ i \geq 0,$$

which are abstract left A-modules. Then, for every such F and every $i \geq 0$, there exists a natural infinite sum structure on the abstract left A-module (2), so that (2) becomes a C-complete left A-module.

And, with $\varprojlim_{d \in \mathscr{D}}^{i} F(d)$ thus being regarded as an object in the category $\mathscr{C}_A$ (for every F in $\mathscr{C}_A$ and every $i \geq 0$), we have that the functors: $\varprojlim_{d \in \mathscr{D}}^{i}$, $i \geq 0$, form a system of right derived functors from the abelian category $\mathscr{A}$ into the abelian category $\mathscr{C}_A$.

Notes. 1. "Right derived functors" in the sense of Remark 2 above.

2. In fact, the proof will show that a stronger conclusion than [COC], Introduction, Chapter 1, §10, Exercise 1, p. 453, Axiom (3), holds; namely, "Given any object F in the category $\mathscr{A}$, there exists an object F' and a monomorphism $f : F \to F'$, such that

$$\varprojlim_{d \in \mathscr{D}}^{i} F'(d) = 0$$

for all integers $i \geq 1$.

4.3   Completion of the C-completion and of the Higher C-completions of an Abstract Left A-module

Theorem 4.3.1. Let A be a topological ring such that the topology is given by denumerably many right ideals $I_1 \supset I_2 \supset I_2 \supset I_3 \supset \ldots \supset I_n \supset \ldots$. Then for every abstract left A-module M, we have the short exact sequences of abelian groups

$$(1): 0 \to \left(\varprojlim_{j \geq 1}^{1} \mathrm{Tor}_{i+1}^{A}(A/I_j, M) \right) \to C_i(M) \to \left(\varprojlim_{j \geq 1} \mathrm{Tor}_{i}^{A}(A/I_j, M) \right) \to 0,$$

for all integers $i \geq 0$.

Note: If the right ideals I_j are two-sided ideals for all integers j (— e.g., this is so if A is *commutative* —) then the short exact sequence (1) of abelian groups is a short exact sequence *of C-complete left A-modules.*

Proof. Let P_* be an acyclic homological projective resolution of the abstract left A-module M. Then

$$\widehat{P_*} = \varprojlim_{j \geq 1} \left((A/I_j) \underset{A}{\otimes} P_* \right),$$

an inverse limit of an inverse system of chain complexes and epimorphisms. Since C_i, $i \geq 0$, resp.: $\text{Tor}_i^A((A/I_j), \quad)$, $i \geq 0$ are the derived functors of "A-adic completion," resp.: of $((A/I_j)\underset{A}{\otimes}, \quad)$, it follows that

$$H_i(\widehat{P_*}) = C_i(M), \ i \geq 0, \text{ resp.: that}$$

$$H_i((A/I_j) \underset{A}{\otimes} P_*) = \text{Tor}_i^A(A/I_j, M),$$

for all integers $i, j \geq 0$. Equation (1) then follows from [COC], Chapter 3, Corollary 1.1, pp. 535 and 536 (with the usual notation shift $C_j^n = [(A/I_j)\underset{A}{\otimes}P_{-n}]$ for turning a non-negative chain complex into a non-negative cochain complex).

If the right ideals I are two-sided ideals, then $(A/I_j) \underset{A}{\otimes} P_n$ is an abstract left A-module. Since this A-module is annihilated by I_j, it is discrete for the A-adic topology, and therefore A-adically complete. Therefore, it has a natural structre as a C-complete left A-module. (Since it is its own "$\widehat{\ }$", and the "$\widehat{\ }$" of any abstract left A-module has a natural sturcture as a C-complete left A-module.) It follows from [COC], — with the abelian category $\mathscr{A}$ being the category of C-complete left A-modules — that we then obtain the short exact sequences (1) in the category of C-complete left A-modules. (And, also note, in this case, that the underlying short exact sequences of abstract left A-modules of these sequences of C-complete left A-modules are the ones constructed in the preceding paragraph. This follows from the fact that the functor "underlying abstract left A-module" from the category of C-completes into the category of abstract left A-modules is exact, and preserves direct products (Corollary 4.2.3).) $\square$

Before stating several useful Corollaries of Theorem 4.3.1, let us note a generalization of Theorem 4.3.1 (that is perhaps a bit "too general" to be useful). We make no use of this more general result in this book, and resist the temptation to present a proof.

Proposition 4.3.2. *Let A be a standard topological ring, and let D be a cofinal subset of the set of open right ideals in A. Then, for every abstract left A-module M, if P_* is a projective resolution of M, then there are induced two homological spectral sequences of abelian groups with*

$$(2.1): \ {}^IE_{p,q}^2 = \varprojlim_{I \in D}{}^{-p} \text{Tor}_q^A(A/I, M) \text{ and}$$

$$(2.2): \ {}^{II}E_{p,q}^2 = H_p(\varprojlim_{I \in D}{}^{-q} (P_*/IP_*)).$$

The first of these is confined to the second quadrant, and the second to the fourth quadrant. The second of these spectral sequences (2.2) comes with an abutment K_n, n an integer, such that the associated graded of K_n is $E_{p,q}^\infty$; and such that the filtration on K_n is discrete, and K_n is the union of its filtered pieces, for $n \geq 0$.

Notes. 1. If the right ideals $I \in D$ are two-sided ideals, then the spectral sequences (2.1) and (2.2), together with their abutments, are spectral sequences in the category of C-complete left A-modules.

2. If there exists a positive integer n such that $\varprojlim_{I \in D}^{n} \equiv 0$, then the first spectral sequence (2.1) also abuts at the same K_n, n an integer, as does (2.2), but for in general a different set of filtrations. In this case, the filtrations induced on the K_n by each of the spectral sequences is finite.

Note: 1. The reader should notice that, in the spectral sequence (2.2) of Proposition 4.3.2, we have that

$$'E_{p,0}^2 = C_p(M), \text{ for all integers } p \geq 0.$$

2. We will not make use of Proposition 4.3.2 in this book; however, we will make extensive use of the more special (but still, really, extremely general) Theorem 4.3.1. We state some corollaries.

Corollary 4.3.3. *Let A be a standard topological ring such that the topology is given by denumerably many ideals $I_1 \supset I_2 \supset \ldots I_n \supset \ldots$. Then there is induced a short exact sequence of C-complete left A-modules:*

$$(2): \quad 0 \to \left(\varprojlim_{j \geq 1}^1 \mathrm{Tor}_1^A(A/I_j, M) \right) \to C(M) \overset{\rho_M}{\to} \widehat{M} \to 0.$$

where $\rho_M : C(M) \to \widehat{M}$ is the natural infinitely linear function discussed in Corollary 2.5.7.

Proof. This is the sequence (1) of Theorem 4.3.1, in the case $i = 0$. (Notice that a conclusion of Corollary 4.3.3 is that the group on the left of the short exact sequence (1) possesses a natural structure as a C-complete left A-module — by virtue of being the kernel of ρ_M.) □

Theorem 4.3.4. *Let A be a topological ring such that the topology is given by denumerably many right ideals $I_1 \supset I_2 \supset I_3 \supset \ldots I_j \ldots$. Suppose also that A is admissible. Then for every abstract left A-module M, we have that*

$$div(C(M)) \approx \varprojlim_{j \geq 1}^1 \mathrm{Tor}_1^A(A/I_j, M).$$

Proof. In Chapter 3, §2, Corollary 3.2.18, equation (2), we have proven that if A is as in the hypotheses of the Corollary, then $div(C(M)) \approx \mathrm{Ker}(\rho_M)$. □

Remark 1. Let A be as in the hypotheses of Theorem 4.3.4. Then for every abstract left A-module M, we have the short exact sequences (1) of Theorem 4.3.1, for all integers $i \geq 0$. By Theorem 4.3.4, we have *for $i = 0$*, that the divisible part of $C_i(M)$ is the leftmost group in the short exact sequence (1). One might ask, "can this result be generalized to integers $i \geq 1$?" It can be shown that the answer is: very definitely not, not even for $i = 1$, and if A

is, say, a regular local ring of dimension ≥ 2 with the topology given by powers of the maximal ideal. See Chapter 5, §6, Remark 1 and Example 1 following Corollary 5.6.5, below.

Remark 2. Let A be a topological ring obeying the hypotheses of Theorem 4.3.4. Then by Chapter 1, §3, Remark 1, A is strongly admissible. Therefore by Chapter 1, §3, Theoerm 1.3.7, conditions [1] and [4], we have that the inverse limit, over any directed set, of A-adically complete left A-modules is A-adically complete.

Suppose now that the topological ring A also admits a sequence $I_1 \supset I_2 \supset I_3 \supset \ldots$ of open *two-sided* ideals that give the topology of A. (E.g., this is the case if A is commutative.) Then by the Note to Theorem 4.3.1 the short exact sequences (1) of Theorem 4.3.1 occur in the category of C-complete left A-modules. The left A-modules on the right of equation (1) are the inverse limit, over the directed set of positive integers, of A-adically discrete (and therefore A-adically complete) abstract left A-modules — therefore, as we have just observed, from Chapter 1, §3, Theorem 1.3.7, we have that the groups on the right side of equation (1) are A-adically complete. Therefore the divisible part $div(C_i(M))$ of the middle group $C_i(M)$ of equation (1) must map into zero in the rightmost group of equation (1). Hence in this case the left side of equation (1) necessarily always contains the divisible part $div(C_i(M))$ of $C_i(M)$, for all integers $i \geq 0$. In Theorem 4.3.4, we have shown that for $i = 0$, the group on the left of equation (1) of Theorem 4.3.1 actually coincides with $div(C(M))$. However, as we have noted in Remark 1 above, for $i \geq 1$, even if the ring A is commutative, Noetherian (even regular of dimension ≥ 2 with topology given by powers of the maximal ideal), then there are examples, e.g., for $i = 1$, where the group on the left side of equation (1) is *strictly bigger* than the divisible part of $C_i(M)$.

Corollary 4.3.5. *Let A be as in Theorem 4.3.4, and let M be a C-complete left A-module. Then there is induced a canonical isomorphism of abelian groups*

$$div(M) \approx \varprojlim_{j \geq 0}{}^1 \mathrm{Tor}_1^A(A/I_j, M).$$

Note: If the ideals I_j are two-sided for $j \geq 1$, then this isomorphism is an isomorphism of C-complete left A-modules.

Proof. Since A is admissible and admits a denumerable neighborhood base at zero, by Chapter 3, §1, Theorem 3.1.6, we have that A is C-O.K. Therefore by Chapter 3, §1, Definition 3 and Chapter 3, §1, Theorem 3.1.5, condition [3] we have that $M = C(M)$. Therefore Theorem 4.3.4 for M becomes Corollary 4.3.5. □

Let us study some of the above theorems and corollaries in an important special case.

Definition 1. An element t of a ring A is a *left non-divisor of zero* iff $x \in A$, $tx = 0$ implies $x = 0$ — i.e., iff the function, left multiplication by t, from A into A, is injective.

Definition 2. A *right t-adic ring* (respectively: a *proper right t-adic ring*) is a topological ring A, such that there exists an element $t \in A$, (respectively: a left non-divisor of zero $t \in A$) such that

(1) $\{t^n A : n \geq 1\}$ is an open base for the neighborhood system at zero.

We sometimes use the briefer phrase "*t-adic ring*" (respectively: "proper *t*-adic ring") for "right *t*-adic ring" (respectively: "proper right *t*-adic ring").

Example 1. Let A be a ring and let $t \in A$. Then we have the *right t-adic topology* (or more briefly the *t*-adic topology) on A, which makes A into a *topological group*, such that a complete system of neighborhoods of zero is $\{t^n A : n \geq 1\}$. Then by Chapter 1, Example 3 (with $F_i = \{t\}$ for all $i \geq 1$), we have that the *t*-adic topology on A will make A into a *topological ring* iff

(2) $a \in A$ implies that there exists an integer $n \geq 1$ (in general depending on a) such that

$$at^n \in tA.$$

Thus if $t \in A$ obeys condition (2), then the right *t*-adic topology on A makes A into a topological ring; and then A together with this topology is a *right t-adic ring* in the sense of Definition 2 (And, of course, every right *t*-adic ring can be so described.)

Remark 3. In [COC], in Chapter IV, one often considers the condition on element $t \in A$, that

$$At \subset tA.$$

Clearly if $t \in A$ obeys this condition, then t obeys condition (2) above; and therefore A together with the right *t*-adic topology is a *t*-adic ring. (As noted in [COC], the condition "$At \subset tA$" is equivalent to: "The right ideal tA is a two-sided ideal." And this condition implies that $t^n A$ is a two-sided ideal for all integers $n \geq 1$.)

Example 2. Let A be a commutative ring and let $t \in A$ be any element (respectively: any non-zero divisor). Then the *t*-adic topology on A makes A into a *t*-adic ring (respectively: a proper *t*-adic ring). (Clearly, Example 2 describes all *t*-adic rings (respectively: all proper *t*-adic rings) such that the underlying abstract ring of A is commutative.)

Example 3. The polynomial ring $\mathbf{Z}[T]$ in one variable T over $\mathbf{Z}$ is a proper *t*-adic ring.

Example 4. Let $\mathcal{O}$ be a discrete valuation ring not a field. Then $\mathcal{O}$ together with the topology given by powers of the maximal ideal is a proper *t*-adic ring. One can take any non-zero element of the maximal ideal for t — e.g., a uniformizing parameter.

Example 5. Let A be any Noetherian (respectively: Noetherian Cohen-Macauley) commutative local ring of dimension 1. Then A together with the topology given by powers of the maximal ideal is a *t*-adic (respectively: a proper *t*-adic) ring.

Remark 4. Clearly, Example 4 is a special case of Example 5; Examples 3 and 5 are special cases of Example 2; and Example 2 is a special case of Example 1; and Example 1 describes the most general *t*-adic ring (i.e., every right *t*-adic ring is a special case of Example 1).

A right t-adic ring (in the sense of Definition 2 above) obeys the hypotheses of Chapter 1, Example 3 (with $F_i = \{t\}$ for all $i \geq 1$). Therefore by that Example a t-adic ring is always admissible. Since also a t-adic ring clearly has a denumerable neighborhood base at zero, we also have by Chapter 3, §1, Theorem 3.1.6 that a t-adic ring is always C-O.K.

If A is a proper right t-adic ring, then we will sometimes use the phrase a *parameter* t, for any element $t \in A$ that is a non-left zero divisor and such that the topology of A is the right t-adic topology. (This terminology is consistent with the usual terminology in the case of Example 5.)

Let us also borrow some terminology from [COC]. Namely,

Definition 3. If M is an abstract left A-module and $t \in A$, then the *precise t^i-torsion in* M denotes $\{x \in M : t^i \cdot x = 0$ in $M\}$ — i.e., the kernel of the endomorphism "left multiplication by t^i" of the abelian group M. (Notice that, if the element t is in the center of the ring A, then this is a left submodule of M. But if t is not in the center of A, then this may no longer be the case.) The t-torsion in M is, as usual, $\{x \in M : \exists i \geq 0$ such that $t^i \cdot x = 0\}$ — i.e., the union of the precise t^i-torsion for $i \geq 0$.

If A is a ring, $t \in A$ and M is an abstract left A-module, then we say that the *t-torsion is bounded below* in M iff there exists an integer $j \geq 0$ such that every t-torsion element in M is a precise t^j-torsion element. (That is, iff there exists an integer $j \geq 0$ such that $x \in M$, $i \geq 0$, $t^i \cdot x = 0$ implies that $t^j \cdot x = 0$.)

Lemma 4.3.6. *Let A be a proper right t-adic ring with parameter t. Then for every abstract left A-module M and every integer $n \geq 0$ we have that*

$$(1): \ (A/t^n A) \underset{A}{\otimes} M = M/(t^n M).$$

$$(2): \ \mathrm{Tor}_1^A(A/t^n A, M) \approx \ (\text{precise } t^n\text{-torsion in } M)$$

canonically as abstract abelian groups (or as abstract left A-modules if $At^n \subset t^n A$). And

$$(3): \ \mathrm{Tor}_i^A(A/t^n A, M) = 0 \ \text{for } i \geq 2.$$

Proof. Use the acyclic homological free resolution:

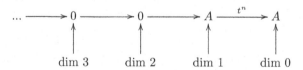

$$\dim 3 \qquad \dim 2 \qquad \dim 1 \qquad \dim 0$$

of the abstract right A-module $A/t^n A$, where "t^n" denotes "left multiplication by t^n." $\qquad \square$

Proposition 4.3.7. *Let A be a proper right t-adic ring with parameter t and let M be an abstract left A-module. Then there are induced canonical isomorphisms of abelian groups*

$$(1): \ div(C(M)) = \varprojlim_{i \geq 1}^1 \ (\text{precise } t^i\text{-torsion in } M).$$

$$(2): \quad C_1(M) = \varprojlim_{i \geq 1} \text{ (precise } t^i\text{-torsion in } M).$$

$$(3): \quad C_i(M) = 0 \text{ for } i \geq 2.$$

Note 1. The right sides of equations (1) and (2) involve "$\varprojlim^1$" and "$\varprojlim$" of the same inverse system (precise t^i-torsion in $M)_{i \geq 0}$. In this inverse system, the map:

$$(\text{precise } t^{i+1}\text{-torsion in } M) \rightarrow (\text{precise } t^i\text{-torsion in } M)$$

is the restriction of "left multiplication by t": $M \rightarrow M$.

Note 2. If $At \subset tA$ (i.e., if tA is a two-sided ideal), then the natural isomorphisms (1) and (2) are isomorphisms of C-complete left A-modules.

Proof. Equation (1) (respectively: Equations (2) and (3)) follow from Theorem 4.3.4 (respectively: Theorem 4.3.1) and from Lemma 4.3.6.   □

The reader should at this point skip briefly to Remark 9 below, for a careful definition of the terms "t-divisible" and "infinitely t-divisible" and a proof that for a right t-adic ring A "t-divisible" (resp.: "infinitely t-divisible") is equivalent to "A-divisible" (resp.: "infinitely A-divisible") as defined earlier in Chapter 1, §2, Definition 1 (resp: Definition 2).

Remark 5. Under the hypotheses of Proposition 4.3.7 we have that $C_1(M) = \{0\}$ iff M has no non-zero infinitely t-divisible t-torsion elements. This follows immediately from equation (2) of the Proposition.

Remark 6. The hypotheses being as in Proposition 4.3.7, suppose also that $At \subset tA$ — so that the isomorphisms (1) and (2) of Proposition 4.3.7 are isomorphisms of C-complete left A-modules. Then

(4): $C_1(M)$ is A-adically complete, and

(5): $C_1(M)$ has no non-zero t-torsion (i.e., left multiplication by $t : C_1(M) \rightarrow C_1(M)$ is injective).

Proof. The abstract left A-module (t^i-torsion in M) is annihilated by the open right (= two-sided) ideal $t^i A$, and is therefore A-adically discrete, and therefore also A-adically complete, $i \geq 1$. As observed immediately following Remark 4 above, a right t-adic ring is always admissible, and therefore by Chapter 3, §1, Theorem 3.1.6, is always strongly admissible. Therefore by Chapter 1, §3, Theorem 1.3.7, condition [4], it follows that the inverse limit $C_1(M)$ is A-adically complete.

Suppose $x = (x_i)_{i \geq 1}$ is an element of the right side of equation (2), and that $tx = 0$. Then $tx_i = 0$ in M, $i \geq 1$. By Note 1 to Proposition 4.3.7, $tx_{i+1} = x_i$. Therefore $x_i = tx_{i+1} = 0$, all $i \geq 0$, and $x = 0$.   □

Remark 7. Consider a special case of Example 5 — namely, the case in which the topological ring A is a discrete valuation ring $\mathcal{O}$ (that is not a field), together with the topology given by powers of the maximal ideal. Let t be a non-zero element of the maximal ideal. Then it is easy to see that an abstract

$\mathcal{O}$-module that is both $\mathcal{O}$-adically complete and that has no non-zero t-torsion is necessarily isomorphic to the completion of a free module. Therefore, in this case, by Remark 6 above, it follows that:

For every abstract $\mathcal{O}$-module M, we have that $C_1(M)$ is isomorphic as a C-complete left $\mathcal{O}$-module to the $\mathcal{O}$-adic completion of a free $\mathcal{O}$-module. Another obvious immediate consequence of Proposition 4.3.7 is:

Corollary 4.3.8. *Let A be a right t-adic ring such that the t-torsion is bounded below, and let t be an element such that the topology of A is the right t-adic topology. Consider the ring homomorphism*

$$g : \mathbf{Z}[T] \to A$$

that maps T into t, where $\mathbf{Z}[T]$ is as in Example 3. Then if we regard M as a $\mathbf{Z}[T]$-module by means of g, we have that the natural mapping is an isomorphism (of $C^{\mathbf{Z}[T]}$-complete left $C^{\mathbf{Z}[T]}$-modules)

$$C_i^{\mathbf{Z}[T]}(M) \stackrel{\approx}{\to} C_i^A(M), \ i \geq 0.$$

Proof. The functors:

$$(1) : M \to C_i^A(M),$$

$i \geq 0$, considered as functors from the category of abstract left A-modules into the category of left $\mathbf{Z}[T]$-modules, are the derived functor of the functor: $M \to \widehat{M^t}$. The functors:

$$(2) : M \to C_i^{\mathbf{Z}[T]}(M),$$

$i \geq 0$, also considered as functors from the category of abstract left A-modules into the category of left $\mathbf{Z}[T]$-modules, are a homological exact connected sequence of functors, the zeroth one being the functor: $M \to C^{\mathbf{Z}[T]}(M)$. If F is a free left A-module, then $F \approx A^{(S)}$, for some set S. Since the t-torsion is bounded below in A, therefore the t-torsion is bounded below in F. In particular, F has no infinitely t-divisible t-torsion elements. Hence, by Remark 5 following Proposition 4.3.7 applied to the proper t-adic ring $\mathbf{Z}[T]$, we have that $C_1^{\mathbf{Z}[T]}(F) = \{0\}$. And by Proposition 4.3.7, equation (3), we have that $C_i^{\mathbf{Z}[T]}(F) = \{0\}$, for $i \geq 2$. Therefore the homological exact connected sequence of functors from the category of abstract left A-modules into the category of abstract left $\mathbf{Z}[T]$-modules in equation (2) are a system of derived functors of the functor:

$$(3) : M \to C^{\mathbf{Z}[T]}(M).$$

Therefore to complete the proof, it is necessary and sufficient to prove that the functor (3) is the zeroth derived functor of the functor $M \to \widehat{M^t}$, from the category of abstract left A-modules into the category of abstract left $\mathbf{Z}[T]$-modules.

Since the functor in equation (3) is right exact, it suffices to show that, for every free left A-module F, that

$$(4): C^{\mathbf{Z}[T]}(F) = \widehat{F}^{T}$$

(since $\widehat{F}^{t} = \widehat{F}^{T}$). Since by hypothesis the t-torsion is bounded below in A, and since $F \approx A^{(S)}$ for some set S, the t-torsion is bounded below in the left A-module F. Therefore there is an integer $n \geq 0$ such that every t-torsion element in F is a precise t^{n}-torsion element. Therefore, in the inverse system:

$$(\text{precise } t^{i}\text{-torsion in } F)_{i \geq 0},$$

(in which the map from the $(i+1)$th group into the ith group is "multiplication by t"), we have that the map from the $(i+n)$th group into the nth group is zero. Therefore $\varprojlim_{i \geq 1}^{1}(\text{precise } t^{i}\text{-torsion in } F) = 0$. Hence, by Proposition 4.3.7, conclusion (1), we have that

$$div(C^{\mathbf{Z}[T]}(F)) = 0.$$

By Chapter 3, §2, Corollary 3.2.18, conclusion (2), we have the short exact sequence:

$$0 \to div(C^{\mathbf{Z}[T]}(F)) \to C^{\mathbf{Z}[T]}(F) \to \widehat{F}^{T} \to 0.$$

Hence

$$(4): C^{\mathbf{Z}[T]}(F) = \widehat{F}^{T}.$$

$\square$

If A is a t-adic ring and if M is an abstract left A-module, then if we regard M as being an abstract left $\mathbf{Z}[T]$-module as in Corollary 4.3.8 above, then for every $i \geq 1$, the t^{i}-torsion in M regarded as left A-module coincides with the T^{i}-torsion in M regarded as $\mathbf{Z}[T]$-module. Therefore, by Corollary 4.3.8, the hypothesis in Proposition 4.3.7 that "the right t-adic ring A is a *proper* right t-adic ring", can be weakened to the less restrictive condition, that "the t-torsion in A is bounded below". We record this result as:

Theorem 4.3.9. *Let A be a right t-adic ring such that the t-torsion is bounded below, and let M be an abstract left A-module. Then there are induced canonical isomorphisms of abelian groups*

$$(1): div(C(M)) = \varprojlim_{i \geq 1}^{1} (\text{precise } t^{i}\text{-torsion in } M).$$

$$(2): C_{1}(M) = \varprojlim_{i \geq 1} (\text{precise } t^{i}\text{-torsion in } M).$$

$$(3): C_{i}(M) = 0 \text{ for } i \geq 2.$$

Again by Corollary 4.3.8, the hypothesis in Remarks 5 and 6 above, that "the right t-adic ring A is a *proper* right t-adic ring", can be weakened to the less restrictive condition, that "the t-torsion in A is bounded below":

Note 1: *Remark 5 generalized*: Under the hypotheses of Theorem 4.3.9, we have that $C_1(M) = \{0\}$ iff M has no non-zero infinitely t-divisible t-torsion elements.

Note 2: *Remark 6 generalized*: The hypotheses being as in Theorem 4.3.9, suppose also that the right ideal tA is a two-sided ideal (i.e., that $At \subset tA$). Then the isomorphisms of abelian groups (1) and (2) of Theorem 4.3.9 are isomorphisms of C-complete left A-modules, and

(4): $C_1(M)$ is A-adically complete, and

(5): $C_1(M)$ has no non-zero t-torsion (i.e., left multiplication by $t : C_1(M) \to C_1(M)$ is injective).

Notice that Corollary 4.3.8 above implies that the C-completion and higher C-completions of an abstract left A-module over a right t-adic ring A such that the t-torsion is bounded below are essentially invariant under changing the ring (if one keeps the same parameter). Actually, a far more general such theorem is true, which is proved in the next section.

Let us continue with a few more special results, to illustrate, e.g., some applications to [COC].

Proposition 4.3.10. *Let A be a standard topological ring that is C-O.K., and let M be an abstract left A-module such that M is A-adically Hausdorff. Suppose also that $C_1(\widehat{M}) = 0$. Then there is induced a canonical isomorphism of C-complete left A-modules:*

$$div(C(M)) \approx C_1(\widehat{M}/M).$$

Note: If we delete the hypothesis that "$C_1(\widehat{M}) = 0$", then the proof of the Proposition shows that

$$div(C(M)) \approx \mathtt{Cok}(C_1(\widehat{M}) \to C_1(\widehat{M}/M)).$$

Proof. Apply the sequence of left derived functors C_i, $i \geq 0$, to the short exact sequence

$$0 \to M \to \widehat{M} \to \widehat{M}/M \to 0.$$

A portion of the long exact sequence obtained is:

$$\ldots \to C_1(\widehat{M}) \to C_1(\widehat{M}/M) \xrightarrow{d_1} C(M) \to C(\widehat{M}) \to \ldots$$

Since A is C-O.K. and $\widehat{M}$ is A-adically complete and therefore C-complete, we have that $C(\widehat{M}) = \widehat{M}$. By hypothesis, $C_1(\widehat{M}) = 0$. Therefore the above exact sequence tells us that

$$C_1(\widehat{M}/M) \approx \mathtt{Ker}(C(M) \to \widehat{M}).$$

Since the topological ring A is C-O.K., it is admissible (Chapter 3, §1, Corollary 3.1.7). Therefore by the Note to Chapter 3, §2, Corollary 3.2.18, conclusion $(2')$, we have that

$$\text{Ker}(C(M) \to \widehat{M}) = div(C(M)).$$

The proof of the Note is similar.   □

Corollary 4.3.11. *Let A be a t-adic ring such that the t-torsion is bounded below, and let M be an abstract left A-module such that M is A-adically Hausdorff. Then there is induced a canonical isomorphism of abelian groups (or of C-complete left A-modules if the right ideal tA is a two-sided ideal)*

$$div(C(M)) \approx \varprojlim_{i \geq 1}(precise \ t^i\text{-}torsion \ in \ \widehat{M}/M).$$

(And, if tA is a two-sided ideal, then by Note 2 of Theorem 4.3.9, $div(C(M))$ is A-adically complete and has no non-zero t-torsion.)

Note: The hypothesis in Corollary 4.3.11 that "M is A-adically Hausdorff" can be weakened to "Every t-divisible element in M is infinitely t-divisible."

Proof. $\widehat{M}$ is A-adically Hausdorff, and therefore has no non-zero t-divisible elements, and therefore also has no non-zero infinitely t-divisible elements. Therefore by Note 1 of Theorem 4.3.9, we have that $C_1(\widehat{M}) = 0$. But then the hypotheses of Proposition 4.3.10 are satisfied. The conclusion follows from Proposition 4.3.10 (applied to M), and from Theorem 4.3.9, conclusion (2) (applied to $\widehat{M}/M$). This proves the Corollary. Let us now prove the Note.

Let $M' = M/(div(M))$. Then clearly

$$(1): \ \widehat{M'}/M' = \widehat{M}/M.$$

On the other hand, since by hypothesis every t-divisible element in M is infinitely t-divisible, we have that $(div(M))$ is an A-divisible module. Therefore by Lemma 4.3.12 below, we have that

$$(2): \ C(M) = C(M').$$

M' obeys the hypotheses of the Corollary. But then, by equations (1) and (2), and the conclusion of the Corollary for M', we deduce the conclusion of the Corollary for M.   □

Lemma 4.3.12. *Let A be a topological ring such that the topology is given by denumerably many right ideals. Let M be an abstract left A-module. Let D be an abstract left A-submodule of M, such that D is A-divisible. Then*

$$C(M) = C(M/D).$$

Proof. Since D is A-divisible, by Chapter 3, §2, Corollary 3.2.4, we have that $C(D) = 0$. Therefore applying the right exact functor C to the short exact sequence:

$$0 \to D \to M \to M/D \to 0$$

gives the conclusion of the Lemma $\qquad\square$

The reader will note that, Proposition 5 of Chapter 4 of [COC], is precisely Corollary 4.3.11 above for proper t-adic rings (after substituting the right side of equation (1) of Proposition 4.3.7 for the left side). In fact, the way that I originally proved Proposition 5, Chapter 4 of [COC], was to use the C-completion and C_1 (which I already knew about at that time, for the case of proper t-adic rings). Similarly, Theorem 6 of Chapter 4 of [COC] was also proved, by first using the C-completion and C_1, and then translating the proof into (messier) terminology not directly using C and C_1.

It will be important, later in this book, for induction arguments (involving much more complicated topological rings than just t-adic rings), to know, for a left A-module over a t-adic ring, when we have that $C(M) = \widehat{M}$.

Corollary 4.3.13. *Let A be a t-adic ring such that the t-torsion is bounded below and let M be an abstract left A-module. Then the following conditions are equivalent:*

[1] $\rho_M : C(M) \to \widehat{M}$ is an isomorphism.
[2] $C(M)$ is A-adically Hausdorff.
[3] $div(C(M)) = 0$.
[4] $\varprojlim_{i\geq 1}^1 (precise\ t^i-torsion\ in\ M) = 0$.

Proof. [2] iff [3] is a tautology. [1] iff [3] follows from Chapter 3, §2, Corollary 3.2.18, conclusion (2). [3] iff [4] follows from Theorem 4.3.9, conclusion (1). $\qquad\square$

Example 6. Let A be a t-adic ring such that the t-torsion is bounded below, and let M be an abstract left A-module such that the t-torsion is bounded below. Then

$$C_i(M) = \begin{cases} \widehat{M}, & i = 0 \\ 0, & i > 0. \end{cases}$$

Proof. Consider the inverse system (precise t^i-torsion in M)$_{i\geq 1}$ of Theorem 4.3.9. Let n be a non-negative integer such that every t-torsion element in M is a precise t^n-torsion element. The map $\pi_{i+n,i}$ from the $(i + n)$th group into the ith group in this inverse systerm is (left multiplication by t^n). But since $t^n \cdot x = 0$ for all t-torsion elements x, it follows that $\pi_{i+n,i} = 0$ for $i \geq 0$. Therefore both $\varprojlim_{i\geq 0}$ and $\varprojlim_{i\geq 0}^1$, of this inverse system vanish. Corollary 4.3.13 and Theorem 4.3.9 complete the proof. $\qquad\square$

Proposition 4.3.14. *Let A be a t-adic ring such that the t-torsion is bounded below and let M be an abstract left A-module. Then the kernel of the natural mapping*

$$\iota_M : M \to C(M)$$

is {infinitely t-divisible elements of M}.

Proof. The set H of all infinitely t-divisible elements of M is an abstract left A-submodule of M, and is an A-divisible abstract left submodule of M (for a proof of this, see Remark 9 below). Therefore by Chapter 3, §2, Corollary 3.2.4, $C(H) = 0$.

(An alternative proof that $C(H) = 0$ using Theorem 4.3.9: By Theorem 4.3.9 equation (1) applied to H, we have $\text{div}(C(H)) = \varprojlim\limits_{i \geq 1}^1 (\text{precise } t^i\text{-torsion in } H)$. Since every element of H is infinitely t-divisible, the maps in this inverse system are surjective. Therefore $\text{div}(C(H)) = 0$. Therefore, by Corollary 4.3.13 $C(H) \approx \widehat{H}$. And since every element in H is infinitely t-divisible, $\widehat{H} = 0$. Therefore $C(H) = 0$.)

Therefore $H \subset \text{Ker}(\iota_M)$. It remains to show that $\text{Ker}(\iota_M)$ is an A-divisible module. Let $K = \text{Ker}(\iota_M)$ and $M' = \text{Im}(\iota_M)$. Then M' is an abstract left A-submodule of the C-complete left A-module $C(M)$. By Chapter 3, §2, Proposition 3.2.1, we have that $C(M)$ has no non-zero infinitely t-divisible elements (see also Remark 9 below). Therefore likewise M' has no non-zero infinitely t-divisible elements. Therefore by Note 1 to Theorem 4.3.9,

$$(1): \ C_1(M') = 0.$$

We have the short exact sequence

$$0 \to K \xrightarrow{j} M \to M' \to 0,$$

which gives rise to the exact sequence

$$(2): \ \ldots \to C_1(M') \xrightarrow{d_1} C(K) \xrightarrow{C(j)} C(M) \to C(M') \to 0.$$

And we have the commutative diagram

$$
\begin{array}{ccc}
C(K) & \xrightarrow{\ C(j)\ } & C(M) \\[2pt]
{\scriptstyle \iota_K}\big\uparrow & & \big\uparrow{\scriptstyle \iota_M} \\[2pt]
K & \xrightarrow{\ \ j\ \ } & M
\end{array}
$$

Since $K = \text{Ker}(\iota_M)$, the composite $\iota_M \circ j : K \to C(M)$ is zero. But the map $C(j) : C(K) \to C(M)$ is by definition the unique infinitely linear function such that the composite $K \xrightarrow{\iota_k} C(K) \xrightarrow{C(j)} C(M)$ is $\iota_M \circ j$. Therefore $C(j) = 0$.

(An alternative proof that $C(j) = 0$: Since $K = \text{Ker}(\iota_M : M \to C(M))$, by the universal mapping property definition of $C(M/K)$, we deduce that

$C(M/K) = C(M)$, i.e., that $C(M) = C(M')$. From the exact sequence (2), it follows that $C(j) = 0$.)

Therefore, in the exact sequence (2), $C(j) = 0$, and by equation (1), $C_1(M') = 0$. Therefore $C(K) = 0$. Therefore by Chapter 3, §2, Corollary 3.2.4, K is infinitely A-divisible.

(Note that the implication: "$C(K) = 0$ implies K is infinitely divisible" is actually very easy. Namely, since the map: $C(K) \to \widehat{K}$ is an epimorphism — e.g., by (the very elementary result) Chapter 3, §2, Corollary 3.2.20 — this implies "$\widehat{K} = 0$," which of course is equivalent to "K is A-divisible.")

Thus, K is infinitely t-divisible, completing the proof. □

Remark 8. Actually, Proposition 4.3.14 shows that the kernel of ι_M is {infinitely A-divisible elements of M} – but, as we shall see in Remark 9 below, in every t-adic ring, this is the same as {infinitely t-divisible elements of M}. (This is especially easy to prove when, as hypothesized in Proposition 4.3.14, the t-torsion in A is bounded below – see Remark 10 below.)

Corollary 4.3.15. *Let A be a t-adic ring such that the t-torsion is bounded below and let M be an abstract left A-module. Then a necessary condition for the four equivalent conditions of Corollary 4.3.13 to hold is that every t-divisible element in M be infinitely t-divisible.*

Proof. If condition [3] of Corollary 4.3.13 holds, then let $x \in M$ be a t-divisible element of M. Then $\iota_M(x)$ is a t-divisible element of $C(M)$ — i.e., $\iota_M(x) \in \text{div}(C(M))$. By condition [3] of Corollary 4.3.13, $\iota_M(x) = 0$. But then by Proposition 4.3.14, we have that the element $x \in M$ is infinitely t-divisible in M. □

Example 7. Let A be a t-adic ring such that the t-torsion is bounded below and let M be an abstract left A-module. Then necessary and sufficient conditions for the four equivalent conditions of Corollary 4.3.13 to hold, is that every t-divisible element of M be infinitely t-divisible, and that

[5] $x \in \widehat{M}$, tx in the image of the natural map: $j_M : M \to \widehat{M}$, implies x is also in the image of j_M.

Proof. By Corollary 4.3.15 we can assume that every t-divisible element of M is infinitely t-divisible. (Then the assertion is actually equivalent to [COC], Chapter 4, Corollary 5.1, p. 555. We recall the easy deduction from Corollary 4.3.11 above.)

Since A is strongly admissible, $\widehat{M}/(t^n \widehat{M}) = M/(t^n M)$, $n \geq 0$. Therefore $(A/(t^n A)) \underset{A}{\otimes} (\widehat{M}/M) = 0$, $n \geq 0$ — i.e., $\widehat{M}/M$ is divisible. Therefore every element of $\widehat{M}/M$ is infinitely t-divisible.

$\widehat{M}/M$ has no non-zero infinitely t-divisible t-torsion elements iff

$$\varprojlim_{i \geq 1}(\text{precise } t^i\text{-torsion in } \widehat{M}/M) = 0.$$

And by the Note to Corollary 4.3.11, this latter condition is equivalent to $div(C(M)) = 0$, which is condition [3] of Corollary 4.3.13. Therefore the four equivalent conditions of Corollary 4.3.13 hold iff $\widehat{M}/M$ has no non-zero infinitely t-divisible t-torsion elements. Since every element of $\widehat{M}/M$ is infinitely t-divisible, it is equivalent to say that $\widehat{M}/M$ has no t-torsion elements – i.e., that condition [5] above holds.   □

Example 8. Let A be a ring and let M be an abstract left A-module. Then for each integer $i \geq 1$ and any $t \in A$, introduce a *topology* on $G_i = $ (precise t^i-torsion in M) so that it becomes a topological group, and such that a complete system of neighborhoods of 0 in G_i is $\{t^n \cdot(\text{precise } t^{i+n}\text{-torsion in } M)\colon n \geq 0\}$. Then, if A is a proper t-adic ring with parameter t,

(1) A sufficient condition for the four conditions of Corollary 4.3.13 to hold is that there exist a smallest open neighborhood of zero in G_i, for all $i \geq 1$.

(2) An alternative sufficient condition for the four conditions of Corollary 4.3.13 to hold is that G_i be a complete topological group, for all $i \geq 1$.

(3) Necessary and sufficient conditions for the four conditions of Corollary 4.3.13 to hold is that both

(3a) Every t-torsion element in M that is t-divisible, is also infinitely t-divisible, *and*

(3b) G_i is "complete but not Hausdorff" — that is, that G_i modulo the closure of zero (with the quotient topology) be a complete topological group — for all integers $i \geq 1$.

Proof. Sufficiency of (2) and necessity of (3b) above follow from Corollaries 1.1 and 2.1 of [FLI], respectively. Let us prove (3). In fact, by Corollary 4.3.15, condition (3a) is necessary for the four equivalent conditions of Corollary 4.3.13 to hold. Therefore we may assume that M obeys condition (3a). In the terminology of [FLI], condition (3a) is equivalent to: "The inverse system (precise t^i-torsion in $M)_{i \geq 1}$ is non-deviant." But then, by Theorem 1 of [FLI], it follows that $\varprojlim_{i \geq 1}^1$ of this inverse system is zero iff $\widehat{G}_i/G_i = 0$ for all $i \geq 1$ — i.e., iff G_i is "complete but not Hausdorff," for all $i \geq 1$. This proves (3).

Finally, suppose (1). Then the inverse system of the Note to Proposition 4.3.7 is such that the images stabilize (in the terminology of [FLI]). Therefore, by Example 1 in [FLI], this inverse system is non-deviant. Therefore, by Theorem 1 of [FLI], $\varprojlim_{i \geq 1}^1$ of this inverse system is zero iff $\widehat{G}_i/G_i = 0$. But, since G_i has a smallest neighborhood of zero, $\widehat{G}_i$ is a discrete quotient group of G_i, and in particular $G_i \to \widehat{G}_i$ is surjective, so that $\widehat{G}_i/G_i = 0$, $i \geq 1$.   □

By contrast to Proposition 4.3.10, it is amusing that

Proposition 4.3.16. *Let A be a standard topological ring that is C-O.K., and let M be an abstract left A-module. If*

$$C_1(C(M)) = 0$$

then

$$C_1(C(M)/M) = 0.$$

Note 1. Of course, by "$C(M)/M$" we mean, more precisely, $\mathrm{Cok}(\iota_M : M \to C(M))$.

Note 2. If we delete the hypothesis "$C_1(C(M)) = 0$" then the proof shows that

$$C_1(C(M)) \to C_1(C(M)/M)$$

is surjective.

Proof. Let M' be the image of M in $C(M)$. Then by the universal mapping property definition of $C(M')$, we have that $C(M) = C(M')$. Therefore replacing M by M' if necessary, we can assume that $\iota_M : M \to C(M)$ is injective. Then applying the sequence of left derived functors to the short exact sequence

$$0 \to M \to C(M) \to C(M)/M \to 0,$$

we obtain the long exact sequence a portion of which is the exact sequence:

$$\ldots \to C_1(C(M)) \to C_1(C(M)/M) \overset{d_1}{\to} C(M) \overset{C(\iota_M)}{\to} C(C(M)) \to \ldots$$

Since A is C-O.K., $C(\iota_M)$ is an isomorphism. Also by hypothesis $C_1(C(M)) = 0$. Therefore $C_1(C(M)/M) = 0$. The proof of Note 2 is similar. □

Also by contrast to Corollary 4.3.11,

Corollary 4.3.17. *Let A be a t-adic ring such that the t-torsion is bounded below and let M be an abstract left A-module. Then the additive function "left multiplication by t" of the abstract left A-module*

$$\mathrm{Cok}(\iota_M : M \to C(M))$$

into itself is bijective.

Proof. By Chapter 1, §3, Example 5.1, the topological ring A is strongly admissible, and hence is admissible. Therefore by Theorem 3.2.11 the map $\iota_M : M \to C(M)$ induces an isomorphism (as topological left $\hat{A}$-modules for the uniform topologies) on the A-adic completions. Equivalently, for all integers $n \geq 1$, we have that

$$(\iota_M \underset{A}{\otimes} A/t^n A) : M/(t^n M) \to C(M)/(t^n C(M))$$

is a bijection, and therefore in particular is an epimorphism. Since the functor: $N \rightsquigarrow N/t^n N$ is right-exact, this latter is equivalent to the assertion that

$$(A/t^n A) \underset{A}{\otimes} \mathrm{Cok}(\iota_M : M \to C(M))$$

is zero, for $n \geq 1$. That is, the abstract left A-module $\mathrm{Cok}(\iota_M)$ is A-divisible — or, equivalently, the function "left multiplication by t": $\mathrm{Cok}(\iota_M) \to \mathrm{Cok}(\iota_M)$ is surjective. It remains to show that $\text{Cok}(\iota_M)$ has no non-zero t-torsion. Since we have shown that $\text{Cok}(\iota_M)$ is an A-divisible module, by Note 1 to Theorem 4.3.9, it is equivalent to show that

$$C_1(\text{Cok}(\iota_M)) = 0.$$

$C(M)$ is a C-complete left A-module. Therefore by Chapter 3, §2, Proposition 3.2.1, $C(M)$ has no non-zero infinitely t-divisible elements. Therefore, by Note 1 to Theorem 4.3.9, $C_1(C(M)) = 0$. Therefore the hypotheses of Proposition 4.3.16 are satisfied. And therefore

$$C_1(\text{Cok}(\iota_M)) = 0.$$

$$\square$$

Remark 9. In several of the above arguments, we have used the terms "t-divisible element" and "infinitely t-divisible element"; and occasionally the fact that: Let A be a t-adic ring, and let $t \in A$ be any element such that the topology of A is the t-adic. Then for any abstract left A-module M, {t-divisible elements of M} = {A-divisible elements of M} as defined in Chapter 1, §2, Definition 1. And {infinitely t-divisible elements of M} = {infinitely A-divisible elements of M} as defined Chapter 1, §2, Definition 2. These facts are not both immediately obvious. Let us prove them; but first let us make clear the terminology.

Let M be an abstract left A-module where A is a ring and let $t \in A$. An element $x \in M$ is t-*divisible* iff $n \geq 1$ implies there exists $y \in M$ such that $t^n y = x$. The element $x \in M$ is *infinitely t-divisible* iff there exist elements y_i, $i \geq 1$, in M, such that $ty_{i+1} = y_i$, $i \geq 1$, and $ty_1 = x$.

If A is a topological ring and t is an element of A such that the topology of A is the right t-adic, then:

(1) $x \in M$ is t-divisible iff $x \in M$ is A-divisible.

(2) $x \in M$ is infinitely t-divisible iff $x \in M$ is infinitely A-divisible.

Proof of (1). x is t-divisible iff $x \in t^n M$, for all $n \geq 1$. A complete system of neighborhoods of zero in M for the A-adic topology is {$t^n M : n \geq 1$}. Therefore an equivalent statement is: $x \in$ (closure of {0} for the A-adic topology of M) — i.e., that x is an A-divisible element of M as defined in Chapter 1, §2, Definition 1. $\square$

Proof of (2). ($\Leftarrow$): By definition, $x \in M$ is infinitely A-divisible iff there exists an abstract left A-submodule N of M, such that N is A-divisible, and $x \in N$. Since N is A-divisible, and tA is an open right ideal in A, we have that $N = (tA) \cdot N$, i.e., $N = tN$. Therefore by induction on $i \geq 1$, choose $y_i \in N$ such that $ty_1 = x$, $ty_{i+1} = y_i$, $i \geq 1$.

($\Rightarrow$): Suppose we can show that the set $N = $ {infinitely t-divisible elements of M} is an abstract left A-submodule of M. Clearly $tN = N$. Therefore $t^n N = N$, $n \geq 1$. Since {$t^n A : n \geq 1$} are a complete system of neighborhoods of zero in A, it follows that the A-module N is A-divisible. Therefore, every infinitely t-divisible element is contained in the A-divisible A-submodule N of M, and therefore is A-divisible.

Therefore, to complete the proof of this Remark, it is necessary and sufficient to show that $N = \{$infinitely t-divisible elements of $M\}$ is a left A-submodule of M.

Proof. It is immediate that N is closed under sums. A is a t-adic ring. Therefore by condition (2) of Example 1, for every $b \in A$, there exist an integer $n(b) \geq 1$ and an element $c(b) \in A$ such that

(0): $b \cdot t^{n(b)} = tc(b)$.

I claim that, if N is any subset of M such that

(1): $tN = N$, then

(2): $A \cdot N = t \cdot A \cdot N$.

(Here, the products mean the set products; e.g., $t \cdot A \cdot N = \{t \cdot a \cdot n : a \in A, n \in N\}$.)

Proof. The inclusion $t \cdot A \cdot N \subset A \cdot N$ is obvious. To show that $A \cdot N \subset t \cdot A \cdot N$. In fact, if $a \in A$ and $x \in N$, then, since $tN = N$, by induction there exists a sequence y_i, $i \geq 1$, of elements of N, such that $ty_1 = x$, and $ty_{i+1} = y_i$ for $i \geq 1$. Since A is a t-adic ring, by equation (0) above, there exist an integer $n \geq 1$ and an element $c \in A$ such that

$$a \cdot t^n = tc.$$

Then

$$a \cdot x = a \cdot t^n y_n = tc \cdot y_n \in t \cdot A \cdot N,$$

completing the proof of (2).

Applying equation (2) to the set N of all infinitely t-divisible elements of M, we obtain that $t \cdot (A \cdot N) = (A \cdot N)$. Therefore every element of the set $A \cdot N$ is infinitely t-divisible. I.e., $A \cdot N \subset N$. It follows that the set N is closed under left scalar multiplication — i.e., N is a left A-submodule of M. □

Remark 10. By Proposition 4.3.14, if A is a t-adic ring such that the t-torsion is bounded below, then for every abstract left A-module M the kernel of the natural mapping

$$\iota_M : M \to C(M)$$

is $\{$infinitely t-divisible elements of $M\}$. If one is careful, this can be used to give an alternative proof, in the case that the t-adic ring A is such that the t-torsion is bounded below, that $\{$infinitely t-divisible elements of $M\}$ is an abstract left A-submodule of M. (Namely, the proof of Proposition 4.3.14 actually shows that $\mathrm{Ker}(\iota_M) = $ (the largest A-divisible left submodule of M), which is not at first glance, (without using Remark 9 above) obviously equal to $\{$infinitely t-divisible elements of $M\}$. However, by Corollary 4.3.8, the statement: $\mathrm{Ker}(\iota_M) = \{$infinitely t-divisible elements of $M\}$ is true for A iff it is true, in the case $A = \mathbf{Z}[T]$, $t = T$. And then since T is in the center of the ring it is clear that the set of infinitely T-divisible elements of M is a submodule of M. This gives another proof, other than the direct one in Remark 9 above (in the case that the t-adic ring A is such that the t-torsion is bounded below), that the set of infinitely t-divisible elements in M is a left A-submodule of M.

(Note: It is probably possible to prove that every t-adic ring is a quotient ring of a proper t-adic ring. If so, then the proof given in this Remark, that {infinitely t-divisible elements} is a left submodule of M, if the t-adic ring A is *proper* — could easily be extended to arbitrary t-adic rings; thus giving an *alternate* proof to the one given in Remark 9 above, that "infinitely t-divisible" and "infinitely A-divisible" are synonomous for all t-adic rings A, whenever $t \in A$ is such that the topology is the t-adic.)

Let us now return to the general situation.

Remark 11. Let A be a standard topological ring. Then we have seen that the category of C-complete left A-modules is an abelian category. If the topological ring A is *admissible*, and if M is any abstract left A-module, then we have seen (Chapter 3, §2, Corollary 3.2.17) that $\operatorname{div} M$ is a C-complete left A-submodule of M. Therefore, if A is an admissible topological ring, we have the additive functor div from the category $\mathscr{C}_A$ of C-complete left A-modules into itself, a subfunctor of the identity functor.

(Also, if A is admissible and admits a denumerable neighborhood base at zero, then, (Chapter 3, §2, Corollary 3.2.18), we have that

$$M/div(M) \approx \widehat{M}$$

for all C-complete left A-modules M.)

The question obviously arises: If we put some sufficiently stringent hypothesis on an admissible topological ring A, then might one hope that, there exists an integer $k \geq 1$ such that

$$div^k(N) = 0$$

for all abstract left A-modules N (where $\operatorname{div}^k = \operatorname{div} \circ \dots \operatorname{div}$ denotes the kth iterate of div)? In fact, we show that this is "about as false as can be" in general: Even in essentially the simplest case — namely, the case in which A is a discrete valuation ring not a field (or, more generally, any proper t-adic ring), then we construct a C-complete left A-module N, such that $\operatorname{div}^k N \neq 0$ for all integers $k \geq 0$.

This is done in chain of Examples 9.1 - 9.6.

Example 9.1. Let A be a proper t-adic ring with parameter t. Then we define an abstract left A-module $M = M_{A,t}$ as follows. M is given by the generators x and y_i for $i \geq 1$; and the relations

$$t^i \cdot y_i = x, i \geq 1$$

Let us assume for simplicity that the element t is in the center of the ring A. Then we have the localization of the ring A at the element t — denote this by $t^{-\infty}A$. (It is the direct limit of the sequence of abstract left A-modules: $A \xrightarrow{t} A \xrightarrow{t} A \to \dots$.) We have $A \subset t^{-\infty}A$, and $t^{-\infty}A = \{t^{-n}a : a \in A, n \geq 0\}$.

Then we have the epimorphism: $\theta : M \to t^{-\infty}A$ that sends x into 1 and y_i into t^{-i}, $i \geq 1$. And the A-homomorphism: $\iota : A \to M$ that sends 1 into x. Then $\theta \circ \iota = $ the inclusion: $A \to t^{-\infty}A$. Therefore $\iota : A \to M$ is injective.

Otherwise stated:

(1) : The left A-submodule $A \cdot x$ of M generated by x is a free left

A-submodule of rank one generated by x.

In particular, $x \in M$ is not a t-torsion element.
Since $x = t^i y_i$, $i \geq 1$, we have that x is t-divisible. In fact, I claim that

(2) : $div(M) = A \cdot x$.

(*Proof.* Let $R = M/(A \cdot x)$. Then R is given by the generators y_i, $i \geq 1$, and the relations $t^i \cdot y_i = 0$. Therefore

$$R \approx \bigoplus_{i \geq 0} A/t^i A,$$

and this abstract left A-module is A-adically Hausdorff. (Since if an element is t-divisible, then the ith coordinate is t-divisible, and therefore a multiple of t^i and therefore zero). Since $R = M/(A \cdot x)$ is A-adically Hausdorff, $divM \subset Ax$, proving (2).)

Thus, we have an example of an abstract left A-module M, such that

$$div(M) \neq 0$$

(in fact, such that $div(M) \approx A$), and such that M has no non-zero infinitely t-divisible elements. Because of this latter, by Proposition 4.3.14, we have that $M \subset C(M)$. Therefore $C(M)$ is a C-complete left A-module such that $div(C(M)) \neq 0$.

Example 9.2. Let A, t be as in Example 9.1. Then we construct an abstract left A-modle $M^{(1)} = M_{A,t}^{(1)}$. Namely, let $M^{(1)}$ be the abstract left A-module given by the generators x, y_i, for $i \geq 1$, and z_{ij} for $i, j \geq 1$, and the relations:

$$t^i y_i = x, \ i \geq 1$$

$$t^j z_{ij} = y_i, \ i, j \geq 1.$$

Then, by the same methods as in Example 9.1, one easily verifies that

(3) : $div(M^{(1)}) = M$.

Then
(4): $M^{(1)}$ has no non-zero infinitely t-divisible elements.
(*Proof.* Let N be an abstract left submodule of $M^{(1)}$ that is A-divisible. Then $N \subset div(M^{(1)})$. Therefore, by equation (3), $N \subset M$. But by Example 9.1, M has no non-zero infinitely divisible elements. Therefore $N = \{0\}$.)

And, by equation (3) of this Example and equation (1) of Example 9.1,

(5) : $div(div(M)) = A \cdot x \approx A$.

In particular, $\text{div}(\text{div}(M)) \neq \{0\}$.

Example 9.3. Let A and t be as in Example 9.1 above. Let $M^{(1)}$ be the abstract left A module constructed in Example 9.2. Let $N = C(M^{(1)})$. Then I claim that:

(6): N is a C-complete left A-module such that

$$div(N) \neq 0 \text{ and}$$

$$div(div(N)) \neq 0.$$

And in fact, $\text{div}(\text{div}(N))$ contains a free submodule of rank one (the one generated by $\iota_{M^{(1)}}(x)$).

Proof. By equation (4) of Example 9.2, and by Proposition 4.3.14,

$$\iota_{M^{(1)}} : M^{(1)} \to N \text{ is injective; i.e.,}$$

$$M^{(1)} \text{ is an abstract left } A\text{-submodule of } N.$$

Therefore

$$\text{div}(M^{(1)}) \text{ is an abstract left } A\text{-submodule of div}(N).$$

$$\text{div}(\text{div}(M^{(1)})) \text{ is an abstract left } A\text{-submodule of div}(\text{div}(N)).$$

Equation (5) of Example 9.2 completes the proof. $\qquad\square$

Example 9.4. Let A and t be as in Example 9.1. Then, for each integer $k \geq 0$, we construct the abstract left A-module $M^{(k)} = M_{A,t}^{(k)}$ as follows. $M^{(k)}$ is given by the (denumerably many) generators x, x_{i_1} for $i_1 \geq 1$, x_{i_1,i_2} for $i_1, i_2 \geq 1, \ldots,$ $x_{i_1,i_2,\ldots,i_k}$ for $i_1, i_2, \ldots, i_k \geq 1$, and the (denumerably many) relations:

$$
\begin{aligned}
&t^{i_1} \cdot x_{i_1} = x, && i_1 \geq 1 \\
&t^{i_2} \cdot x_{i_1,i_2} = x_{i_1}, && i_1, i_2 \geq 1 \\
&\quad\cdot \\
&\quad\cdot \\
&\quad\cdot \\
&t^{i_r} x_{i_1,i_2,\ldots,i_{r-1},i_r} = x_{i_1,i_2,\ldots,i_{r-1}}, && i_1, i_2, \ldots, i_r \geq 1 \\
&\quad\cdot \\
&\quad\cdot \\
&\quad\cdot \\
&t^{i_k} x_{i_1,i_2,\ldots,i_k} = x_{i_1,i_2,\ldots,i_{k-1}}, && i_1, i_2, \ldots, i_k \geq 1.
\end{aligned}
$$

Then $M^{(0)} \approx M$, the module of Example 9.1 (under the left A-isomorphism $(x \longleftrightarrow x, x_i \longleftrightarrow y_i, i \geq 1)$, and $M^{(1)} \approx$ the module of Example 9.2 (under the left A-isomorphism $x \longleftrightarrow x, x_i \longleftrightarrow y_i, i \geq 1, x_{ij} \longleftrightarrow z_{ij}, i,j \geq 1$). And the analogous argument to that in Example 9.1 shows that

$$(7): \ div(M^{(k+1)}) = M^{(k)}, \text{ for } k \geq 0.$$

Since, as we have seen in Example 9.1,

$$(8): \ div(M^{(0)}) = A \cdot x \approx A,$$

it follows, by induction on k, by the Proof of equation (4) of Example 9.2, that, for every integer $k \geq 0$,

$$(9): \ M^{(k)} \text{ has no non-zero infinitely } t\text{-divisible elements.}$$

Example 9.5. Let A, t be as in Example 9.1. For every integer $k \geq 0$, let

$$N^{(k)} = C(M^{(k)}).$$

Then, exactly as in Example 9.3, we see that
(10): N is a C-complete left A-module such that

$$div(N) \neq 0$$

$$div(div(N)) \neq 0$$

.

.

.

$$div^{k+1}(N) \neq 0$$

(where "div^{k+1}" means the $(k+1)$th iterate of the operator "div" with itself). And in fact, $div^{k+1}(N)$ contains a free abstract left A-module (namely, the one generated by $\iota_{M^{(k)}}(x)$).

Example 9.6. Let A, t be as in Example 9.1, and let $N^{(k)}$, $k \geq 0$, be as in Example 9.5. Define

$$N = \prod_{k \geq 0} N^{(k)}.$$

Then I claim that
(11): N is a C-complete left A-module, such that, for every integer $k \geq 1$, we have that

$$div^k(N) \neq \{0\}$$

— and, in fact, such that $div^k(N)$ contains a free left A-submodule of rank one.

Proof. In general, for any set I, any right t-adic ring A (or even any topological ring A having an open neighborhood base at zero consisting of right finitely generated right ideals), and any family $(M_i)_{i \in I}$ of abstract left A-modules,

$$div \left(\prod_{i \in I} M_i \right) = \prod_{i \in I} (div(M_i)).$$

Equation (11) therefore follows from equation (10) of Example 9.5. $\square$

Remark 12. Let A be a t-adic ring such that the t-torsion is bounded below, and let M be an abstract left A-module. Then there exists a C-complete left A-module N such that

(1) : $M/$(infinitely t-divisible elements of M) $\subset N$.

Namely, by Proposition 4.3.14 above, $N = C(M)$ is such a C-complete left A-module.

This observation is essentially equivalent to [COC], Chapter 4, Example 2, pp. 599 and 600. (In fact, the construction of C^* such that equation (1) of [COC], p. 600, holds, is equivalent (once one knows Chapter 3, §1, Proposition 3.1.13), to the weaker statement, that

$$div(M)/(\text{infinitely } t\text{-divisible elements of } M)$$

is isomorphic to an abstract left A-submodule of a C-complete left A-module.)

Remark 13. In fact, truthfully, this author made the construction of [COC], Chapter 4, Example 2, by using the C-completion and equation (1) of Remark 12 above. The C-completion, for the case of t-adic rings, was discovered by the author while making up Chapter IV of [COC] at Pennsylvania State University in 1971-72. Many of the additional constructions of [COC], Chapter 4, that were done after that discovery, were greatly assisted by this knowledge.

The next corollary of Theorem 4.3.1 is important enough to state as a theorem.

Theorem 4.3.18. *Let A be a standard topological ring such that there exists a denumerable neighborhood base at zero. Then, for every abstract left A-module M that is left flat, we have that*

$$C(M) = \widehat{M},$$

$$C_i(M) = 0, i \geq 1.$$

Note: The hypothesis that "M is left flat" can be weakened to: "There exists a sequence $I_1 \supset I_2 \supset I_3 \supset \ldots \supset I_n \supset$ of open right ideals in A that are a complete system of neighborhoods of zero, such that

$$\text{Tor}_i(A/I_j, M) = 0 \text{ for } i, j \geq 1.\text{"}$$

Proof. If the indicated Tor's vanish, then, for $i \geq 1$, the short exact sequences (1) of Theorem 4.3.1 tell us that $C_i(M) = 0$, $i \geq 1$. And Corollary 4.3.3 tells us that $C(M) = \widehat{M}$. □

Corollary 4.3.19. *Let A be a standard topological ring such that there exists a denumerable neighborhood base at zero. Let M be an arbitrary abstract left A-module. Let B_* be an acyclic flat homological resolution of M. Then there are induced natural isomorphisms of C-complete left A-modules*

$$C_i(M) \approx H_i(\widehat{B_*}),$$

for all integers $i \geq 0$.

Note: The hypothesis that B_* be a *flat* resolution — i.e., that B_p be left flat, for all $p \geq 0$ — can be replaced by the weaker hypothesis, that $\mathrm{Tor}_i^A(A/I_j, B_p) = 0$ for $i, j \geq 1$, $p \geq 0$, for some complete system of neighborhoods $I_1 \supset I_2 \supset I_3 \supset \dots I_n \supset \dots$ of zero that are right ideals in the topological ring A.

Proof. By Theorem 4.3.18, we have that

$$C(B_j) = \widehat{B_j}, j \geq 0.$$

$$C_i(B_j) = 0, \quad i \geq 1, \quad j \geq 0.$$

The Corollary follows from §1, Proposition 4.1.5. □

By Corollary 3.2.18, conclusion (2), if A is a t-adic ring, and if M is an abstract left A-module, then we have the short exact sequence: $0 \to div(C(M)) \to C(M) \to \widehat{M} \to 0$. And, by Theorem 4.3.9, conclusion (1), if the t-torsion in M is bounded below, then $div(C(M)) = 0$. Hence, if the t-torsion in M is bounded below, then $C(M) = \widehat{M}$. We conclude this chapter by recalling a Lemma proved in [COC] that will be useful in Chapter 5:

Lemma 4.3.20. *Let A be a right t-adic ring and let M be an abstract left A-module. If M has no non-zero t-torsion, then the t-adic completion, $M^{\wedge^t}$ of M for the right t-adic topology, also has no non-zero t-torsion.*

Proof 1. Chapter 4, Lemma 1.1.1, p. 540 of [COC]. □

Another, more elementary, proof, follows by first noting that an equivalent statement to Lemma 4.3.20 is:

Lemma 4.3.21. *Let M be an abelian group and let $t : M \to M$ be an endomorphism of M as abelian group. Regard M as topological abelian group such that $\{t^i M\}$, $i \geq 1$, is a complete system of neighborhoods of zero. If t is injective, then the mapping induced by $t : M^{\wedge^t} \to M^{\wedge^t}$ is injective.*

Proof 2. Since the mapping $t : M \to M$ is injective, it follows that the mapping $t : M/t^i M \to M/t^{i+1}M$ is injective for $i \geq 1$. Since inverse limits over directed sets is left-exact in the category of abelian groups, passing to the inverse limit for $i \geq 1$, we have that the mapping induced by t:

$$\varprojlim_{i \geq 1}(M/t^i M) \to \varprojlim_{i \geq 1}(M/t^{i+1}M)$$

is injective — i.e., that the induced mapping: $M^{\wedge^t} \to M^{\wedge^t}$ is injective. □

Corollary 4.3.22. *The hypotheses being as in Lemma 4.3.20, we have that the cokernel:*

$$\mathrm{Cok}(M \to \widehat{M})$$

has no non-zero t-torsion.

Proof. Using the ring homomorphism $\mathbf{Z}[T] \to A$ that sends T into t, we can regard every abstract right A-module M as being a $\mathbf{Z}[T]$-module. Then $\widehat{M}^t = \widehat{M}^T$. Hence replacing A with $\mathbf{Z}[T]$ and t with T if necessary, we can and do assume that A is a proper right t-adic ring.

If M' is the image of M in $\widehat{M}$, then $\widehat{M}$ satisfies the universal mapping property for the completion $\widehat{M'}$. Hence $\widehat{M} = \widehat{M'}$. Therefore

$$\mathrm{Cok}(M \to \widehat{M}) \approx \mathrm{Cok}(M' \to \widehat{M'}).$$

Hence replacing M with M', we can assume that $M \subset \widehat{M}$; so we have the short exact sequence:

$$(1) : 0 \to M \to \widehat{M} \to \widehat{M}/M \to 0.$$

By Lemma 4.3.6, for every left A-module N,

$$(2) : \quad \mathrm{Tor}_1^A(A/tA, N) \approx \text{(precise } t\text{-torsion in } N\text{)}.$$

From the short exact sequence (1) we have the long exact sequence of Tor's, a portion of which is:

$$(3) : \dots \mathrm{Tor}_1^A(A/tA, \widehat{M}) \to \mathrm{Tor}_1^A(A/tA, \widehat{M}/M) \to M/tM \to \widehat{M}/t\widehat{M} \to \dots$$

(Here we're using that $\mathrm{Tor}_0^A(A/tA, N) = N/tN$, for all left A-modules N.) By Lemma 4.3.20, $\widehat{M}$ has no non-zero t-torion. Hence $\mathrm{Tor}_1^A(A/tA, \widehat{M}) = 0$. Since also the natural map is an isomorphism: $M/tM \approx \widehat{M}/t\widehat{M}$, from the exact sequence (3) we deduce that $\mathrm{Tor}_1^A(A/tA, \widehat{M}/M) = 0$ — i.e., that $\widehat{M}/M$ has no non-zero t-torsion. □

Corollary 4.3.23. *Let A be a right t-adic ring such that the t-torsion is bounded below, and let M be an abstract left A-module such that the t-torsion is bounded below. Then*

$$(1) : C(M) = \widehat{M}$$

$$(2) : M \text{ and } \widehat{M} \text{ have the same } t\text{-torsion subgroup.}$$

Note: Hence the t-torsion is bounded below in $\widehat{M}$.

Proof. By Corollary 4.3.8, it suffices to prove the Corollary in the case in which A is the polynomial ring in one variable X over $\mathbf{Z}$, $A = \mathbf{Z}[X]$, and $t = X$. Therefore we can, and do, assume that the t-torsion subgroup of M is a right A-submodule, call it T, of M. Let M' be the quotient module, which has no non-zero t-torsion. Then we have the short exact sequence:

$$(3) : 0 \to T \to M \to M' \to 0.$$

By Chapter 3, §2, equation (2) of Corollary 3.2.18, we have the short exact sequence

$$0 \to div(C(M)) \to C(M) \to \widehat{M} \to 0.$$

And, by Theorem 4.3.9, conclusion (1), since the t-torsion in M is bounded below, $div(C(M)) = 0$. Hence, if the t-torsion in M is bounded below, then $C(M) = \widehat{M}$, proving conclusion (1) above. Since T is the t-torsion in M, the t-torsion in T is bounded below. Therfore likewise $C(T) = \widehat{T}$. Since M' is t-torion free, the t-torsion in M' is likewise bounded below; therefore also $C(M') = \widehat{M'}$. Since the t-adic topology of T is the discrete topology, $\widehat{T} = T$. Therefore $C(T) = \widehat{T} = T$. Hence, the image of the short exact sequence (3) under the exact connected sequence of functors C_* yields the exact sequence

$$C_1(M') \to T \to \widehat{M} \to \widehat{M'} \to 0.$$

By conclusion (2) of Theorem 4.3.9 applied to the t-torsion free left A-module M', $C_1(M') = 0$. Therefore this last exact sequence is the short exact sequence

$$(4): 0 \to T \to \widehat{M} \to \widehat{M'} \to 0.$$

Since M' has no non-zero t-torsion, by Lemma 4.3.20, $\widehat{M'}$ has no non-zero t-torsion. Therefore equation (4) implies that T is the t-torsion submodule of $\widehat{M}$. Since T is the t-torsion submodule of M, it follows that M and $\widehat{M}$ have the same t-torsion submodule, proving conclusion (2). □

Chapter 5

The Direct Sum and Direct Limit of C-complete Left A-modules

5.1 Direct Sum of C-complete Left A-modules

Theorem 5.1.1. *Let A be a C-O.K. topological ring. Let I be a set, and for every $i \in I$, let M_i be a C-complete left A-module. Then the C-complete left A-module*

$$C\left(\bigoplus_{i \in I} M_i\right)$$

is the direct sum of the family $(M_i)_{i \in I}$ in the category of C-complete left A-modules.

Proof. We verify the universal mapping property. Let D be a C-complete left A-module, and let $\phi_i : M_i \to D$ be maps of C-complete left A-modules for all $i \in I$. Then $\phi_i : M_i \to D$ is also a homomorphism of abstract left A-modules. Therefore there exists a unique homomorphism

$$\phi : \bigoplus_{i \in I} M_i \to D$$

such that $\phi \circ (\text{inclusion: } M_i \to \bigoplus_{i \in I} M_i) = \phi_i$. Then, by the universal mapping property definition of $C(\bigoplus_{i \in I} M_i)$, it follows that there exists a unique infinitely linear function

$$I : C\left(\bigoplus_{i \in I} M_i\right) \to D$$

such that $I \circ \left(\text{natural map } \bigoplus_{i \in I} M_i \to C\left(\bigoplus_{i \in I} M_i\right)\right) = \phi.$ $\qquad\square$

<u>Remark 1</u>. Of course, the canonical injections: $M_j \to C\left(\underset{i \in I}{\oplus} M_i\right)$ for $j \in J$,

are the composites: $M_j \to \underset{i \in I}{\oplus} M_i \to C\left(\underset{i \in I}{\oplus} M_i\right)$, for all $j \in I$.

<u>Remark 2</u>. Notice that the proof of Theorem 5.1.1 used that A was C-O.K. — namely, we used the fact that the "stripping functor" from C-complete left A-modules into abstract left A-modules is a *full* functor. (Explicitly, we used that the composites described in Remark 1 are infinitely linear). And, (Chapter 3, §1, Definition 3) a standard topological ring A has this property iff A is C-O.K.

However, as the next Remark shows, this hypothesis is unnecessary to construct the direct sum of C-complete left A-modules. (The substance of Theorem 5.1.1 is that, in the case in which the topological ring A is C-O.K., we have a nice explicit construction of this direct sum.)

<u>Remark 3</u>. Suppose that A is a standard topological ring that is *not necessarily* C-O.K. Let I be a set, and let $(M_i)_{i \in I}$ be a family of C-complete left A-modules. Then as in Chapter 2, §4, Theorem 2.4.13, we have seen that the direct sum of the family $(M_i)_{i \in I}$ exists in the category of C-complete left A-modules. Let us present an alternative explicit construction:

In fact, let $\underset{i \in I}{\oplus} M_i$ denote the direct sum in the category of abstract left A-modules. Let $j_k : M_k \to \underset{i \in I}{\oplus} M_i$ be the inclusion, for all $k \in I$. Let $\alpha : \underset{i \in I}{\oplus} M_i \to C\left(\underset{i \in I}{\oplus} M_i\right)$ be the natural map. If S is a set, $(a_k)_{k \in S}$ is a zero-bound family of elements of $\widehat{A}$, if $i \in I$, and if $(x_k)_{k \in S}$ is a family of elements of M_i, then we have

$$\sum_{k \in S} a_k x_k \in M_i,$$

using the infinite sum structure in M_i. On the other hand, we have likewise

$$\sum_{k \in S} a_k(\alpha(j_i(x_k)) \in C\left(\underset{i \in I}{\oplus} M_i\right)$$

using the infinite sum structure in $C\left(\underset{i \in I}{\oplus} M_i\right)$. Let H be the C-complete left A-submodule of $C\left(\underset{i \in I}{\oplus} M_i\right)$ generated by the elements:

$$\alpha\left(j_i\left(\sum_{k \in S} a_k \cdot x_k\right)\right) - \sum_{k \in S} a_k(\alpha(j_i(x_k))) \in C\left(\underset{i \in I}{\oplus} M_i\right),$$

for all sets S, all $i \in I$, all families $(a_k)_{k \in S}$ of elements of $\widehat{A}$ that are zero-bound and all families $(x_k)_{k \in S}$ of elements of M_i. Then it is easy to show, by the universal mapping property definition, that the quotient C-complete left A-module

$$C\left(\underset{i \in I}{\oplus} M_i\right)/H,$$

is the direct sum of the objects M_i, $i \in I$, in the category of C-complete left A-modules.

Remark 4. Another way of making the construction of Remark 3 is as follows: Let A be a standard topological ring and let M_i, $i \in I$, be C-complete left A-modules. For each $i \in I$, let G_i be a set of generators and let R_i be a set of relations for M_i as a C-complete left A-module. (Thus, every element of R_i is an element of $\widehat{A^{(G_i)}}$.) Let G be the disjoint union of the sets G_i, $i \in I$, and let g_i be the inclusion from G_i into G, for all $i \in I$. Let $\widehat{g}_i$ denote the induced infinitely linear map: $\widehat{A^{(G_i)}} \to \widehat{A^{(G)}}$. The direct sum of the M_i, $i \in I$, in the category of C-complete left A-modules, is the C-complete left A-module, with the set of generators the disjoint union G of the G_i, for $i \in I$, and with set of relations the union of the relations $\widehat{g}_i(R_i) \subset \widehat{A^{(G)}}$, $i \in I$.

Theorem 5.1.2. *(Direct limits in the category of C-completes). Let A be a standard topological ring that is C-O.K. Let D be a directed set (resp.: let $\mathscr{D}$ be a category that is a set), and let $(M_i, \alpha_{ij})_{i,j \in D}$ be a direct system in (resp.: and let F be a covariant functor from $\mathscr{D}$ into) the category of C-complete left A-modules. Then the C-complete left A-module*

$$C\left(\varinjlim_{i \in D} M_i\right)$$

$$\left(\text{respectively: } C\left(\varinjlim_{d \in \mathscr{D}} F(d)\right)\right)$$

is the direct limit of the direct system $(M_i, \alpha_{ij})_{i,j \in D}$ (resp.: of the covariant functor F) in the category of C-complete left A-modules.

Proof. Exactly the same as Theorem 5.1.1. □

Remark 5. Suppose that we have all the hypotheses of Theorem 5.1.2, except that the standard topological ring A is not necessarily C-O.K. Then in Chapter 2, §4, Corollary 2.4.14 we have constructed the direct limit of the direct system $(M_i, \alpha_{ij})_{i,j \in D}$ (resp.: of the covariant functor F). Let us present this construction, very explicitly.

Explicitly, the direct limit in the category of C-complete left A-modules is the C-complete quotient-module of the C-complete left A-module

$$C\left(\varinjlim_{i \in D} M_i\right)$$

$$\left(\text{resp.: } C\left(\varinjlim_{d \in \mathscr{D}} F(d)\right)\right)$$

by the C-complete left submodule H, where H is the C-complete left submodule generated by the subset consisting of all elements of the form

$$\alpha j_i \left(\sum_{k \in S} a_k x_k \right) - \sum_{k \in S} a_k \cdot (\alpha j_i(x_k))$$

where $i \in D$ (resp.: i is an arbitrary object of $\mathscr{D}$), S is a set, $(a_k)_{k \in S}$ is an arbitrary family of elements of $\widehat{A}$ that is zero-bound, and $(x_k)_{k \in S}$ is an arbitrary family of elements of M_i (resp.: of elements of $F(i)$), where $j_i : M_i \to \varinjlim_{j \in D} M_j$ (resp.: $j_i : F(i) \to \varinjlim_{d \in \mathscr{D}} F(d)$) is the ith canonical injection, and where $\alpha :$ $\varinjlim_{j \in D} M_j \to C \left(\varinjlim_{j \in D} M_j \right)$, $\left(\text{resp.: } \alpha : \varinjlim_{d \in \mathscr{D}} F(d) \to C \left(\varinjlim_{d \in \mathscr{D}} F(d) \right) \right)$ is the canonical map.

Notation: Let A be a standard topological ring. Then if I is a set and $(M_i)_{i \in I}$ are an indexed family of C-complete left A-modules, then by Theorem 5.1.1 (or Remark 2), we have that the direct sum in the category of C-complete left A-modules exists, and is, almost always, quite different from the direct sum as abstract left A-modules. We therefore wish to use a different notation than "$\underset{i \in I}{\oplus}$" for the direct sum in the category of C-complete left A-modules; namely, we use the notation $\int_{i \in I} M_i$ for the direct sum of the C-complete left A-modules M_i, $i \in I$, in the category $\mathscr{C}_A$. Then, by Theorem 5.1.1, if A is C-O.K., then

$$(1): \quad \int_{i \in I} M_i = C \left(\underset{i \in I}{\oplus} M_i \right).$$

And, if A is an arbitrary standard topological ring, by Remark 3, $\int_{i \in I} M_i$ is then a certain quotient of the right side of equation (1). Similarly, if $(M_i, \alpha_{ij})_{i,j \in D}$ is a direct system in (resp.: if $\mathscr{D}$ is a category that is a set and if F is a covariant functor from $\mathscr{D}$ into) the category of C-complete left A-modules, where A is any standard topological ring, then once again the direct limit exists in the category of C-complete left A-modules, but again is in general quite different from the direct limit as abstract left A-modules. So, we use the notation

$$\varinjlim_{i \in D}^{C} M_i$$

$$\left(\text{resp.: } \varinjlim_{d \in \mathscr{D}}^{C} F(d) \right).$$

So, if the standard topological ring A is C-O.K., then

$$(2): \quad \varinjlim_{i \in D}^{C} M_i = C \left(\varinjlim_{i \in D} M_i \right)$$

$$\left(\text{resp.: (2)}: \varinjlim_{d \in \mathscr{D}}^{C} F(d) = C\left(\varinjlim_{d \in \mathscr{D}} F(d)\right)\right).$$

And, by Remark 5, if the standard topological ring A is not C-O.K., then the left side of (2) is a certain quotient of the right side of (2).

Remark 6. As we have seen in Chapter 4, §2, *direct products* and *inverse limits* in the category of C-completes — and even higher inverse limits — coincide with the corresponding construction for abstract left A-modules. Therefore a different notation is not needed for those constructions (we use the familiar "$\prod_{i \in I} M_i$," or "$\varprojlim_{i \in D} M_i$," or "$\varprojlim_{i \in \mathscr{D}} F(d)$").

Proposition 5.1.3. *Let A be a standard topological ring.*

(1) Then if I is a set and $(M_i)_{i \in I}$ is an indexed family of abstract left A-modules, we have a canonical isomorphism of C-complete left A-modules:

$$C\left(\bigoplus_{i \in I} M_i\right) \approx \int_{i \in I} C(M_i).$$

(2) If D is a directed set (resp.: If $\mathscr{D}$ is a category that is a set), and if $(M_i, \alpha_{ij})_{i,j \in D}$ is a direct system of abstract left A-modules indexed by the directed set D (resp.: and if F is a covariant functor from $\mathscr{D}$ into the category of abstract left A-modules), then we have the canonical isomorphism of C-complete left A-modules

$$C\left(\varinjlim_{i \in D} M_i\right) \approx \varinjlim_{i \in D}^{C} C(M_i)$$

$$\left(resp.: C\left(\varinjlim_{d \in \mathscr{D}} F(d)\right) \approx \varinjlim_{d \in \mathscr{D}}^{C} C(F(d))\right).$$

Proof. Follows immediately from the universal mapping properties. □

Remark 7. Another way of phrasing Proposition 5.1.3 is: "The functor C, from the category of abstract left A-modules into the category of C-complete left A-modules, preserves arbitrary direct limits indexed by arbitrary categories that are sets": We have already observed this (by the same method of proof) in Chapter 2, §5, Corollary 2.5.3.

5.2 The Image of Direct Sum in the Direct Product

Let A be a standard topological ring. If I is a set, and if $(M_i)_{i \in I}$ is an indexed family of C-complete left A-modules, then we have (§1) the direct sum

$$\int_{i \in I} M_i,$$

and also (Chapter 2, §3, Theorem 2.3.9) the direct product

$$\prod_{i \in I} M_i$$

of the family $(M_i)_{i \in I}$ in the category of C-complete left A-modules. Then we have the canonical map:

$$(1): \quad \theta : \int_{i \in I} M_i \to \prod_{i \in I} M_i.$$

This is the unique infinitely linear functor θ such that, for every $j, k \in I$, the composite:

$$M_j \to \int_{i \in I} M_i \stackrel{\theta}{\to} \prod_{i \in I} M_i \to M_k$$

is

$$\begin{cases} 0, & \text{if } j \neq k \\ \text{identity of } M_k, & \text{if } j = k. \end{cases}$$

Definition 1. We use the notation

$$(2): \quad \mathcal{Z}_{i \in I} M_i$$

for the image of the map θ.

Thus, given any family $(M_i)_{i \in I}$ of C-complete left A-modules, we have that $\mathcal{Z}_{i \in I} M_i$ is a C-complete left A-submodule of $\prod_{i \in I} M_i$. And, of course, the assignment: $(M_i)_{i \in I} \rightsquigarrow \mathcal{Z}_{i \in I} M_i$ is a covariant functor (from the category of I-tuples of C-complete left A-modules and I-tuples of infinitely linear functions into the category of C-complete left A-modules).

Example 1. If $M_i = \widehat{A}$, for all $i \in I$, then $\mathcal{Z}_{i \in I} M_i$ consists of all families $(a_i)_{i \in I}$ of elements of $\widehat{A}$ that are zero-bound.

Example 2. If $M_i = M$, for all $i \in I$, then we sometimes use the notation

$$M^{<I>} \text{ for } \mathcal{Z}_{i \in I} M_i.$$

Example 3. In the notations of Example 2, Example 1 can be written:

$$\widehat{A}^{<I>} = \widehat{A^{(I)}}, \text{ for all sets } I.$$

Example 4. Of course, if I is a finite set, then trivially

$$\int_{i \in I} M_i = \mathcal{Z}_{i \in I} M_i = \prod_{i \in I} M_i,$$

$$M^{(I)} = M^{<I>} = M^I.$$

Lemma 5.2.1. *Let A be a standard topological ring, let I be a set, and let $f_i : M_i \to N_i$ be epimorphisms of C-complete left A-modules. Then*

$$\underset{i \in I}{\mathcal{Z}} \, f_i : \underset{i \in I}{\mathcal{Z}} \, M_i \to \underset{i \in I}{\mathcal{Z}} \, N_i$$

is an epimorphism.

Proof. The functor: $(M_i)_{i \in I} \rightsquigarrow \int_{i \in I} M_i$ is right-exact. And the map: $\int_{i \in I} N_i \to \underset{i \in I}{\mathcal{Z}} \, N_i$ is an epimorphism. The result follows from the commutative diagram:

$$
\begin{array}{ccc}
\underset{i \in I}{\mathcal{Z}} \, M_i & \xrightarrow{\underset{i \in I}{\mathcal{Z}} \, f_i} & \underset{i \in I}{\mathcal{Z}} \, N_i \\
\uparrow & & \uparrow \\
\underset{i \in I}{\int} M_i & \xrightarrow{\ \delta\ } & \underset{i \in I}{\int} N_i
\end{array}
$$

where $\delta = \int_{i \in I} f_i$. $\qquad\square$

Proposition 5.2.2. *Let A be a standard topological ring. Let I be a set, and let M_i be a C-complete left A-module for all $i \in I$. For each $i \in I$, let F_i be a projective abstract left A-module and let*

$$\phi_i : \widehat{F}_i \to M_i$$

be an epimorphism of C-complete left A-modules. Then $\underset{i \in I}{\mathcal{Z}} \, M_i$ is the image of the induced mapping

$$\widehat{\underset{i \in I}{\oplus} F_i} \to \prod_{i \in I} M_i.$$

Note: The hypothesis that "F_i is projective for all $i \in I$," can be weakened to:

$$C \left(\underset{i \in I}{\oplus} F_i \right) = \widehat{\underset{i \in I}{\oplus} F_i}.$$

(E.g, by Theorem 4.3.18, in the special case that the topological ring *admits a denumerable neighborhood base at zero*, it suffices that F_i be left flat, for all $i \in I$.)

Proof. We have the commutative diagram

$$
\begin{array}{ccc}
\underset{i \in I}{\int} \widehat{F}_i & \xrightarrow{\ \rho\ } & \prod_{i \in I} M_i \\
{\scriptstyle \gamma}\downarrow & \nearrow{\scriptstyle \theta} & \\
\underset{i \in I}{\int} M_i & &
\end{array}
$$

(where $\gamma = \int_{i \in I} \phi_i$). Since the direct sum is a right exact functor, the map γ is an epimorphism. By definition, we have that $\underset{i \in I}{Z} M_i = \text{Im}(\theta)$. Therefore from the commutative diagram

$$\underset{i \in I}{Z} M_i = \text{Im}(\rho).$$

But, since $\widehat{F_i} = C(F_i)$, for all $i \in I$, and since $\widehat{\underset{i \in I}{\oplus} F_i} = C\left(\underset{i \in I}{\oplus} F_i\right)$, the proposition follows from §1, Proposition 5.1.3.   □

Proof of the Note. The argument is exactly the same; except that we must show that the condition

$$C\left(\underset{i \in I}{\oplus} F_i\right) = \widehat{\underset{i \in I}{\oplus} F_i}$$

implies that $C(F_j) = \widehat{F_j}$, for all $j \in I$. In fact, if $j \in I$, then $C\left(\underset{i \in I}{\oplus} F_i\right) = C(F_j) \oplus C\left(\underset{i \neq j,\, i \in I}{\oplus} F_i\right)$, and $\widehat{\underset{i \in I}{\oplus} F_i} = F_j \oplus \left(\widehat{\underset{i \neq j,\, i \in I}{\oplus} F_i}\right)$. Therefore, if $C\left(\underset{i \in I}{\oplus} F_i\right) = \widehat{\underset{i \in I}{\oplus} F_i}$, then $C(F_j) = \widehat{F_j}$.   □

Remark. Notice, in the course of proving the Note to Proposition 5.1.3, that we have also shown that: "If $C\left(\underset{i \in I}{\oplus} F_i\right) = \widehat{\underset{i \in I}{\oplus} F_i}$, then $C(F_i) = \widehat{F_i}$, for all $i \in I$". One might wonder if the converse is true. It is not, as will be shown in Example 9 below.

Theorem 5.2.3. *Let A be a standard topological ring that is admissible and that admits a denumerable neighborhood base at zero. Then for any set I, and any family $(M_i)_{i \in I}$ of C-complete left A-modules, we have that*

$$(3): \underset{i \in I}{Z} M_i = \{(x_i)_{i \in I} \in \prod_{i \in I} M_i : \text{For every open right ideal } J \subset A,$$

we have that $x_i \in J \cdot M_i$ for all but finitely many $i \in I\}$.

Proof. Let $A = J_0 \supset J_2 \supset J_3 \supset \ldots$ be a sequence of right ideals that gives the topology of A.

For a fixed set I, and for every family $(M_i)_{i \in i}$ of C-complete left A-modules, let $G((M_i)_{i \in I})$ denote the right side of equation (3). Then G is a functor from the category $\mathscr{C}_A^I$ into the category of abstract left A-modules, where $\mathscr{C}_A$ is the category of C-complete left A-modules.

I claim that, if $f_i : N_i \to M_i$ are epimorphisms of C-complete left A-modules, for all $i \in I$, then

$$G((f_i)_{i \in I}) : G((N_i)_{i \in I}) \to G((M_i)_{i \in I})$$

is an epimorphism of abstract left A-modules. For, if $(x_i)_{i \in I} \in G((M_i)_{i \in I})$, then for every $i \in I$, let $n(i) \geq 0$ be defined by

$$
\begin{cases}
\text{the integer } n \text{ such that } x_i \in J_n \cdot M \text{ but } x_i \notin J_{n+1} \cdot M, & \text{if there exists such} \\
& \text{an integer } n \geq 0; \\
\infty, & \text{if } x_i \in J_n \cdot M \\
& \text{for all integers } n \geq 0.
\end{cases}
$$

Then the assertion that "$(x_i)_{i \in I} \in G((M_i)_{i \in I})$" is equivalent to:

$$(4): \ \lim_{i \in I} n(i) = +\infty.$$

(I.e., for every $N \geq 0$, there are only finitely many $i \in I$ such that $n(i) < N$.) By definition of $n(i)$, we have that

$$x_i \in J_{n(i)} \cdot M_i.$$

Since the homomorphism $f_i : N_i \to M_i$ is surjective, it follows that there exists

$$(5): \ y_i \in J_{n(i)} \cdot N_i \text{ such that}$$

$$(6): \ f_i(y_i) = x_i.$$

Then, equations (4) and (5) imply that $(y_i)_{i \in I} \in G((N_i)_{i \in I})$. And equation (6) implies that $G((f_i)_{i \in I})((y_i)_{i \in I}) = ((x_i)_{i \in I})$.

Therefore the functor G maps families $(f_i)_{i \in I}$ of epimorphisms into epimorphisms, as asserted. But by Lemma 5.2.1, so does the functor $\underset{i \in I}{Z}$. Therefore, I claim that to prove the Theorem in general, it suffices to prove it in the case that the M_i are completions of the same free module F, for all $i \in I$. For, suppose that the Theorem is known in that case. Then choose F a fixed free module (of large enough rank), and $\phi_i : \widehat{F} \to M_i$ epimorphisms of C-complete left A-modules. Then, since

$$G((M_i)_{i \in I}) = \left(\prod_{i \in I} \phi_i \right) (G((\widehat{F})_{i \in I}))$$

and

$$\underset{i \in I}{Z} M_i = \left(\prod_{i \in I} \phi_i \right) \left(\underset{i \in I}{Z} (\widehat{F}) \right),$$

if we have that

$$(7): \ \underset{i \in I}{Z} \widehat{F} = G((\widehat{F})_{i \in I}),$$

then the Theorem is established in general. We therefore need only prove equation (3), when $M_i = F$, all $i \in I$, where F is the free abstract left A-module $A^{(S)}$ for some fixed set S.

First, observe that, since F is a free abstract left A-module, and since the topological ring A is by hypothesis admissible, we have (Chapter 1, section 3, Definition 1, condition (11)) for every open right ideal J in A, that

$$(8): \quad J \cdot \widehat{A^{(S)}} = (J^S \cap \widehat{A^{(S)}}) \text{ in } A^S.$$

Otherwise stated

$$(9): \text{ If } a = (a_i)_{i \in S} \in \widehat{A^{(S)}} = \widehat{F},$$

and if J is any open right ideal in A, then $a \in J \cdot \widehat{F}$ iff $a_i \in J$, for all $i \in S$.

We now complete the proof of the Theorem, by proving equation (3). Both sides of equation (3) are subgroups of $\prod_{i \in I} \widehat{F} = \prod_{i \in I} \widehat{A^{(S)}}$. We now show they are the same subset.

Since F is free, we have $C(F) = \widehat{F}$. Then by §1, Proposition 5.1.3, we have that

$$\int_{i \in I} \widehat{F} = \int_{i \in I} C(F) = C(F^{(I)}).$$

Since F is free, $F^{(I)}$ is free, so $C(F^{(I)}) = \widehat{F^{(I)}}$. But then

$$\int_{i \in I} \widehat{F} = \widehat{F^{(I)}} \subset \prod_{i \in I} \widehat{F}.$$

Thus, in this case, the natural map θ (from the direct sum in the category of C-completes into the direct product) is injective. It follows (in the definition of $\mathcal{Z}_{i \in I} \widehat{F}$ as the image of θ), that

$$\mathcal{Z}_{i \in I} \widehat{F} = \int_{i \in I} \widehat{F} = \widehat{F^{(I)}}$$

$$(10): \quad \mathcal{Z}_{i \in I} \widehat{F} = \widehat{F^{(I)}}.$$

$$(11): \quad \mathcal{Z}_{i \in I} \widehat{F} = \widehat{A^{(S \times I)}}.$$

Suppose now that $\chi = (\chi_i)_{i \in I} \in \mathcal{Z}_{i \in I} \widehat{F} = \widehat{F^{(I)}} = \widehat{A^{(S \times I)}}$. We have $\chi_i = (a_{ji})_{j \in S}$, for all $i \in I$. Then since $\chi \in \widehat{A^{(S \times I)}}$, this means that the family $(a_{ji})_{(j,i) \in S \times I}$ of elements of A converges to zero. Therefore, for every open right ideal J in A, we have that $a_{ji} \in J$, for all but finitely many pairs $(j,i)_{S \times I}$. Therefore, for all but finitely many $i \in I$ (namely, one avoids those $i \in I$ such that there exists $j \in S$ such that $a_{ji} \notin J$), we have that $a_{ji} \in J$ for all $j \in S$. But then, by equation (9) applied to the element

$$\chi_i = (a_{ji})_{j \in S} \in \widehat{F},$$

we have that $\chi_i \in J \cdot \widehat{F}$, for all but finitely many $i \in I$. This being true for every open right ideal $J \subset A$, it follows (by definition of G), that $\chi = (\chi_i)_{i \in I} \in G((\widehat{F})_{i \in I})$. Thus, the left side of equation (3) is contained in the right side.

Conversely, suppose that $\chi = (\chi_i)_{i \in I} \in G((\widehat{F})_{i \in I})$. Then $\chi_i \in \widehat{F}$, for all $i \in I$ — say $\chi_i = (a_{ji})_{j \in S} \in \widehat{A^{(S)}}$, for all $i \in I$ — and we know that, for every open right ideal $J \in A$, that we have that $\chi_i \in J \cdot \widehat{F}$ for all but finitely many $i \in I$. For any given $i \in I$, the condition "$\chi_i \in J \cdot \widehat{F}$" clearly implies that "$a_{ji} \in J$ for all $j \in S$." Therefore, the element

$$\chi = (a_{ji})_{(j,i) \in S \times I} \in \prod_{i \in I} \widehat{A^{(S)}}$$

is such that: For all but finitely many i, we have $a_{ji} \in J$ for all $j \in S$." Now, for each of those finitely many i, we have that $\chi_i = (a_{ji})_{j \in S} \in \widehat{A^{(S)}}$ — i.e., that the family $(a_{ji})_{j \in S}$ of elements of A converge to zero. Therefore, for every such i, we have that for all but finitely many j, that $a_{ji} \in J$. It follows that $a_{ji} \in J$ for all but finitely many pairs $(j,i) \in S \times I$. I.e., $(a_{ji})_{(j,i) \in S \times I} \in \widehat{A^{(S \times I)}}$. But then by equation (11), $\chi = (a_{ji})_{(j,i) \in S \times I} \in \underset{i \in I}{\mathcal{Z}} \widehat{F}$.

Thus the right side of equation (3) is contained on the left side. $\square$

Remark 1. Notice that, under the hypothesis of Theorem 5.2.3, we have (using the right side of equation (3)) that

$$\prod_{i \in I} (div \, M_i) \subset \underset{i \in I}{\mathcal{Z}} M_i.$$

Definition 2. Let us introduce some terminology. If A is a standard topological ring, if I is a set and if M_i are C-complete left A-modules for all $i \in I$, then given an indexed family of elements $(\chi_i)_{i \in I} \in \prod_{i \in I} M_i$, let us say that the family $(\chi_i)_{i \in I}$ is *zero-bound* iff for every open right ideal $J \subset A$, we have that $\chi_i \in J \cdot M_i$ for all but finitely many $i \in I$.

Example 5. If the set I is infinite and if $M_i = M$ for all $i \in I$, then the family $(\chi_i)_{i \in I}$ is zero-bound iff $(\chi_i)_{i \in I}$ converges to zero in M for the A-adic topology.

Example 6. If the set I is finite, then $(\chi_i)_{i \in I}$ is zero-bound for all $(\chi_i)_{i \in I} \in \prod_{i \in I} M_i$.

Example 7. If the set I is infinite, and if $\iota_j : M_j \to \prod_{i \in I} M_i$ is the inclusion for all $j \in I$, then a family $(\chi_i)_{i \in I} \in \prod_{i \in I} M_i$ is zero-bound iff the family $(\iota_i(\chi_i))_{i \in I}$ of elements of $\prod_{i \in I} M_i$ converges to zero in $\prod_{i \in I} M_i$ for the A-adic topology.

Example 8. If the A-adic and uniform topologies coincide in $\widehat{A}$ (e.g., it suffices that *either* A is admissible *or* A is complete) then let $M_i = \widehat{A}$ for all $i \in I$. Then the above definition of "zero-bound" coincides with that given at the beginning of Chapter 2, §1. In particular, a family $(a_i)_{i \in I} \in \widehat{A}^I$ is zero-bound iff $(a_i)_{i \in I} \in \widehat{A^{(I)}}$.

In the terminology of Definition 2, Theorem 5.2.3 can be stated as follows:

Corollary 5.2.4. *Let A be a complete topological ring that is admissible and such that A admits a denumerable neighborhood base at zero. Then for any set I, and any family $(M_i)_{i \in I}$ of C-complete left A-modules, we have that*

$$\mathop{\mathcal{Z}}_{i \in I} M_i = \{ (\chi_i)_{i \in I} \in \prod_{i \in I} M_i : (\chi_i)_{i \in I} \text{ is zero-bound} \}.$$

Remark 2. Corollary 5.2.4 motivates the notation "$\mathop{\mathcal{Z}}_{i \in I} M_i$" ("$\mathcal{Z}$" for "zero-bound").

Lemma 5.2.5. *Let A be a standard topological ring. Let I be a set and let M_i be an abstract left A-module for all $i \in I$. Suppose that either condition (1) or condition (2) holds:*
Either (1) M_i is a projective abstract left A-module for all $i \in I$.
Or (2) The topological ring A admits a denumerable neighborhood base at zero, and the abstract left A-module M_i is left-flat, for all $i \in I$.
Then the natural map:

$$(1): \quad \theta : \int_{i \in I} \widehat{M_i} \to \prod_{i \in I} \widehat{M_i}$$

is injective. And

$$(2): \quad \widehat{\bigoplus_{i \in I} M_i} = C\left(\bigoplus_{i \in I} M_i \right) = \int_{i \in I} M_i = \mathop{\mathcal{Z}}_{i \in I} M_i.$$

Note: Another condition more general than each of conditions (1) and (2) above, that suffices for the conclusions of the Lemma, is:

$$(3): \quad C\left(\bigoplus_{i \in I} M_i \right) = \widehat{\bigoplus_{i \in I} M_i}.$$

Proof. First notice that, if condition (1) holds, then $\bigoplus_{i \in I} M_i$ is projective, and therefore condition (3) holds. Also, if condition (2) holds, then $\bigoplus_{i \in I} M_i$ is left-flat, so once again by Chapter 4, §3, Theorem 4.3.18, we have that condition (3) holds. Therefore in all cases we have that condition (3) holds.

But for all $j \in I$, the natural map:

$$C\left(\bigoplus_{i \in I} M_i \right) \to \widehat{\bigoplus_{i \in I} M_i}$$

is the direct sum of the two maps:

$$C(M_j) \to \widehat{M_j}$$

and

$$C\left(\bigoplus_{i \in I - \{j\}} M_i\right) \to \left[\bigoplus_{i \in I - \{j\}} M_i\right]^{\widehat{\ }}.$$

(This is because the functors: C, and A-adic completion, are additive functors, and therefore preserve finite direct sums of objects — see [COC], Introduction, Chapter 1, §3, pp. 41 and 42, Remark following Definition 7.) Therefore the natural mapping:

$$C(M_j) \to \widehat{M}_j$$

must be an isomorphism for all $j \in I$. Therefore

$$\int_{i \in I} \widehat{M}_i = \int_{i \in I} C(M_i) = C\left(\bigoplus_{i \in I} M_i\right) = \widehat{\bigoplus_{i \in I} M_i}$$

so that

$$(1): \quad \int_{i \in I} \widehat{M}_i = \widehat{\bigoplus_{i \in I} M_i}.$$

By Chapter 1, §3, Lemma 1.3.4, we have that $\widehat{\bigoplus_{i \in I} M_i} \subset \prod_{i \in I} \widehat{M}_i$. Therefore equation (1) implies that the map θ is injective. Therefore $\int_{i \in I} \widehat{M}_i = \mathcal{Z}_{i \in I} M_i$. This and equation (1) complete the proof. $\qquad\square$

Theorem 5.2.6. (Computation of the direct sum in the category of C-complete left A-modules). *Let A be a standard topological ring. Let I be a set and let M_i be a C-complete left A-module, for all $i \in I$. For every $i \in I$, suppose that we have the short exact sequence of C-complete left A-modules:*

$$(1): \quad 0 \to K_i \to \widehat{F}_i \to M_i \to 0$$

where F_i is an abstract left A-module, and $\widehat{F}_i$ is the A-adic completion of F_i, for all $i \in I$. Suppose also that

$$(2): \quad C\left(\bigoplus_{i \in I} F_i\right) = \widehat{\bigoplus_{i \in I} F_i}.$$

Then the following short exact sequence of C-complete left A-modules is induced:

$$(3): \quad 0 \to \mathcal{Z}_{i \in I} K_i \to \mathcal{Z}_{i \in I} \widehat{F}_i \to \int_{i \in I} M_i \to 0.$$

<u>Note</u>: In the special case that the standard topological ring A is C-O.K., then by §1, Theorem 5.1.1, the C-complete left A-module on the right side of equation (3) is

$$\int_{i \in I} M_i = C\left(\bigoplus_{i \in I} M_i\right).$$

Proof. For all $i \in I$, let G_i be a free abstract left A-module and let $\widehat{G}_i \to K_i$ be an epimorphism. Then we have the exact sequence of C-complete left A-modules

$$\widehat{G}_i \to \widehat{F}_i \to M_i \to 0.$$

Throwing this sequence through the right-exact functor " $\int\limits_{i \in I}$," and using Lemma 5.2.5 and the Note to the Lemma, we obtain an exact sequence

$$(4): \quad \underset{i \in I}{\mathcal{Z}} \, \widehat{G}_i \to \underset{i \in I}{\mathcal{Z}} \, \widehat{F}_i \to \int\limits_{i \in I} M_i \to 0.$$

By Lemma 5.2.1, the image of $\underset{i \in I}{\mathcal{Z}} \, \widehat{G}_i$ in $\underset{i \in I}{\mathcal{Z}} \, \widehat{F}_i$ is $\underset{i \in I}{\mathcal{Z}} \, K_i$. Therefore we have the induced short exact sequence

$$(3): \quad 0 \to \underset{i \in I}{\mathcal{Z}} \, K_i \to \underset{i \in I}{\mathcal{Z}} \, \widehat{F}_i \to \int\limits_{i \in I} M_i \to 0$$

as claimed.   □

Note: Under the hypotheses of Theorem 5.2.6 we also have an exact sequence of C-complete left A-modules:

$$(5): \quad 0 \to \underset{i \in I}{\mathcal{Z}} \, K_i \to \left(\left(\prod_{i \in I} K_i \right) \cap \left(\underset{i \in I}{\mathcal{Z}} \, \widehat{F}_i \right) \right) \to \int\limits_{i \in I} M_i \overset{\theta}{\to} \prod_{i \in I} M_i.$$

(In this exact sequence, the intersection in the second-from-the-left term occurs in $\prod\limits_{i \in I} \widehat{F}_i$, which contains both $\prod\limits_{i \in I} K_i$ and $\underset{i \in I}{\mathcal{Z}} \, \widehat{F}_i$.) An equivalent way of writing the exact sequence (5) is as a short exact sequence:

$$(6): \quad 0 \to \underset{i \in I}{\mathcal{Z}} \, K_i \to \left(\left(\prod_{i \in I} K_i \right) \cap \left(\underset{i \in I}{\mathcal{Z}} \, \widehat{F}_i \right) \right) \to \mathrm{Ker}\left(\int\limits_{i \in I} M_i \overset{\theta}{\to} \prod_{i \in I} M_i \right) \to 0.$$

The equivalence of (5) and (6) is clear.

Proof of the Note. From the short exact sequence (3) of Theorem 5.2.6, if

$$H = \mathrm{Ker}\left(\int\limits_{i \in I} M_i \to \prod_{i \in I} M_i \right),$$

then we have a short exact sequence

$$(7): \quad 0 \to \underset{i \in I}{\mathcal{Z}} \, K_i \to \left(\left(\underset{i \in I}{\mathcal{Z}} \, \widehat{F}_i \right) \cap L \right) \to H \to 0,$$

where L is the kernel of the map:

$$\prod_{i \in I} \widehat{F}_i \to \prod_{i \in I} M_i.$$

But by exactness of the sequence (1), $L = \prod_{i \in I} K_i.$ $\qquad\qquad\square$

Corollary 5.2.7. *Let A be a standard topological ring that is admissible, and such that A admits a denumerable neighborhood base at zero. Let I be a set, and let $(M_i)_{i \in I}$ be an indexed family of C-complete left A-modules. For every $i \in I$, suppose that we have a short exact sequence of abstract left A-modules*

$$(0): \quad 0 \to K_i \to \widehat{F}_i \to M_i \to 0,$$

where F_i is an abstract left A-module that is left-flat, and where $\widehat{F}_i$ is the A-adic completion of F_i, for all $i \in I$. Then the following three conditions are equivalent.

(1) The natural map: $\int_{i \in I} M_i \to \prod_{i \in I} M_i$ is injective.

(2) The natural epimorphism: $\int_{i \in I} M_i \to \underset{i \in I}{\mathcal{Z}} M_i$ is an isomorphism.

(3) If $(\chi_i)_{i \in I} \in \prod_{i \in I} K_i$ are a family of elements, such that $(\chi_i)_{i \in I}$ is zero-bound in $\prod_{i \in I} \widehat{F}_i$, then the family $(\chi_i)_{i \in I}$ is zero-bound in $\prod_{i \in I} K_i$.

Note: The hypothesis in the Corollary, that the F_i are left-flat, for all $i \in I$, can be replaced by the more general condition

$$C\left(\underset{i \in I}{\oplus} F_i \right) = \widehat{\underset{i \in I}{\oplus} F_i}.$$

(That the latter condition is more general than the former follows from Chapter 4, §3, Theorem 4.3.18, applied to the left-flat A-module, $\underset{i \in I}{\oplus} F_i$.)

Proof. By definition of $\underset{i \in I}{\mathcal{Z}} M_i$, the map in (2) is the co-image of the map in (1). Therefore the map in (1) is injective iff the map in (2) is an isomorphism. On the other hand, we have that the hypotheses of Theorem 5.2.6 hold. Therefore we have the short exact sequence (3) of the conclusion of Theorem 5.2.6. But by that short exact sequence we have that condition (1) of the corollary holds iff the sequence:

$$(4): \quad 0 \to \underset{i \in I}{\mathcal{Z}} K_i \to \underset{i \in I}{\mathcal{Z}} \widehat{F}_i \to \prod_{i \in I} M_i$$

is exact at the middle spot. By Corollary 5.2.4, the last assertion is equivalent to:

"If $(f_i)_{i \in I} \in \prod_{i \in I} \widehat{F}_i$, and $(f_i)_{i \in I}$ is zero-bound in $\prod_{i \in I} \widehat{F}_i$, and if f_i maps into zero in M_i for all $i \in I$; then $f_i \in K_i$ for all $i \in I$, and $(f_i)_{i \in I}$ is zero-bound in $\prod_{i \in I} K_i$." But, since the sequences

$$0 \to K_i \to \widehat{F}_i \to M_i \to 0$$

are exact for all $i \in I$ this statement simplifies to:
"If $(f_i)_{i \in I}$ is a family of elements in $\prod_{i \in I} K_i$ such that $(f_i)_{i \in I}$ is zero-bound in $\prod_{i \in I} \widehat{F}_i$, then $(f_i)_{i \in I}$ is zero-bound in $\prod_{i \in I} K_i$." Hence condition (1) holds iff condition (3) holds. $\Box$

Note: Let $\mathbf{Z}[T]$ be regarded, as usual, as topological ring with the T-adic topology. Then if A is any right t-adic ring such that the t-torison is bounded below, then by Chapter 4, §3, Corollary 4.3.8, we have that $C_i^A(N) = C_i^{\mathbf{Z}[T]}(N)$, $i \geq 0$, for all abstract left A-modules N. Since also the topological ring A is C-O.K., by Theorem 5.1.1 we have that, for any indexed family $(M_i)_{i \in I}$ of C-complete left A-modules,

$$\int_{i \in I} M_i \approx C(\bigoplus_{i \in I} M_i).$$

And by Theorem 5.1.2, for any direct system: $(M_i)_{i \in D}$ of (resp.: any covariant functor F from a category that is a set $\mathscr{D}$ into the category of) C-complete left A-modules, we have that

$$\varinjlim_{d \in D}^{C} M_d \approx C(\varinjlim_{d \in D} M_d),$$

$$\left(\text{resp: } \varinjlim_{d \in \mathscr{D}}^{C} F(d) \approx C(\varinjlim_{d \in \mathscr{D}} F(d)) \right).$$

It follows that, if I is a set and if M_i are C-complete left modules for all $i \in I$, then

$$\int_{i \in I}^{A} M_i = \int_{i \in I}^{\mathbf{Z}[T]} M_i,$$

and that, whenever $(M_d)_{d \in D}$ is a direct system of C-complete left A-modules, (resp.: whenever F is a covariant functor from a category that is a set into the category of C-complete left A-modules), that

$$\varinjlim_{d \in D}^{C \ A} M_d = \varinjlim_{d \in D}^{C \ \mathbf{Z}[T]} M_d,$$

$$\left(\text{resp.: } \varinjlim_{d \in \mathscr{D}}^{C \ A} F(d) = \varinjlim_{d \in \mathscr{D}}^{C \ \mathbf{Z}[T]} F(d) \right).$$

Corollary 5.2.8. *Let A be a t-adic ring, such that there exists an element $t \in A$ such that the topology of A is the right t-adic, and such that the t-torsion in the ring A is bounded below. Let I be a set, and let M_i be a C-complete left A-module, for all $i \in I$. Then each of the following two conditions is itself equivalent to the three equivalent conditions of Corollary 5.2.7.*

(4) There exist a finite subset F of I, and a non-negative integer n, such that $i \in I - F$, $x \in M_i$, x is a t-torsion element, implies that $t^n x = 0$.

(5) There exists a finite subset F of I, such that the t-torsion in $\underset{i \in I-F}{\oplus} M_i$ is bounded below.

Proof. As in the Note immediately preceding this Corollary, let $\phi : \mathbf{Z}[T] \to A$ be the unique ring homomorphism such that $\phi(T) = t$. Then, since $\int\limits_{i \in I} M_i = C\left(\underset{i \in I}{\oplus} M_i \right)$, by Corollary 4.3.8, replacing A by $\mathbf{Z}[T]$ and t by T if necessary, we can assume that $A = \mathbf{Z}[T]$. Therefore, we can and do assume that A is a proper t-adic ring and that t is a parameter.

(4) implies (5): Take F as in (4). Then the t-torsion in $\underset{i \in I-F}{\oplus} M_i$ is bounded below (by the integer n of condition (4)).

(5) implies (1): By Chapter 4, §3, Example 6, since the t-torsion is bounded below in $M = \underset{i \in I-F}{\oplus} M_i$,

$$C\left(\underset{i \in I-F}{\oplus} M_i \right) = \widehat{\underset{i \in I-F}{\oplus} M_i}.$$

But the latter is contained in $\underset{i \in I-F}{\prod} M_i$ (by Chapter 1, §3, Lemma 1.3.4) Therefore

$$(1) : C\left(\underset{i \in I-F}{\oplus} M_i \right) \subset \prod_{i \in I-F} M_i.$$

Since the functor C is an additive functor, and since F is a finite set,

$$C\left(\underset{i \in I}{\oplus} M_i \right) = \underset{i \in F}{\oplus} C(M_i) \oplus C\left(\underset{i \in I-F}{\oplus} M_i \right).$$

Since M_i is C-complete, (and since A is C-O.K.), $C(M_i) = M_i$, for all $i \in F$. So this last equation becomes:

$$C\left(\underset{i \in I}{\oplus} M_i \right) = \underset{i \in F}{\oplus} M_i \oplus C\left(\underset{i \in I-F}{\oplus} M_i \right).$$

Combining with equation (1), we have that

$$C\left(\underset{i \in I}{\oplus} M_i \right) \subset \prod_{i \in I} M_i.$$

But, since A is C-O.K.,

$$\int\limits_{i \in I} M_i = C \left(\bigoplus_{i \in I} M_i \right).$$

Thus

$$\int\limits_{i \in I} M_i \subset \prod_{i \in I} M_i.$$

(3) implies (4): Choose short exact sequences (0) as in Corollary 5.2.7, with F_i left flat, for all $i \in I$. Then $\operatorname{Tor}_A^i(A/tA, F_i) = 0$, for $i \geq 1$, so by Lemma 4.3.6, F_i has no non-zero t-torsion, for all $i \in I$. Therefore, by Lemma 4.3.20, the t-adic completion $\widehat{F_i}$ has no non-zero t-torsion.

Suppose that (4) were false. Then there exists a sequence $i_1, i_2, i_3, \ldots$ of distinct elements of I, such that M_{i_j} has a precise t^j-torsion element that is not a precise t^{j-1}-torsion element, $j \geq 1$. Let $\overline{y_j} \in M_{i_j}$ be such a precise t^j-torsion element, $j \geq 1$. Let $y_j \in \widehat{F_{i_j}}$ represent $\overline{y_j} \in M_{i_j}$, $j \geq 1$. Then

(2): $t^j \cdot y_j \in K_{i_j}$, $t^{j-1} \cdot y_j \notin K_{i_j}$, $j \geq 1$.

If $t^j \cdot y_j \in t \cdot K_{i_j}$, say $t^j y_j = tr$, $r \in K_{i_j}$, then, since K_{i_j} is a submodule of $\widehat{F_{i_j}}$, and since $\widehat{F_{i_j}}$ has no non-zero t-torsion, from the relation $t(t^{j-1}y_j - r) = 0$, we would deduce $t^{j-1}y_j = r$, whence $t^{j-1}y_j \in K_{i_j}$, a contradiction to equation (2). Therefore $t^j \cdot y_j \notin t \cdot K_{i_j}$, for all $j \geq 1$. Let $(\chi_i)_{i \in I} \in \prod\limits_{i \in I} K_i$ be the family of elements such that $\chi_{i_j} = t^j y_j$, for all $j \geq 1$, and $\chi_i = 0$, for all $i \in I - \{i_1, i_2, i_3, \ldots\}$. Then, since $\chi_{i_j} = t^j \cdot y_j \notin t \cdot K_{i_j}$, $j \geq 1$, the family $(\chi_i)_{i \in I}$ is not zero-bound in $\prod\limits_{i \in I} K_i$. But since $\chi_{i_j} = t^j \cdot y_{i_j}$, and $y_{i_j} \in \widehat{F_{i_j}}$, the family $(\chi_i)_{i \in I}$ is zero-bound in $\prod\limits_{i \in I} \widehat{F_i}$. Thus condition (3) of Corollary 5.2.7 fails, a contradiction. Therefore (4) must be true.   □

Exercise 1: Let A be a standard topological ring, let I be a set, let $(N_i)_{i \in I}$ be an indexed family of C-complete left A-modules and let $(\chi_i)_{i \in I} \in \prod\limits_{i \in I} N_i$. Then show that $(\chi_i)_{i \in I}$ is zero-bound iff for every denumerable subset D of I, then family $(\chi_i)_{i \in D} \in \prod\limits_{i \in D} N_i$ is zero-bound.

Exercise 2: Hypotheses as in Exercise 1, suppose that A has a denumerable neighborhood base at zero. Suppose that K_i is a C-complete left A-submodule of N_i, for all $i \in I$. Then show that condition (3) of Corollary 5.2.7 holds (with N_i replacing $\widehat{F_i}$) iff

(3′) for every open right ideal $J \subset A$, there exists an open right ideal $J' \subset A$ such that $J \cdot K_i \supset K_i \cap (J' \cdot N_i)$ for all but finitely many $i \in I$.

Remark 3. Given A an arbitrary standard topological ring, a set I and C-complete left A-modules M_i, for all $i \in I$, it is very easy to construct short exact sequences in the category of C-complete left A-modules

$$0 \to K_i \to \widehat{F_i} \to M_i \to 0, \ i \in I,$$

obeying all the hypotheses of Theorem 5.2.6.

(1) Let F_i be abstract projective left A-modules and $F_i \to M_i$ epimorphisms of abstract left A-modules. Then (by the universal mapping property definition of $C(F_i) = \widehat{F_i}$) there exist unique infinitely linear functions

$$\widehat{F_i} \to M_i$$

extending the function $F_i \to M_i$, for all $i \in I$. Or,

(2) Let S_i be a set of generators for M_i as C-complete left A-module as defined in Chapter 2, §4. (I.e., every element of M_i can be written as an infinite linear combination of elements of S_i.) Then we have the induced epimorphism of C-complete left A-modules

$$\widehat{A^{(S_i)}} \to M_i,$$

for all $i \in I$. Let $F_i = A^{(S_i)}$, $i \in I$.

Whether one makes the construction (1) or (2), define $K_i = \text{Ker}(\widehat{F_i} \to M_i)$, for all $i \in I$. And then all the hypotheses of Theorem 5.2.6 hold. (Since $\underset{i \in I}{\oplus} F_i$

is projective, and therefore $C \left(\underset{i \in I}{\oplus} F_i \right) = \widehat{\underset{i \in I}{\oplus} F_i}$.)

Remark 4. And, of course, given A, I and $(M_i)_{i \in I}$ as in Corollary 5.2.7, it is again easy to choose short exact sequences (0) as in the hypotheses of Corollary 5.2.7. E.g., the constructions (1) or (2) of Remark 3 are adequate (and in construction (1) of Remark 3, since the standard topological ring A admits a denumerable neighborhood base at zero, by Chapter 4, §3, Theorem 4.3.18, one only needs to take F_i to be left flat rather than the more stringent condition of projective).

Corollary 5.2.9. *The hypotheses and notations being as in Corollary 5.2.7, suppose in addition that the abstract left A-modules M_i are A-adically complete, for all $i \in I$. Then the following two conditions are each equivalent to each of the conditions (1), (2), (3) of Corollary 5.2.7:*

(4) : $\int_{i \in I} M_i$ *is A-adically Hausdorff*

(5) : $\int_{i \in I} M_i$ *is A-adically complete.*

Proof. Since A has a denumerable neighborhood base at zero and is admissible, by Chapter 3, §1, Theorem 3.1.6, A is C-O.K. Therefore by §1, Theorem 5.1.1,

$$(6) : \int_{i \in I} M_i = C \left(\underset{i \in I}{\oplus} M_i \right).$$

By Chapter 3, §1, Theorem 3.1.6, A is admissible. Therefore, by Theorem 3.2.11, for every abstract left A-module M,

$$\widehat{C(M)} \approx \widehat{M}.$$

In particular, taking $M = \bigoplus\limits_{i \in I} M_i$,

$$(7): \quad \left[C(\bigoplus_{i \in I} M_i) \right]^{\displaystyle\frown} = \widehat{\bigoplus_{i \in I} M_i}.$$

The A-divisible part of $\int\limits_{i \in I} M_i = C(\bigoplus\limits_{i \in I} M_i)$ is the kernel of the mapping into its A-adic completion $[C(\bigoplus\limits_{i \in I} M_i)]^{\displaystyle\frown}$. Therefore by equation (7),

$$(8): \quad div \left(\int_{i \in I} M_i \right) = \mathtt{Ker} \left(\int_{i \in I} M_i \to \widehat{\bigoplus_{i \in I} M_i} \right).$$

By Chapter 1, §3, Lemma 1.3.4, we have that the natural map:

$$(9): \quad \widehat{\bigoplus_{i \in I} M_i} \subset \prod_{i \in I} \widehat{M_i}$$

is a monomorphism. Since by hypothesis M_i is A-adically complete, $\widehat{M_i} = M_i$, for all $i \in I$. Combining with equation (8), we obtain

$$(10): \quad div \left(\int_{i \in I} M_i \right) = \mathtt{Ker} \left(\int_{i \in I} M_i \to \prod_{i \in I} M_i \right).$$

Therefore $\int\limits_{i \in I} M_i$ is A-adically Hausdorff iff the natural map $\int\limits_{i \in I} M_i \to \prod\limits_{i \in I} M_i$ is injective. I.e., condition (4) of this corollary is equivalent to condition (1) of Corollary 5.2.7. Finally, by Chapter 3, §2, Corollary 3.2.23, we have that conditions (4) and (5) of this corollary are equivalent.   □

A consequence deduced in the proof of Corollary 5.2.9 is worth recording separately:

Corollary 5.2.10. *Under the hypotheses of Corollary 5.2.7, we have that*

$$(11): \quad div \left(\int_{i \in I} M_i \right) = \mathtt{Ker} \left(\int_{i \in I} M_i \xrightarrow{\theta} \prod_{i \in I} \widehat{M_i} \right).$$

Proof. Follows from equations (8) and (9) deduced in the proof of Corollary 5.2.9.
□

Proposition 5.2.11. *Let A be a topological ring that is admissible and such that A admits a denumerable neighborhood base at zero. Let M be an arbitrary C-complete left A-module. Then the following two conditions are equivalent:*

(1) For all (or equivalently: There exists an) infinite set(s) I, such that the natural map:

$$\int_{i \in I} M \to \prod_{i \in I} M = M^I$$

is injective.

(2) For all (or equivalently: There exists a) left-flat abstract left A-module F, and an epimorphism

$$\phi : \widehat{F} \to M$$

of abstract left A-modules, such that if $K = \mathtt{Ker}(\phi)$, then the A-adic topology of K is the topology induced from the A-adic topology of $\widehat{F}$.

Note: By §1, Theorem 5.1.1, since by Chapter 3, §1, Theorem 3.1.6, A is C-O.K., another way of wording condition (1), is that: The natural map

$$C(M^{(I)}) \to M^I$$

is injective.

Proof. Let I be an infinite set, and let ϕ, F and K be as in condition (2) of the proposition. Then Corollary 5.2.7 applies, with $M_i = M$, $F_i = F$, and $K_i = K$, for all $i \in I$. But then, the equivalence of conditions (1) and (3) of Corollary 5.2.7 tells us that condition (1) of the proposition for the fixed infinite set I is equivalent to:

(3′) If χ_i are elements of K, for all $i \in I$, such that $(\chi_i)_{i \in I}$ converges to zero A-adially in $\widehat{F}$, then $(\chi_i)_{i \in I}$ converges to zero A-adically in K.

Clearly, if condition (2) of this proposition holds — that is, if the A-adic topology of K is induced from the A-adic topology of $\widehat{F}$ — then condition (3′) above holds. Conversely, suppose that (3′) above holds.

If I_j, $j \geq 0$, are a complete system of open right ideals for the topology of A, then for any abstract left A-module N, if $(\chi_i)_{i \in I} \in N^I$ converges to zero A-adically in N, then for every integer $j \geq 1$ there is a finite subset F_j of I such that $\chi_i \in I_j \cdot N$ for all $i \notin F_j$. The union D of the finite sets F_j for $j \geq 0$ is a denumerable subset D of I, such that χ_i is in the closure of $\{0\}$ in N for all $i \notin D$. Hence a family $(\chi_i)_{i \in I}$ of elements of N converges to 0 in N iff $(\chi_i)_{i \in D}$ converges to 0 in N for some denumerable subset of D of I, and also χ_i is in the closure of $\{0\}$ in N for all $i \notin D$.

The left A-module $\widehat{F}$ is A-adically complete, and therefore the closure of $\{0\}$ is $\{0\}$. Hence a family of elements of $(\chi_i)_{i \in I}$ in $\widehat{F}$ converges to 0 in $\widehat{F}$ iff there is a denumerable subset D of I such that $\chi_i = 0$ for $i \notin D$, and such that $(\chi_i)_{i \in D}$ converges to zero in $\widehat{F}$. Therefore condition (3′) for the infinite set I holds iff condition (3′) holds for every denumerable subset D of I. Hence condition (3′) holds for all infinite sets I iff it holds for any single denumerable set D.

But, since the A-adic topology in the topological groups K and $\widehat{F}$ admit denumerable neighborhood bases at zero, condition (3′) for D is clearly equivalent to: The A-adic topology of K is induced from the A-adic topology of F. Therefore (3′) implies (2). $\qquad\square$

Corollary 5.2.12. *Let A be a t-adic ring, such that there exists an element $t \in A$, such that the topology of A is the right t-adic, and such that the t-torsion in the ring A is bounded below. Let M be a C-complete left A-module. Then the two equivalent conditions of Proposition 5.2.11 are also equivalent to:*
(3) The t-torsion in the left A-module M is bounded below.

Proof. If I is any infinite set, then let $M_i = M$ for all $i \in I$. Then the hypotheses of Corollary 5.2.8 are satisfied. And then condition (1) of Proposition 5.2.11 is identical to condition (1) of Corollary 5.2.7; and condition (3) of the corollary is clearly equivalent to condition (4) of Corollary 5.2.8. Therefore the corollary follows from Corollary 5.2.8. □

Remark 5. The following examples, in a very concrete case, illustrate some of the above theorems, corollaries and propositions.

Let $\mathscr{O}$ be a fixed complete discrete valuation ring that is not a field. We regard $\mathscr{O}$ as a topological ring with the $\mathscr{M}$-adic topology, where $\mathscr{M}$ is the maximal ideal of $\mathscr{O}$. Let t be a generator of $\mathscr{M}$. Then of course, $\mathscr{O}$ is admissible, has a denumerable neighborhood base at zero, etc.

Example 9. Let $M_i = \mathscr{O}/t^i\mathscr{O}$, $i \geq 1$. Then M_i is discrete, and therefore also $\mathscr{O}$-adically complete — and a fortiori C-complete. Let us apply Corollaries 5.2.7 and 5.2.9 to determine whether or not the C-complete $\mathscr{O}$-module:

$$\int_{i \geq 1} (\mathscr{O}/t^i\mathscr{O}) = C \left(\bigoplus_{i \geq 1} \mathscr{O}/t^i\mathscr{O} \right)$$

is Hausdorff. By Corollaries 5.2.7 and 5.2.9, this is so iff condition (3) of Corollary 5.2.7 holds. We have the short exact sequences:

$$(0): \quad 0 \to t^i\mathscr{O} \hookrightarrow \mathscr{O} \to \mathscr{O}/t^i\mathscr{O} \to 0,$$

which obey the hypotheses of Corollary 5.2.7. The sequence: $(t^i)_{i \in I} \in \prod_{i \in I} t^i\mathscr{O}$ is not zero-bound in $\prod_{i \geq 1} t^i\mathscr{O}$ (since in fact t^i is not even in $t(t^i\mathscr{O})$, for any $i \geq 1$). But the sequence $(t^i)_{i \geq 1} \in \prod_{i \geq 1} \mathscr{O}$ is zero-bound in $\prod_{i \in I} \mathscr{O}$. Therefore, by Corollaries 5.2.7 and 5.2.9,

$$\int_{i \geq 1} (\mathscr{O}/t^i\mathscr{O}) = C \left(\bigoplus_{i \geq 1} \mathscr{O}/t^i\mathscr{O} \right)$$

is not Hausdorff. It is an example of a C-complete $\mathscr{O}$-module that is not Hausdorff.

This is, perhaps, the simplest example of a C-complete module that is not complete. Since the module is C-complete, given any family $(\chi_i)_{i \in I}$ of elements of the module, we have the infinite sum: $\sum_{i \geq 1} t^i \chi_i$. And infinite sums *cannot* be defined using the $\mathscr{O}$-adic topology, since the module is not Hausdorff. Also, (since $\mathscr{O}$ is obviously admissible), the infinite sum structure on this module can be characterized as being the unique such structure obeying the axioms of Chapter 2.

Perhaps we should construct this module explicitly. A free resolution of $\underset{i\geq 1}{\oplus}\mathscr{O}/t^i\mathscr{O}$ is the chain complex

$$\mathscr{O}^{(\omega)} \xrightarrow{d_1} \mathscr{O}^{(\omega)}$$

where $d_1((\chi_i)_{i\geq 1}) = (t^i\chi_i)_{i\geq 1}$, for all $(\chi_i)_{i\geq 1} \in \mathscr{O}^{(\omega)}$. The A-adic completion of this chain complex is the chain complex

$$(1): \ \widehat{\mathscr{O}^{(\omega)}} \xrightarrow{\hat{d}_1} \widehat{\mathscr{O}^{(\omega)}}$$

where again $(\hat{d}_1)((\chi_i)_{i\geq 1}) = (t^i\chi_i)_{i\geq 1}$, for all $(\chi_i)_{i\geq 1} \in \widehat{\mathscr{O}^{(\omega)}}$ (i.e., all $(\chi_i)_{i\geq 1} \in \mathscr{O}^\omega$ such that $(\chi_i)_{i\geq 1}$ converges to zero in $\mathscr{O}$). By definition, $C_n\left(\underset{i\geq 1}{\oplus}\mathscr{O}/t^i\mathscr{O}\right)$ is the nth homology group of the chain complex (1) $n \geq 0$. Clearly $\hat{d}_1$ is injective. Therefore

$$(2): C_n\left(\underset{i\geq 1}{\oplus}\mathscr{O}/t^i\mathscr{O}\right) = 0, \ n \geq 1.$$

If we let K denote the image of the monomorphism $\hat{d}_1$, then K is the completion of the free submodule of $\widehat{\mathscr{O}^{(\omega)}}$ generated by the elements $u_j = (\delta_{ij}t^i)_{i\geq 1} \in \widehat{\mathscr{O}^{(\omega)}}$.

Remark 6. Clearly, K does not have the induced topology from $\widehat{\mathscr{O}^{(\omega)}}$ (since, e.g., $u_j \to 0$ in $\widehat{\mathscr{O}^{(\omega)}}$, but $(u_j)_{j\geq 1}$ does not converge in K). Therefore, by Proposition 5.2.11, if $M = C\left(\underset{i\geq 1}{\oplus}\mathscr{O}/t^i\mathscr{O}\right)$, then the natural map

$$\int_{i\geq 1} M \to \prod_{i\geq 1} M$$

is not injective.

Remark 7. Given a family $(F_i)_{i\in I}$ as in Theorem 5.2.6, a hypothesis of Theorem 5.2.6 on the family is that

$$(a): \ C\left(\underset{i\in I}{\oplus}F_i\right) = \widehat{\underset{i\in I}{\oplus}F_i}.$$

As noted in the Proof of Theorem 5.2.6, this condition implies that

$$(b): \ C(F_i) = \widehat{F_i}, \ \text{for all } i \in I.$$

One might wonder, whether conversely condition (b) implies condition (a)? The answer is "no." E.g., let $A = \mathscr{O}$ as in Example 9, $I = \{\text{positive integers}\}$, $F_i = \mathscr{O}/t^i\mathscr{O}$, $i \geq 1$. Then we have seen that $C\left(\underset{i\geq 1}{\oplus}F_i\right)$ possesses divisible elements.

To continue with Example 9: From the short exact sequence

$$(3): \quad 0 \to K \to \widehat{\mathscr{O}^{(\omega)}} \to C\left(\bigoplus_{i \geq 1} \mathscr{O}/t^i\mathscr{O}\right) \to 0,$$

it follows by topology that the divisible part of $C\left(\bigoplus_{i \geq 1} \mathscr{O}/t^i\mathscr{O}\right)$ is the image of the closure $\overline{K}$ of K in $\widehat{\mathscr{O}^{(\omega)}}$ (since as observed in Chapter 1 the A-adic topology on a quotient left module coincides with the quotient topology from the A-adic topology of the module — for any standard topological ring and any abstract left module and abstract left quotient module). But K itself is generated topologically by the elements $u_j \in K$, and $u_j \to 0$ in $\mathscr{O}^{(\omega)}$ (in fact, $u_j \in t^j \cdot \mathscr{O}^{(\omega)}$, $j \geq 1$). Therefore for *any* family $(a_i)_{i \in I} \in \mathscr{O}^\omega$ (i.e., any family a_i, $i \in I$, of elements of $\mathscr{O}$), we have that the series: $\sum_{j \geq 1} a_j u_j$ converges in $\widehat{\mathscr{O}^{(\omega)}}$. Therefore $\overline{K} \approx \mathscr{O}^\omega$ (the direct product of $\aleph_0$ copies of $\mathscr{O}$). But the divisible part of $C\left(\bigoplus_{i \geq 1} \mathscr{O}/t^i\mathscr{O}\right)$ is isomorphic to $\overline{K}/K$. Therefore

$$(4): \quad div\left(C\left(\bigoplus_{i \geq 1} \mathscr{O}/t^i\mathscr{O}\right)\right) \approx \mathscr{O}^\omega/\widehat{\mathscr{O}^{(\omega)}}.$$

Notice that this is an $\mathscr{O}$-adically complete $\mathscr{O}$-module. Thus, if $M = C\left(\bigoplus_{i \geq 1} \mathscr{O}/t^i\mathscr{O}\right)$, then M is C-complete,

$$(5): \quad M/div(M) \approx \widehat{[\bigoplus_{i \in I} \mathscr{O}/t^i\mathscr{O}]}$$

and $div(M)$ (is A-adically complete and) is given by equation (4).

It might be instructive to study some infinite linear combinations of divisible elements in $M = C\left(\bigoplus_{i \geq 1} \mathscr{O}/t^i\mathscr{O}\right)$. Let $e_j = (\delta_{ij})_{i \geq 1} \in \widehat{\mathscr{O}^{(\omega)}}$, for $j \geq 1$. Then $\widehat{\mathscr{O}^{(\omega)}}$ is the completion of the free module generated by e_j, $j \geq 1$. For every $v \in \widehat{\mathscr{O}^{(\omega)}}$, let $\overline{v}$ denote the image in $M = C\left(\bigoplus_{i \geq 1} \mathscr{O}/t^i\mathscr{O}\right)$ (under the natural epimorphism in the short exact sequence (3)). Then since $t^j e_j = u_j \in K$, we have that $t^i \cdot \overline{e_j} = 0$ in M, for all $j \geq 1$. Nevertheless, the infinite linear combination in M:

$$(6): \quad \sum_{j \geq 1} t^j \overline{e_j}$$

is *not* zero — namely, it is the image $\overline{\sum_{j \geq 1} t^j e_j}$ of $\sum_{j \geq 1} t^j e_j \in \widehat{\mathscr{O}^{(\omega)}}$ under the natural epimorphism into M. And this image is *not* zero, since $\sum_{j \geq 1} t^i e_j = (t^j)_{j \geq 1}$, and this latter is *not* an element of K (but is an element of $\overline{K}$). Since $\sum_{j\geq 1} t^j e_j$ is in $\overline{K}$, it follows that the infinite linear combination (6) in M is a divisible element of M — in fact, under the isomorphism (4), the element (6) corresponds to the coset of the tuple $(1,1,1,\ldots,1,\ldots) \in \mathcal{O}^\omega$.

More generally, given any family $(a_j)_{j\geq 1}$ of elements of $\mathcal{O}$, such that $a_j \in t^j \cdot \mathcal{O}$, for $j \geq 1$, then the infinite linear combination

$$(7): \quad \sum_{j\geq 1} a_j \overline{e_j} \in M$$

is also a divisible element of M. Given another such family $(b_j)_{j\geq 1}$, it will determine the same divisible element as in (7) iff the sequence $t^{-j}(b_j - a_j)$, $j \geq 1$ of elements of $\mathcal{O}$ converges to zero in $\mathcal{O}$.

Notice that, if $(a_j)_{j\geq 1}$ is as above (i.e., $a_j \in t^j \mathcal{O}$, $j \geq 1$), then

$$a_j \cdot \overline{e_j} = 0 \text{ in } M$$

(since $a_j \cdot \overline{e_j} = (t^{-j} a_j) u_j \in K$). Therefore each of the finite sums

$$\sum_{1\leq j\leq n} a_j \overline{e_j} = 0 \text{ in } M.$$

Nevertheless, the *infinite* sum (7) is *not* zero in M, e.g., if $a_j = t^j$, for all $j \geq 1$. Therefore, the "infinite linear combination (7)" *must* be understood in the sense of the definitions of Chapter 2 — e.g., it certainly is *not* a topological, or any other kind, of limit of the sequence

$$\sum_{1\leq j\leq n} a_j \overline{e_j}, \ n \geq 1$$

(since this sequence is always the constant sequence $0, 0, 0, \ldots$ if $a_j \in t^j \mathcal{O}$, $j \geq 1$).

What about infinite linear combinations of the $\overline{e_j}$, $j \geq 1$, that represent other than divisible elements of M? If $(a_i)_{i\geq 1}$ is an arbitrary family of elements of $\mathcal{O}$ that converges to zero in $\mathcal{O}$, then we have the infinite linear combination:

$$(8): \quad \sum_{j\geq 1} a_j \overline{e_j} \in M.$$

Every element of M can be written in this form. This element (8) will be a divisible element of M iff $a_j \in t^j \mathcal{O}$, for all $j \geq 1$. If also $b_j \in \mathcal{O}$ and $b_j \to 0$ in $\mathcal{O}$, then the infinite linear combination $\sum_{j\geq 1} b_j \overline{e_j} \in M$ will be the same element of M as (8) iff (1): $t^j | (a_j - b_j)$, and also (2): $t^{-j}(a_j - b_j)_{j\geq 1}$, converges to zero in $\mathcal{O}$. Finally, notice also that the image of the element (8) in $M/\mathrm{div}(M) = \widehat{M} = [\bigoplus_{i\geq 1} \mathcal{O}/t^i \mathcal{O}]\,\widehat{\ }$, is the element $(a_i + t^i \mathcal{O})_{i\geq 1} \in [\bigoplus_{i\geq 1} \mathcal{O}/t^i \mathcal{O}]\,\widehat{\ }$ (by Chapter 1, §3, Lemma 1.3.4, the latter is a subset of $\prod_{i\geq 1} \mathcal{O}/t^i \mathcal{O}$).

The above example (Example 9) is perhaps the easiest example of a C-complete module that is not complete (in the "reasonable", Noetherian commutative case, with the I-adic topology). And, it is also not untypical of such examples. Most (but not all) pathologies discussed in general in this and other sections, that can occur in the Noetherian commutative I-adic case, actually already occur in the case $A = \mathscr{O}$, as above.

Note: The $\mathscr{O}$-module $M = C\left(\underset{i \geq 1}{\oplus}\, \mathscr{O}/t^i\mathscr{O}\right)$ just constructed coincides with the $\mathscr{O}$-module G constructed in [COC], Chapter 4, Example 2, pp. 586 - 588. (In fact, the cochain complex D^* used to construct G is the same as the chain complex (1) of Example 9, after re-indexing.)

Example 10. Let $\mathscr{O}$ and t be as in Example 9. Let $N = [\underset{i \geq 1}{\oplus}\, \mathscr{O}/t^i\mathscr{O}]\widehat{}$. (This is $N = M/\mathrm{div}(M)$, where M is the C-complete of Example 9.) Then N is $\mathscr{O}$-adically complete.

Is the natural map

$$\int_{i \geq 1} N \to \prod_{i \geq 1} N$$

injective? (By §1, Theorem 5.1.1, this map is also the map

$$C(N^{(\omega)}) \to N^{\omega}.)$$

By Corollary 5.2.9, another way of asking this question is: Is the C-complete

$$\int_{i \geq 1} N = C(N^{(\omega)})$$

Hausdorff?

From the short exact sequence (3) of Example 9, (and the facts that $N = M/\mathrm{div}(M)$, and that the pre-image of $\mathrm{div}(M)$ in $\mathscr{O}^{(\omega)}$ is $\overline{K}$), we deduce the short exact sequence

$$(9): \quad 0 \to \overline{K} \to \widehat{\mathscr{O}^{(\omega)}} \to N \to 0.$$

As we have seen in Example 9,

$$(10): \quad \overline{K} \approx \mathscr{O}^{\omega},$$

under which isomorphism the inclusion in equation (9) corresponds to the monomorphism

$$\mathscr{O}^{\omega} \to \widehat{\mathscr{O}^{(\omega)}}$$

that maps $(a_i)_{i \geq 1}$ into $\sum_{i \geq 1} a_i t^i e_i = (a_i t^i)_{i \geq 1}$ (where, as in Example 9, $e_i = (\delta_{ij})_{j \geq 1}$). The elements $u_j = (t^j \delta_{ij})_{i \geq 1} = (t^j e_j)$ of $\overline{K}$ (in fact, even of K), correspond under the isomorphism (10) to the elements $(\delta_{ij})_{i \geq 1}$ of $\mathscr{O}^{\omega}$. Therefore

u_j, $j \geq 1$, does not converge to zero in $\overline{K}$. But $u_j = t^j e_j$, $j \geq 1$, does converge to zero in $\widehat{\mathscr{O}^{(\omega)}}$. Therefore, by Proposition 5.2.11, we have that

$$\int_{i \geq 1} N \to \prod_{i \geq 1} M$$

is *not* injective — i.e., that the C-complete $\mathscr{O}$-module $\int_{i \geq 1} N = C(N^{(\omega)})$ is *not* Hausdorff.

Example 11. A similar argument to that in Example 10 (with K replacing $\overline{K}$, but using the same sequence $u_j \in K$, $j \geq 1$) shows similarly that if $M = C\left(\bigoplus_{i \geq 1} M_i\right)$, then the natural map:

$$\int_{i \geq 1} M \to \prod_{i \geq 1} M$$

(i.e., the natural map: $C(M^{(\omega)}) \to M^\omega$) is not injective.

Remark 8. Notice that for the $\mathscr{O}$-module studied in Example 11, we have that

$$\int_{j \geq 1} M = \int_{j \geq 1} C\left(\bigoplus_{i \geq 1} \mathscr{O}/t^i \mathscr{O}\right) = C\left(\bigoplus_{j \geq 1}\left(\bigoplus_{i \geq 1} \mathscr{O}/t^i \mathscr{O}\right)\right).$$

I.e., that

$$(11): \quad C(M^{(\omega)}) = C\left(\bigoplus_{j \geq 1}\left(\bigoplus_{i \geq 1} \mathscr{O}/t^i \mathscr{O}\right)\right).$$

If the valuation ring $\mathscr{O}$ is of mixed characteristic, then the module on the left of this equation is of some interest in p-adic cohomology. E.g., one can show that if one computes the ith p-adic cohomology group of Euclidean n-space over the residue class field $\mathscr{O}$, where $1 \leq i \leq n$, and if one were to use the "full completion" instead of the "†-completion" (see [PPWC]), then one would get the $\mathscr{O}$-module (11) as the answer. And one can show that, the reason for this follows from the spectral sequence of the C-completion, defined in Chapter 3 above. As we have seen, the module (11) is enormous — it has a large divisible part, and modulo divisible elements it has non-zero topological torsion. Perhaps that alone motivates why "†" is better in p-adic cohomology (see [PPWC]).

In a later paper, I hope to introduce a (computable) homological functor "D", related to "†" in essentially the same way as "C" is related to "$\widehat{}$". This functor "D" should, it is hoped, be useful in studying (and computing), the proper p-adic cohomology (with "†" instead of "$\widehat{}$").

Remark 9. All of the results, in Remark 5, Example 9, Remark 6, Remark 7, Example 10 and Example 11 above, of course all generalize to arbitrary proper t-adic rings without any change. (Of course, part of Remark 8 does not similarly generalize — namely, the part that deals with p-adic cohomology.)

5.3   Right Flatness of the Completion of a Free Module

In general, let A be a standard topological ring and B be an abstract ring. Let M be an abstract (A, B)-bimodule [CEHA]. Then the A-adic completion $\widehat{M}$ of M has a natural structure as an $(\widehat{A}, B)$-bimodule.

Proof. For every $b \in B$, let ρ_b be the function "right multiplication by b" from M into itself, $\rho_b(x) = xb$. Then since M is an (A, B)-bimodule, $\rho_b : M \to M$ is an endomorphism of M as an abstract left A-module. Since $(\widehat{M}, unif)$ is the completion of M as a topological left A-module (and the assignment: $M \rightsquigarrow (\widehat{M}, unif)$ is a covariant functor from the category of abstract left A-modules into the category of complete topological $(\widehat{A}, unif)$-modules), therefore we have $\widehat{\rho}_b$, which is an endomorphism of $\widehat{M}$ as an abstract left $\widehat{A}$-module (in fact, even as a topological left module over $(\widehat{A}, unif)$ for the uniform, A-adic, and the $\widehat{A}$-adic topologies. These latter two topologies are of course the same, if the standard topological ring A is strongly admissible). Since $b \to \rho_b$ is a homomorphism of rings from B^o into $\mathrm{Hom}_A(M, M)$, it follows that $b \to \widehat{\rho}_b$ is a homomorphism of rings from B^o into $\mathrm{Hom}_{\widehat{A}}(\widehat{M}, \widehat{M})$. Otherwise stated, $\widehat{M}$ is an $(\widehat{A}, B)$-bimodule.   □

Example 1. If A is a commutative standard topological ring then every abstract A-module M is of course an (A, A)-bimodule. By the above, $\widehat{M}$ becomes an $(\widehat{A}, A)$-bimodule. Of course, the structure of $\widehat{M}$ as a left $\widehat{A}$-module thus obtained is the usual one, and the right A-module structure obtained is identical to the usual structure of $\widehat{M}$ as a left A-module.

Example 2. Let A be a standard topological ring and let I be a set. Then $A^{(I)}$ has a natural structure as (A, A)-bimodule. Therefore by the above $\widehat{A^{(I)}}$ has a natrual structure as an abstract $(\widehat{A}, A)$-bimodule. Explicitly, the structure as abstract right A-module is: if $(a_i)_{i \in I} \in \widehat{A^{(I)}}$ (i.e., if $(a_i)_{i \in I}$ is a zero-bound family of elements of $\widehat{A}$), then for every $a \in A$ we have that
(1): $(a_i)_{i \in I} \cdot a = (a_i \cdot a)_{i \in I}$.
Clearly, in fact, $\widehat{A^{(I)}}$ even has a structure as an abstract $(\widehat{A}, \widehat{A})$-bimodule, by means of the same equation (1).

Exercise 1: Let A be a (right) standard topological ring and let I be a set. Then show that
(1) For the structure of $A^{(I)}$ as an abstract (A, A)-bimodule, endow $A^{(I)}$ with its A-adic topology as a *right* A-module (meaning its A^o-adic topology for its structure as *left* A^o-module). Denote this topology on $A^{(I)}$ by τ. Then for the *right* A-module structure, show that scalar multiplication:

$$(A^{(I)}, \tau) \times A \to (A^{(I)}, \tau)$$

is continuous, so that $(A^{(I)}, \tau)$ is a topological right A-module. Show also that left scalar multiplication:

$$A \times (A^{(I)}, \tau) \to (A^{(I)}, \tau)$$

is likewise continuous, so that $(A^{(I)}, \tau)$ is also a topological left A-module.

(2) Let $(\widehat{A^{(I)}}, unif)$ denote $\widehat{A^{(I)}}$ regarded as a topological left $(\widehat{A}, unif)$-module with the uniform topology (so a complete system of neighborhoods of zero in $(\widehat{A^{(I)}}, unif)$ is $\{J^I \cap \widehat{A^{(I)}} : J$ is an open right ideal in $(\widehat{A}, unif)\}$). Then show that for the structure of $\widehat{A^{(I)}}$ as a *right* $\widehat{A}$-module, described in Example 2 above, that $(\widehat{A^{(I)}}, unif)$ is also a topological *right* $(\widehat{A}, unif)$-module.

Remark 1. The next exercise shows that one should, however, be a little cautious in assuming that bimodule structures are always topological bimodules.

Exercise 2: Let A be a right standard topological ring. Then show that: If the set I is of infinite cardinality $\geq$ card(A); if we regard A^I as a topological left A-module, with the topology $\bar{r}$ for which a complete system of neighborhoods of zero is $\{J^I : J$ is an open right ideal in $A\}$ — then, for the structure of A^J as an (A, A)-bimodule given by equation (1) of Example 2 we have that the following two conditions are equivalent:

(1) $(A^I, \bar{r})$ is a topological right A-module.

(2) There exists a complete system of neighborhoods of zero in A consisting of two-sided ideals.

Remark 2. Let A be an abstract ring such that A is not commutative. Then for I a non-empty set, the structure of $A^{(I)}$ as an (A, A)-bimodule in general *depends on the fixed basis* in $A^{(I)}$. That is: If F is a free abstract left A-module, and if A is a non-commutative ring, then there is in general no natural map to endow F with a natural structure as an (A, A)-bimodule. By Example 2, once we choose a basis for F as a free abstract left A-module, then such an (A, A)-bimodule structure is induced. But in general, if the ring A is not commutative, then different bases for F may yield different such right A-module structures on F. (E.g., if there exists an element $a \in A$ that admits a two-sided inverse, and such that a is not in the center of A, then the right module structure described above, on a free abstract left module, depends on the choice of basis, even for free abstract left A-modules of rank one.)

Remark 3. However, if A is a *commutative* topological ring, then of course the definition above of (A, A)-bimodule on $\widehat{A^{(I)}}$, and of an $(\widehat{A}, \widehat{A})$-bimodule structure on $\widehat{A^{(I)}}$, are independent of any choices of bases — since (see Example 1) the indicated structures as right modules simply coincide with the corresponding structures as left modules.

Theorem 5.3.1. *Let A be a standard topological ring and let I be a set. Then for every abstract left A-module M there is induced a homomorphism of abstract left $\widehat{A}$-modules*

$$(1): \quad \phi_M : \widehat{A^{(I)}} \underset{A}{\otimes} M \to C(M^{(I)}).$$

The homomorphisms ϕ_M are a natural transformation of functors from the category of abstract left A-modules into the category of abstract left $\hat{A}$-modules.

If M is an abstract left A-module of finite presentation, then the homomorphism ϕ_M is an isomorphism.

Proof. Let $\iota_i : M \to M^{(I)}$ be the ith canonical injection, for all $i \in I$, and let $\theta : M^{(I)} \to C(M^{(I)})$ be the natural map. Then consider the function

$$\phi : \widehat{A^{(I)}} \times M \to C(M^{(I)})$$

that maps $((a_i)_{i \in I}, \chi)$ into $\sum_{i \in I} a_i \cdot (\theta(\iota_i(\chi)))$, (where the infinite linear combination is taken relative to the infinite sum structure of the C-complete left A-module $C(M^{(I)})$). The function ϕ is an A-balanced map, and therefore defines a homomorphism

$$\phi_M : \widehat{A^{(I)}} \underset{A}{\otimes} M \to C(M^{(I)})$$

of left $\hat{A}$-modules as in equation (1). It is clear that the ϕ_M, for M an abstract left A-module, together are a homomorphism of functors. Finally let us show, for M of finite presentation, that ϕ_M is an isomorphism.

Since M is of finite presentation, there exists an exact sequence of abstract left A-modules

$$A^n \to A^m \to M \to 0.$$

Since the functors: $M \rightsquigarrow \widehat{A^{(I)}} \underset{A}{\otimes} M$ and $M \rightsquigarrow C(M^{(I)})$ are both right exact, to prove that ϕ_M is an isomorphism it therefore suffices to prove that ϕ_{A^n} and ϕ_{A^m} are isomorphisms. Since the functors: $M \rightsquigarrow \widehat{A^{(I)}} \underset{A}{\otimes} M$ and $M \rightsquigarrow C(M^{(I)})$ are also additive, it therefore suffices to prove the assertion in the case $M = A$. But then

$$\widehat{A^{(I)}} \underset{A}{\otimes} A = \widehat{A^{(I)}} = C(A^{(I)}).$$

$\square$

Corollary 5.3.2. *Let A be a standard topological ring and let M be an abstract left A-module of finite presentation. Then the natural homomorphism is an isomorphism of abstract left $\hat{A}$-modules*

$$\phi_M : \hat{A} \underset{A}{\otimes} M \overset{\approx}{\to} C(M).$$

Proof. Take I to be a set of cardinality one in Theorem 5.3.1. $\square$

Corollary 5.3.3. *Let A be a complete standard topological ring and let M be an abstract left A-module of finite presentation. Then*

(1) $M = C(M)$ in the sense of Chapter 3, §1, Definition 2.

(2) $\iota_M : M \to C(M)$ is an isomorphism of abstract left A-modules.

(3) There exists a unique infinite sum structure on the abstract left A-module M.

Proof. Conclusion (2) follows from Corollary 5.3.2. Conclusion (1) is equivalent to conclusion (2) by Chapter 3, §1, Definition 2. Conclusion (3) follows from Chapter 3, §1, Proposition 3.1.4. □

Remark 4. Conclusion (3) of Corollary 5.3.3 is identical to Chapter 2, §4, Corollary 2.4.12. The proof of Corollary 5.3.3, conclusion (3) is a different proof of the same result as Chapter 2, §4, Corollary 2.4.12.

Remark 5. In the terminology of Chapter 3, §1, Definition 2, the conclusion of Chapter 2, §4, Proposition 2.4.11, can be written:

If A is a complete standard topological ring, and if M is a finitely generated left A-module, then M admits an infinite sum structure iff $M = C(M)$.

Remark 6. By §1, Proposition 5.1.3, if A is a standard topological ring, then the functor C, from the category of abstract left A-modules into the category of C-complete left A-modules, preserves arbitrary direct sums. Therefore:

Let A be a complete standard topological ring and let M be an abstract left A-module of finite presentation (or, more generally, a finitely generated abstract left A-module that admits an infinite sum structure). Then for every set I, we have that

$$C(M^{(I)}) \approx \int_{i \in I} M$$

canonically as C-complete left A-modules.

Proof. By Corollary 5.3.3 (or by Remark 5 above) we have that $C(M) = M$. Therefore by §1, Proposition 5.1.3,

$$C(M^{(I)}) = \int_{i \in I} C(M) = \int_{i \in I} M.$$

□

Remark 7. Therefore, when the standard topological ring A is complete, the conclusion of the second paragraph of Theorem 5.3.1 can also be written:
M of finite presentation as abstract left A-module implies

$$\int_{i \in I} M = \widehat{A^{(I)}} \underset{A}{\otimes} M$$

as abstract left A-modules.

Definition 1. Let A be an abstract ring. Let M be an abstract left A-module and let n be a non-negative integer. Then we say that the abstract left A-module M *has a chain of syzygies of length n* iff there exists an exact sequence

$$F_n \to F_{n-1} \to \cdots \to F_1 \to F_0 \to M \to 0$$

of abstract left A-modules such that F_i is a free abstract left A-module of finite rank, $0 \le i \le n$.

Corollary 5.3.4. *Let A be a standard topological ring, let N be a non-negative integer and let M be an abstract left A-module such that M admits a chain of syzygies of length N. Then for every integer i, $0 \leq i < N$, the abstract left A-module*

$$\operatorname{Tor}_i^A(\widehat{A}, M)$$

has a natural infinite sum structure; and therefore has a natural structure as a C^A-complete left A-module.

Note: The proof also shows that, if M' is another abstract left A-module that admits a chain of syzygies of length N, and if $f : M \to M'$ is any A-linear homomorphism, then

$$\operatorname{Tor}_i^A(\widehat{A}, f) : \operatorname{Tor}_i^A(\widehat{A}, M) \to \operatorname{Tor}_i^A(\widehat{A}, M')$$

is infinitely A-linear, $0 \leq i < N$ for the natural structure of C^A-complete left A-modules so defined on the abstract left $\widehat{A}$-modules $\operatorname{Tor}_i^A(\widehat{A}, M)$ and $\operatorname{Tor}_i^A(\widehat{A}, M')$, $0 \leq i < N$.

Proof. Let F_* be a free acyclic homological resolution of M as abstract left A-module, such that F_i is free of finite rank, $0 \leq i \leq N$. Then

$$(1): \quad \operatorname{Tor}_i^A(\widehat{A}, M) \approx H_i(\widehat{A} \underset{A}{\otimes} F_*)$$

canonically, $0 \leq i < \infty$. Since $\widehat{A} \underset{A}{\otimes} F_i$ is a free $\widehat{A}$-module of finite rank, and in particular is a finitely presented abstract left $\widehat{A}$-module, by Corollary 5.3.3 $\widehat{A} \underset{A}{\otimes} F_i$ has a natural structure as a C^A-compete left A-module, $0 \leq i < N+1$; and, by Chapter 2, §4, Corollary 2.4.12, the boundary map: $\widehat{A} \underset{A}{\otimes} F_i \to \widehat{A} \underset{A}{\otimes} F_{i-1}$, being a map of finitely presented abstract left $\widehat{A}$-modules, is infinitely linear, $1 \leq i < N+1$. Therefore from equation (1), $\operatorname{Tor}_i^A(\widehat{A}, M)$ has a structure as a C^A-complete left A-module, $0 \leq i < N$.

Let M' and $f : M \to M'$ be as in the note, and let F_*' be an acyclic free homological resolution of M' as abstract left A-module such that F_i' is free of finite rank as abstract left A-module, $0 \leq i \leq N$. Regard $\operatorname{Tor}_i^A(\widehat{A}, M')$, $0 \leq i < N$, as C^A-complete left A-module by the isomorphism

$$(1'): \quad \operatorname{Tor}_i^A(\widehat{A}, M') \approx H_i(\widehat{A} \underset{A}{\otimes} F_*').$$

Then we show that $\operatorname{Tor}_i^A(\widehat{A}, f)$ is infinitely A-linear, $0 \leq i < N$.

In fact, let $\overline{f}_* : F_* \to F_*'$ be a map over f. Then by Chapter 2, §4, Corollary 2.4.12, $\widehat{A} \underset{A}{\otimes} \overline{f}_i : \widehat{A} \underset{A}{\otimes} F_i \to \widehat{A} \underset{A}{\otimes} F_i'$ is infinitely A-linear $0 \leq i < N+1$. Therefore by equations (1) and (1') $\operatorname{Tor}_i^A(\widehat{A}, f) = H_i(\widehat{A} \underset{A}{\otimes} \overline{f}_*)$ is infinitely A-linear $0 \leq i < N$.

Finally, we must prove that the infinite sum structure that we have defined on the abstract left $\widehat{A}$-module $\operatorname{Tor}_i^A(\widehat{A}, M)$, $0 \leq i < N$, is independent of the choice of a free acyclic homological resolution F_* of M such that F_i is free of finite rank as abstract left A-module, $0 \le i < N + 1$. In fact, if F'_* is another such, then taking $M' = M$, by the last paragraph we have the composite isomorphism:

$$H_i(\widehat{A} \underset{A}{\otimes} F_*) \approx \operatorname{Tor}_i^A(\widehat{A}, M) \approx H_i(\widehat{A} \underset{A}{\otimes} F'_*)$$

is infinitely A-linear, $0 \le i < N$.   □

Example 1. M has a chain of syzygies of length 0 iff M is finitely generated.
Example 2. M has a chain of syzygies of length 1 iff M is finitely presented.
Example 3. Let

$$0 \to K \to F \to M \to 0$$

be a short exact sequence of abstract left A-modules with F a free abstract left A-module of finite rank. Then for $n \ge 0$, M admits a chain of syzygies of length $n + 1$ iff K admits a chain of syzygies of length n.

Remark 8. Example 3 above is equivalent to:

Proposition 5.3.5. *Let M be an abstract left A-module such that M admits a chain of syzygies of length $n \ge 0$. Let*

$$(*) : \quad F_r \overset{d_r}{\to} F_{r-1} \overset{d_{r-1}}{\to} \cdots \overset{d_2}{\to} F_1 \overset{d_1}{\to} F_0 \to M \to 0$$

be an exact sequence of left A-modules, such that $0 \le r < n$ and such that F_i is free of finite rank. Then the chain (*) can be continued to a chain of syzygies of length n for M — i.e., there exist $F_{r+1}, \ldots, F_n$ free abstract left A-modules of finite rank and homomorphisms $d_n : F_n \to F_{n-1}, \ldots, d_{r+1} : F_{r+1} \to F_r$ such that the sequence

$$F_n \overset{d_n}{\to} F_{n-1} \overset{d_{n-1}}{\to} \cdots \overset{d_{r+2}}{\to} F_{r+1} \overset{d_{r+1}}{\to} F_r \overset{d_r}{\to} \cdots \overset{d_1}{\to} F_0 \to M \to 0$$

is exact.

Proof. We must show that the sequence (*) can be extended one step. If $r = 0$, then the assertion is identical to [COC], Introduction, Chapter 1, §6, Corollary 2.1, p. 69. So assume that $r \ge 1$. Let $K_i = \operatorname{Ker}(d_i)$, $1 \le i \le r$, $K_0 =$ (kernel of the epimorphism: $F_0 \to M$), $K_{-1} = M$. Then we have a short exact sequence:

$$(1) : \quad 0 \to K_r \to F_r \to K_{r-1} \to 0.$$

The sequence (*) can be extended one step iff K_r is finitely generated. By [COC], Chapter 1, §6, Corollary 2.1, p. 69, this latter is so iff K_{r-1} is finitely presented. By [COC], Chapter 1, §6, Corollary 2.2, an A-module N is finitely presented iff for every set I, the natural map:

$$A^I \underset{A}{\otimes} N \to N^I$$

is bijective.

Consider the short exact sequence

$$(2):\ 0 \to K_{r-1} \to F_{r-1} \to K_{r-2} \to 0.$$

Since F_{r-1} and K_{r-2} are of finite presentation, for every set I we have the commutative diagram with exact rows and columns:

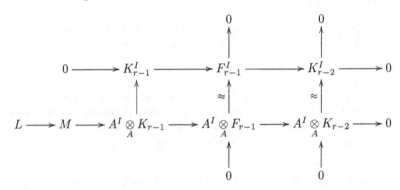

where $L = \mathrm{Tor}_1^A(A^I, F_{r-1})$ and $M = \mathrm{Tor}_1^A(A^I, K_{r-2})$. By diagram chasing (or the Five Lemma), it follows that the map $A^I \underset{A}{\otimes} K_{r-1} \to K_{r-1}^I$ is an epimorphism; and that the map $A^I \underset{A}{\otimes} K_{r-1} \to K_{r-1}^I$ is an isomorphism iff the map $\mathrm{Tor}_1^A(A^I, F_{r-1}) \to \mathrm{Tor}_1^A(A^I, K_{r-2})$ is an epimorphism. Since F_{r-1} is a free left A-module, the latter is equivalent to: $\mathrm{Tor}_1^A(A^I, K_{r-2}) = 0$. Thus, we have shown that:

(3): The chain of syzygies (*) can be extended one step iff K_{r-1} is of finite presentation iff $\mathrm{Tor}_1^A(A^I, K_{r-2}) = 0$ for all sets I.

But, from the short exact sequence

$$0 \to K_i \to F_i \to K_{i-1} \to 0,\ r \geq i \geq 0,$$

and the fact that F_i is free, we deduce that

$$\mathrm{Tor}_1^A(A^I, K_{r-2}) = \mathrm{Tor}_r^A(A^I, K_{-1}).$$

Since $K_{-1} = M$, this reads

$$\mathrm{Tor}_1^A(A^I, K_{r-2}) = \mathrm{Tor}_r^A(A^I, M).$$

Therefore, from (3),

(4): The chain of syzygies (*) can be extended one step iff $\mathrm{Tor}_r^A(A^I, M) = 0$ for all sets I.

However, the right side of equation (4) is independent of the choice of chain (*). Therefore, the chain (*) can be extended one step iff there exists some chain of syzygies for M of length r that can be extended one step — i.e., iff there exists some chain of syzygies for M of length $r + 1$.   □

Corollary 5.3.6. *Let A be an abstract ring, let M be an abstract left A-module and let n be an integer ≥ 0. Then M admits a chain of syzygies of length n iff there exists an exact sequence*

$$F_n \to F_{n-1} \to \cdots \to F_1 \to F_0 \to M \to 0$$

of abstract left A-modules such that F_i is flat and of finite presentation as left A-module.

Corollary 5.3.7. *Let A be a ring, n an integer ≥ 0 and*

$$0 \to K \to F \to M \to 0$$

a short exact sequence of left A-modules. Suppose that F is left flat and of finite presentation as a left A-module. Then M admits a chain of syzygies of length $n+1$ iff K admits a chain of syzygies of length n.

Lemma 5.3.8. *Let A be a ring, let M be an A-module, let n be an integer and suppose that there exists a sequence of length n as in the statement of Corollary 5.3.6. Let r be an integer such that $n > r \geq 0$ and let*

$$F_r \to F_{r-1} \to F_{r-2} \to \cdots \to F_1 \to F_0 \to M$$

be an exact sequence of left A-modules such that F_i is flat of finite presentation, $0 \leq i \leq r$. Then there exist $F_{r+1}, \ldots, F_n$ free left A-modules of finite rank and $F_n \to F_{n-1}, \ldots, F_{r+2} \to F_{r+1}$, $F_{r+1} \to F_r$ homomorphisms of left A-modules such that the sequence

$$F_n \to \cdots \to F_0 \to M$$

is exact.

Proof of Corollary 5.3.6, Corollary 5.3.7, and Lemma 5.3.8. It suffices to prove Lemma 5.3.8. And the proof of Lemma 5.3.8 is essentially identical to the proof of Proposition 5.3.5. □

The proof of Proposition 5.3.5 also shows:

Corollary 5.3.9. *Let A be a ring, let M be a left A-module and let n be an integer ≥ 1. Then M admits a chain of syzygies of length n iff for every set I,*

$$\mathrm{Tor}_i^A(A^I, M) = \begin{cases} M^I, & i = 0 \\ 0, & 1 \leq i \leq n-1, \end{cases}$$

Remark 9. Actually under the hypotheses of Corollary 5.3.9, if I is any fixed set of cardinality greater than or equal to the supremum of card(A) and card(M), then M admits a chain of syzygies of length n iff the Tor-condition of Corollary 5.3.9 holds for the fixed set I. (This follows from [COC], Introduction, Chapter 1, §6, Note to Lemma 2 (pp. 68 and 69), and Remark following Corollary 2.2 (pp. 70 and 71)).

Definition 2. Let A be a ring and let M be an abstract left A-module. Then we say that M *admits a chain of syzygies of infinite length* iff M admits a chain of syzygies of length n for every integer n. By Proposition 5.3.5, we have that

Corollary 5.3.10. *Let A be a ring and let M be an abstract left A-module. Then M admits a chain of syzygies of infinite length iff there exists an acyclic homological resolution F_* of M such that F_i is free of finite rank, $i \geq 0$ (or, in fact, by Corollary 5.3.6, it suffices to have such an F_* with F_i flat of finite presentation as left A-module, for $i \geq 0$).*

Also,

Corollary 5.3.11. *Let A be a ring, let M be an abstract left A-module and let $F_r \to F_{r-1} \to \cdots F_1 \to F_0 \to M \to 0$ be as in Lemma 5.3.8, with $r \geq 0$. Then M admits an infinite chain of syzygies iff there exists F_i, $i \geq r+1$, free of finite rank, and $F_i \to F_{i-1}$, $i \geq r+1$, homomorphisms of left A-modules, such that F_* is an acyclic homological resolution of M.*

The proofs of Corollaries 5.3.10 and 5.3.11 are left as an exercise (from Lemma 5.3.8).

Exercise 3: A ring is *left coherent* iff every finitely generated left ideal is finitely presented. By [COC], Introduction, Chapter 1, §6, Corollary 5.1, p. 75, a ring is left coherent iff the kernel of a mapping of finitely presented left modules is finitely presented. Using this, prove that:

Let A be an abstract ring. Then A is left coherent iff every finitely presented left A-module admits an infinite chain of syzygies. Also

A is left coherent iff, for every left A-module M, the following conditions are equivalent:

(1) M admits a chain of syzygies of length one.

(2) M admits a chain of syzygies of length two.

Theorem 5.3.12. *Let A be a standard topological ring and let M be an abstract left A-module. Let N be an integer ≥ 0. Suppose that M admits a chain of syzygies of length $\geq N+1$. Then for every set I there are induced canonical isomorphisms of abstract left $\widehat{A}$-modules*

$$\operatorname{Tor}_i^A(\widehat{A^{(I)}}, M) \approx C_i(M^{(I)}), \ 0 \leq i \leq N.$$

Proof. Let

$$(1): \ F_{N+1} \to F_N \to \cdots \to F_1 \to F_0 \to M \to 0$$

be a chain of syzygies of length $N+1$ — so that (1) is exact and F_i is free of finite rank, $0 \leq i \leq N+1$. Then by the definition of Tor,

$$(2): \ \operatorname{Tor}_i^A(\widehat{A^{(I)}}, M) = H_i(\widehat{A^{(I)}} \underset{A}{\otimes} F_*), \ 0 \leq i \leq N.$$

Taking the direct sum of the sequence (1) with itself I times, gives an exact sequence

$$(3): \ F_{N+1}^{(I)} \to F_N^{(I)} \to \cdots \to F_1^{(I)} \to F_0^{(I)} \to M^{(I)} \to 0.$$

Then $F_i^{(I)}$ is a free abstract left A-module, for all integers i, $0 \le i \le N + 1$. Since C_i, $i \ge 0$, are the left derived functors of the functor "A-adic completion," from equation (3) we have

$$(4): \ C_i(M^{(I)}) = H_i(\widehat{F_*^{(I)}}), \ 0 \le i \le N.$$

Since F_i is free of finite rank, in particular F_i is of finite presentation as left A-modlue. Therefore by Theorem 5.3.1

$$(5): \ \widehat{A^{(I)}} \underset{A}{\otimes} F_i \approx \widehat{F_i^{(I)}}, \ 0 \le i \le N + 1.$$

Equations (2), (4), and (5) complete the proof. $\qquad\square$

Corollary 5.3.13. *Let the hypotheses be as in Theorem 5.3.12. Then we have the canonical isomorphisms of abstract left $\widehat{A}$-modules*

$$C_i(M) \approx \mathrm{Tor}_i^A(\widehat{A}, M), \ 0 \le i \le N.$$

Proof. Take I to be a set of cardinality one in Theorem 5.3.12. $\qquad\square$

Corollary 5.3.14. *Let A be a complete standard topological ring, and let M be an abstract left A-module that admits a chain of syzygies of length $N + 1$, where $N \ge 0$. Then*

$$C_i(M) = \begin{cases} M, & i = 0 \\ 0, & 1 \le i \le N. \end{cases}$$

Proof. The hypotheses of Corollary 5.3.13 are satisfied. Since $\widehat{A} = A$, we have

$$\mathrm{Tor}_i^A(\widehat{A}, M) = \begin{cases} M, & i = 0 \\ 0, & i \ge 1. \end{cases}$$

$\qquad\square$

Remark 10. Notice that Corollary 5.3.14 in the special case $N = 0$ is another restatement of Chapter 2, §4, Corollary 2.4.12 (or equivalently of conclusion (3) of Corollary 5.3.3 above).

Corollary 5.3.15. *Let A be a standard topological ring such that A is left coherent as an abstract ring. Let M be an abstract left A-module of finite presentation. Then for all sets I, we have the canonical isomorphisms of abstract left A-modules.*

$$\mathrm{Tor}_i^A(\widehat{A^{(I)}}, M) \approx C_i(M^{(I)}), \ i \ge 0.$$

Proof. By Exercise 3, M admits an infinite chain of syzygies. The result follows from Theorem 5.3.12. $\qquad\square$

Corollary 5.3.16. *Under the hypotheses of Theorem 5.3.12, we have that the following two conditions are equivalent:*

$$(1): \ \text{Tor}_i^A(A^{(I)}, M) = \begin{cases} \widehat{M^{(I)}}, & i = 0 \\ 0, & N \geq i \geq 1 \end{cases}$$

$$(2): \ C_i(M^{(I)}) = \begin{cases} \widehat{M^{(I)}}, & i = 0 \\ 0, & N \geq i \geq 1. \end{cases}$$

Proof. Follows immediately from Theorem 5.3.12. $\qquad\qquad\square$

Theorem 5.3.17. *Let A be a standard topological ring such that A is left coherent as an abstract ring. Let I be a set. Then the following four conditions are equivalent:*

(1) $\widehat{A^{(I)}}$ is right flat as a right A-module.

(2) The assignment $M \rightsquigarrow C(M^{(I)})$ is an exact functor, from the category of finitely presented abstract left A-modules into the category of C-complete left A-modules.

(3) $C_i(M^{(I)}) = 0$ for all integers $i \geq 1$ and all finitely presented abstract left A-modules M.

(4) Condition (3) holds for $i = 1$.

Proof. Since the ring A is left coherent, therefore by [COC], Introduction, Chapter 1, Remark 1 following Corollary 5.1, pp. 75 and 76, we have that the category $\mathscr{A}$ with objects all finitely presented abstract left A-modules and all A-homomorphisms, is an abelian category such that the inclusion functor is exact, into the category of all left A-modules. Category $\mathscr{A}$ has enough projectives: namely, the free left modules of finite rank are "enough projectives" in the category $\mathscr{A}$.

The assignments: (1) $M \rightsquigarrow C_i(M^{(I)})$, $i \geq 0$, for M in $\mathscr{A}$, are clearly an exact connected sequence of functors from the abelian category $\mathscr{A}$ into the category of C-complete left A-modules. If F is a free abstract left A-module of finite rank, then $F^{(I)}$ is a free abstract left A-module, whence $C_i(F^{(I)}) = 0$ for $i \geq 0$. Therefore the assignments (1) are a system of derived functors from the abelian category with enough projectives $\mathscr{A}$ into the category of C-complete left A-modules. The equivalence of conditions (2), (3), and (4) follows.

(1) implies (2): By Corollary 5.3.15, the functors: $M \rightsquigarrow C(M^{(I)})$ and $M \rightsquigarrow \widehat{A^{(I)}} \underset{A}{\otimes} M$ are naturally equivalent. If $\widehat{A^{(I)}}$ is right flat, then the latter functor is exact.

(2) implies (1): Let J be a finitely generated left ideal in the ring A. Then since the ring A is left coherent, J is finitely presented. Therefore the inclusion: $J \rightarrow A$ is a map in the category A. By condition (2) it follows that

$$C(J^{(I)}) \rightarrow C(A^{(I)})$$

is injective. Therefore, by Corollary 5.3.15,

$$\widehat{A^{(I)}} \underset{A}{\otimes} J \to \widehat{A^{(I)}} \underset{A}{\otimes} A = \widehat{A^{(I)}}$$

is injective. This being true for all finitely generated left ideals J in A, by [COC], Introduction, Chapter 1, Lemma 4, p. 72, we have that $\widehat{A^{(I)}}$ is right flat as a right A-module. $\qquad\square$

Remark 11. Suppose that we have the hypotheses of Theorem 5.3.17, and that the set I is non-empty. Then a necessary condition for the four equivalent conditions of Theorem 5.3.17 to hold is that $\widehat{A}$ is right flat as a right A-module.

Proof. $\widehat{A}$ is a direct summand of $\widehat{A^{(I)}}$ as a right module. The result follows from condition (1). $\qquad\square$

Lemma 5.3.18. *Let A be a standard topological ring. Then the following two conditions are equivalent.*

(1) $n \geq 0$, N a finitely generated abstract left A-submodule of A^n implies that the A-adic topology of N is the induced topology from the A-adic topology of A^n.

(2) M a finitely presented abstract left A-module, N a finitely generated submodule of M, implies that the A-adic topology of N is the induced topology from the A-adic topology of M.

Proof. (2) implies (1): Obvious.

(1) implies (2): Choose an epimorphism:

$$\rho : A^n \to M.$$

Since M is finitely presented, $K = \texttt{Ker}(\rho)$ is finitely generated. Therefore $N' = \rho^{-1}(N)$ is finitely generated.

(*Proof.* We have the short exact sequence: $0 \to K \to N' \to N \to 0$, with K and N finitely generated. If $x_1, \ldots, x_r$ generate K, and $x_{r+1}, \ldots, x_n \in N'$ are such that $\rho(x_{r+1}), \ldots, \rho(x_n)$ generate N, then one verifies that $x_1, \ldots, x_n$ generate N'.)

But then

$$(1): \quad M \approx A^n/K, \ N \approx N'/K.$$

Since by (1) N' has the induced topology from A^n, it follows easily that N has the induced topology from M.

(*Proof.* Since N' has the induced topology from A^n, for every open right ideal I in A there exists an open right ideal J in A such that

$$IN' \supset (J \cdot A^n) \cap N'.$$

The pre-image of $J \cdot M$ in A^n is $J \cdot A^n + K$. If $y \in (J \cdot M) \cap N$, choose $\chi \in A^n$ such that $\rho(\chi) = y$. Then $\chi \in J \cdot A^n + K$, say $\chi = \chi_0 + k_0$, $\chi_0 \in J \cdot A^n$, $k_0 \in K$. Since $\rho(\chi) = y \in N$, $\chi \in N'$, therefore $\chi_0 = \chi - k_0 \in N' + K = N'$. Thus $\chi_0 \in (J \cdot A^n) \cap N' \subset IN'$. Hence $y = \rho(\chi) = \rho(\chi_0) + \rho(k_0) = \rho(\chi_0) \in I \cdot N$. Thus $(J \cdot M) \cap N \subset I \cdot N$.) $\qquad\square$

In general, given a function $f : Y \to X$ of topological spaces, we say that *the topology of Y is induced from X* iff the topology of Y is $\{f^{-1}(U) : U$ is open in $X\}$.

Lemma 5.3.19. *Let A be a standard topological ring. Then the two conditions of Lemma 5.3.18 are each equivalent to each of the following two conditions:*

(3) $n \geq 0$, N a finitely generated abstract left A-submodule of A^n, implies that the $\widehat{A}$-adic topology of $\widehat{A} \underset{A}{\otimes} N$ is the induced topology from the $\widehat{A}$-adic topology of $\widehat{A}^n$.

(4) M a finitely presented abstract left A-module, N a finitely generated submodule of M, implies that the $\widehat{A}$-adic topology of $\widehat{A} \underset{A}{\otimes} N$ is the induced topology from the $\widehat{A}$-adic topology of $\widehat{A} \underset{A}{\otimes} M$.

Note: In this Lemma, as always unless there is explicit mention otherwise, $\widehat{A}$ is regarded as a topological ring with the *uniform topology*.

Proof. First, I claim that if N is an arbitrary abstract left A-module, then the $\widehat{A}$-adic completion of the abstract left $\widehat{A}$-module $\widehat{A} \underset{A}{\otimes} N$, together with its uniform topology (i.e., with the topology that makes it the completion of $(\widehat{A} \underset{A}{\otimes} N, \widehat{A}$-adic$)$ as a uniform space), is $(\widehat{N}, unif)$.

(*Proof.* For each open right ideal I in A, let $\overline{I}$ be the closure of the image of I in $\overline{A} = (\widehat{A}, unif)$. Then

$$(\widehat{A}/\overline{I}) \underset{\widehat{A}}{\otimes} (\widehat{A} \underset{A}{\otimes} N) = (\widehat{A}/\overline{I}) \underset{A}{\otimes} N = (A/I) \underset{A}{\otimes} N.$$

Taking the inverse limit as topological groups for I an open right ideal in A gives the assertion.)
Therefore
   (4) The $\widehat{A}$-adic topology of $\widehat{A} \underset{A}{\otimes} N$ is induced from the topology of $(\widehat{N}, unif)$.

Since $(\widehat{N}, unif)$ is the completion of $(N, A$-adic$)$ as a topological group,
   (5) The A-adic topology of N is induced from the topology of $(\widehat{N}, unif)$.
Equations (4) and (5), applied to the commuative diagram:

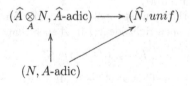

implies that
   (6) The A-adic topology of N is induced from the $\widehat{A}$-adic topology of $\widehat{A} \underset{A}{\otimes} N$.

From equation (6) it is clear that (4) implies (2).

(2) implies (4): Since the A-adic topology of N is induced from the A-adic topology of M, it follows that the topology of $(\widehat{N}, unif)$ is induced from the topology of $(\widehat{M}, unif)$. But then by equation (5), applied to M and N, it follows that the A-adic topology of N is induced from the A-adic topology of M. The equivalence of (1) and (3) is a special case of the equivalence of (2) and (4). □

Remark 12. The proof of Lemma 5.3.19 actually showed a bit more, namely

Corollary 5.3.20. *Let A be a standard topological ring and let $f : N \to M$ be a homomorphism of abstract left A-modules. Then the following three conditions are equivalent:*

(1) The A-adic topology of N is induced from the A-adic topology of M.

(2) The $\widehat{A}$-adic topology of $\widehat{A} \underset{A}{\otimes} N$ is induced from the $\widehat{A}$-adic topology of $\widehat{A} \underset{A}{\otimes} M$.

(3) The topology of $(\widehat{N}, unif)$ is induced from the topology of $(\widehat{M}, unif)$.

Also, in the course of the proof of Lemma 5.3.19, we have observed

Corollary 5.3.21. *Let A be a standard topological ring and let N be an abstract left A-module. Then*

(1) The A-adic topology of N is induced from the $\widehat{A}$-adic topology of $\widehat{A} \underset{A}{\otimes} N$.

And

(2) The completion of the topological left $\widehat{A}$-module $(\widehat{A} \underset{A}{\otimes} N, \widehat{A}$-adic) is the topological left $\widehat{A}$-module $(\widehat{N}, unif)$.

Lemma 5.3.22. *Let A be a standard topological ring, such that $\widehat{A}$ is admissible, and such that $\widehat{A}$ admits a denumerable neighborhood base at zero. Then conditions (5b) and (6b) below are equivalent. Let I be any fixed infinite set. Then the four equivalent conditions (1), (2), (3), (4) of Lemmas 5.3.18 and 5.3.19, imply each of conditions (5b) and (6b) below. Conversely, if $\widehat{A}$ is right flat as an abstract right A-module, then conditions (1), (2), (3), (4), (5b), and (6b) are all equivalent.*

(5b) M is a finitely presented left A-module implies that the natural infinitely linear function

$$\theta : \int_{i \in I} (\widehat{A} \underset{A}{\otimes} M) \to \prod_{i \in I} (\widehat{A} \underset{A}{\otimes} M)$$

is injective.

(6b) M a finitely presented abstract left A-module implies that the natural infinitely linear function

$$C(M^{(I)}) \to (\widehat{A} \underset{A}{\otimes} M)^I$$

is injective.

Proof. Let M be a finitely presented left A-module and let

$$(1):\ 0 \to N \to A^n \to M \to 0$$

be a short exact sequence of abstract left A-modules. Then N is finitely generated. Tensoring (1) over A with $\widehat{A}$ on the left gives the exact sequence

$$(2):\ \widehat{A} \underset{A}{\otimes} N \to (\widehat{A})^n \to (\widehat{A} \underset{A}{\otimes} M) \to 0.$$

Let N' be the image of the map: $\widehat{A} \underset{A}{\otimes} N \to (\widehat{A})^n$. Then we have the short exact sequence of abstract left A^n-modules

$$(3):\ 0 \to N' \to (\widehat{A})^n \to (\widehat{A} \underset{A}{\otimes} M) \to 0.$$

The map: $\widehat{A} \underset{A}{\otimes} N \to (\widehat{A})^n$ factors: $\widehat{A} \underset{A}{\otimes} N \to N' \to (\widehat{A})^n$. Therefore, if condition (3) of Lemma 5.3.19 holds, so that the $\widehat{A}$-adic topology of $\widehat{A} \underset{A}{\otimes} N$ is induced from the $\widehat{A}$-adic topology of $(\widehat{A})^n$, then the $\widehat{A}$-adic topology of N' is induced from the $\widehat{A}$-adic topology of $(\widehat{A})^n$. But then, by §2, Proposition 5.2.11 (with $\widehat{A}$ replacing A, $\widehat{A} \underset{A}{\otimes} M$ replacing M, $(\widehat{A})^n$ replacing F and N' replacing K), we have condition (5b).

Conversely, suppose that (5b) holds and that $\widehat{A}$ is right flat. Let n and N be as in condition (3) of Lemma 5.3.19. Let $M = A^n/N$. Then tensoring on the left with $\widehat{A}$ we have the short exact sequence

$$(4):\ 0 \to \widehat{A} \underset{A}{\otimes} N \to (\widehat{A})^n \to (\widehat{A} \underset{A}{\otimes} M) \to 0$$

of C-complete left A-modules. By §2, Proposition 5.2.11, (with $\widehat{A}$ replacing A), we therefore have that the $\widehat{A}$-adic topology of $\widehat{A} \underset{A}{\otimes} N$ is induced from the $\widehat{A}$-adic topology of $(\widehat{A})^n$, proving (3). Therefore if $\widehat{A}$ is right flat, then (5b) iff (3).

Finally, whether $\widehat{A}$ is right flat or not, by §1, Proposition 5.1.3 and Corollary 5.3.2 above we have that

$$C(M^{(I)}) \approx \int_{i \in I} (\widehat{A} \underset{A}{\otimes} M).$$

Therefore conditions (5b) and (6b) are always equivalent.   □

Remark 13.   In Lemma 5.3.22 above, of course conditions (5b) and (6b) are equivalent for any standard topological ring A (admissible or not, having a denumerable neighborhood base at zero or not) since the left sides of equations (5b) and (6b) are canonically isomorphic, for any standard topological ring A and any finitely presented abstract left A-module M.

Theorem 5.3.23. *Let A be a standard topological ring. Suppose that*
(1) A is left coherent as an abstract ring.
(2) $\widehat{A}$ is admissible.
(3) A has a denumerable neighborhood base at zero.
Then the following six conditions are equivalent:
(1) Condition (1) of Lemma 5.3.18
(2) Condition (2) of Lemma 5.3.18
(3) Condition (3) of Lemma 5.3.19
(4) Condition (4) of Lemma 5.3.19
(5) Conditions (5a) and (5b) below both hold:
(5a) $\widehat{A}$ is right flat as an abstract right A-module.
(5b) Condition (5b) of Lemma 5.3.22 holds.
(6) Conditions (6a) and (6b) below both hold:
(6a) $\widehat{A}$ is right flat as an abstract right A-module.
(6b) Condition (6b) of Lemma 5.3.22 holds.
Moreover, when these six equivalent conditions hold, then the four equivalent conditions of Theorem 5.3.17 all hold.

Corollary 5.3.24. *The hypotheses being as in Theorem 5.3.23, suppose that the six equivalent conditions of Theorem 5.3.23 hold. Then for every left finitely presented abstract A-module M, we have canonical isomorphisms of C-complete left A-modules*

$$\widehat{A} \otimes_A M \approx C(M) \approx \widehat{M}.$$

Proof of Theorem 5.3.23 and Corollary 5.3.24. Let

$$(1): \quad 0 \to N \to M \to Q \to 0$$

be a short exact sequence of finitely presented abstract left A-modules. Then in the short exact sequence (1), the A-adic topology of Q is the quotient topology from the A-adic topology of M. (This is always true for any epimorphism of abstract left A-modules, as we have observed in Chapter 1.) That is,

$$(2): \quad (Q, A\text{-adic}) \approx (M, A\text{-adic})/N$$

as topological groups. By the universal mapping property definition of the completion $(\widehat{Q}, \text{unif})$ of the topological group $(Q, A\text{-adic})$, it follows that

(2) $(\widehat{Q}, unif)$ is the completion of the topological group $(\widehat{M}, unif)/\overline{N}$, where $\overline{N}$ is the closure of the image of N in $(\widehat{M}, unif)$. Since this closure $\overline{N}$ is obviously closed in $(\widehat{M}, \text{unif})$, we have that $(\widehat{M}, unif)/\overline{N}$ is a Hausdorff quotient of the complete topological left $\widehat{A}$-module $(\widehat{M}, unif)$. By hypothesis (3) A admits a denumerable neighborhood base at zero. Therefore there exist open right ideals $J_1, J_2, J_3, \ldots$ in A that are a complete system of neighborhoods of zero in A. And then the closures of the images of $J_1 \cdot M, J_2 \cdot M, J_3 \cdot M, \ldots$ in the complete topological $\widehat{A}$-modoule $(\widehat{M}, unif)$ are a complete systerm of neighborhoods of

zero in $(\widehat{M}, unif)$. Therefore $(\widehat{M}, unif)$ admits a denumerable neighborhood base at zero. But then,

$$(3):\ (\widehat{M}, unif)/\overline{N}$$

is a Hausdorff quotient of a complete topological group $(\widehat{M}, unif)$ such that $(\widehat{M}, unif)$ admits a denumerable neighborhood base at zero. The proof of Chapter 3, §2, Corollary 3.2.22 actually shows that: "The Hausdorff quotient H of complete topological abelian group G, when G admits a denumerable neighborhood base at zero, is such that the quotient topological group H is complete." Therefore the topological group (3) is complete. By equation (2), it follows that

$$(4):\ (\widehat{Q}, unif) = (\widehat{M}, unif)/\overline{N}$$

as topological groups.

We first show that condition (2) of Lemma 5.3.18 implies the conclusion of Corollary 5.3.24. In fact, suppose that we have the short exact sequence (1) of finitely presented abstract left A-modules. Then if condition (2) of Lemma 5.3.18 holds, we have that the A-adic topology of N is the induced topology from the A-adic topology of M. It follows that the closure $\overline{N}$ of the image of N in the completion $(\widehat{M}, unif)$ of the topological group $(M, A$-adic$)$ obeys all the axioms for a completion of the topological group $(N, A$-adic$)$. That is,

$$(5):\ \widehat{N} = \overline{N},$$

where $\overline{N}$ is, as in equation (4), the closure of the image of N in $(\widehat{M}, unif)$. Combining equations (4) and (5), it follows that the sequence

$$(6):\ 0 \to \widehat{N} \to \widehat{M} \to \widehat{Q} \to 0$$

is exact in the category of C-complete left A-modules.

Let $\mathscr{A}$ be the category of finitely presented abstract left A-modules and let $\mathscr{C}_A$ be the category of C-complete left A-modules. Then by equation (6) the functors "A-adic completion" and C are both right exact functors from $\mathscr{A}$ into $\mathscr{C}_A$. We also have the natural transformation of functors

$$\rho_M : C(M) \to \widehat{M}$$

for M in $\mathscr{A}$. If F is free of finite rank, then clearly $C(F) = \widehat{F}$. Given any M in $\mathscr{A}$, since M is finitely presented, choose an exact sequence

$$F \to G \to M \to 0$$

with F and G free of finite rank. Then we have the commutative diagram with exact rows

$$
\begin{array}{ccccccc}
\widehat{F} & \longrightarrow & \widehat{G} & \longrightarrow & \widehat{M} & \longrightarrow & 0 \\
\approx \uparrow{\rho_F} & & \approx \uparrow{\rho_G} & & \uparrow{\rho_M} & & \\
C(F) & \longrightarrow & C(G) & \longrightarrow & C(M) & \longrightarrow & 0
\end{array}
$$

Therefore ρ_M is an isomorphism. We have shown that

$$(7): \quad C(M) \approx \widehat{M}$$

canonically for all M in $\mathscr{A}$. Equation (7), and Corollary 5.3.2, are the conclusion of Corollary 5.3.24.

Condition (2) of Lemma 5.3.18 implies condition (5): If condition (2) holds, then we have observed that, for every short exact sequence (1) in $\mathscr{A}$, the sequence (6) is exact in $\mathscr{C}_A$. That is, the functor "A-adic completion" is exact. By the conclusion of Corollary 5.3.24, it follows that the functor

$$(8): \quad M \rightsquigarrow \widehat{A} \underset{A}{\otimes} M$$

is exact from $\mathscr{A}$ into $\mathscr{C}_A$.

Let I be a finitely generated left ideal in the ring A. Then since the ring A is left coherent, we have that I is finitely presented as an abstract left A-module. Therefore I, and the inclusion $I \to A$, is a map, in fact a monomorphism, in the category $\mathscr{A}$. Since the functor (8) is exact, therefore the image under (8) $\widehat{A} \underset{A}{\otimes} I \rightsquigarrow \widehat{A}$, is a monomorphism in $\mathscr{C}_A$, and is therefore injective. I being an arbitrary finitely generated left ideal in the ring A, by [COC], Introduction, Chapter 1, Lemma 4, p. 72, we have that $\widehat{A}$ is right flat as an abstract right A-module. Therefore condition (2) implies condition (5a). By Lemma 5.3.22, condition (2) implies condition (5b). Therefore condition (2) of Lemma 5.3.18 implies condition (5).

Condition (5) implies condition (2): This is part of the statement of Lemma 5.3.22. By Lemmas 5.3.18 and 5.3.19, conditions (1), (2), (3), and (4) are equivalent. And by Lemma 5.3.22, conditions (5) and (6) are equivalent. Therefore conditions (1), (2), (3), (4), (5), and (6) are equivalent.

It remains to show that, if conditions (1) - (6) hold, then the four equivalent conditions of Theorem 5.3.17 hold. We prove that (6) implies (condition (2) of Theorem 5.3.17): In fact, assume (6). Then in particular we have condition (6a), that $\widehat{A}$ is right-flat as an abstract right A-module. Therefore the functor: $M \rightsquigarrow \widehat{A} \underset{A}{\otimes} M$ is exact from $\mathscr{A}$ into $\mathscr{C}_A$. Since the functor "I-fold direct product" is of course exact, it follows that the composite functor: $M \rightsquigarrow (\widehat{A} \underset{A}{\otimes} M)^I$ is an exact functor from $\mathscr{A}$ into $\mathscr{C}_A$. By condition (6b), the functor: $M \rightsquigarrow C(M^{(I)})$ is a subfunctor. Therefore the functor: $M \rightsquigarrow C(M^{(I)})$ preserves monomorphisms. Since this functor is also obviously right exact (being the composite of the exact functor $M \rightsquigarrow M^{(I)}$ and the right exact functor C), it follows that the functor $M \rightsquigarrow C(M^{(I)})$ from $\mathscr{A}$ into $\mathscr{C}_A$ is exact. And this proves condition (2) of Theorem 5.3.17. $\qquad\square$

Corollary 5.3.25. *Let the hypotheses be as in Theorem 5.3.23, and suppose that the six equivalent conditions of Theorem 5.3.23 hold. Then*

(6b′) For every finitely presented abstract left A-module M and every set I, we have that

$$C(M^{(I)}) = \widehat{M^{(I)}}.$$

Proof. By Corollary 5.3.24, we can assume that I is infinite. We first show that

$$(1): \ C(M^{(I)}) = \{(\chi_i)_{i\in I} \in (\widehat{M})^I : (\chi_i)_{i\in I} \text{ converges to zero in}$$

$$\widehat{M} \text{ for the } \widehat{A}\text{-adic topology}\}.$$

By Corollary 5.3.24, $C(M) = \widehat{A} \underset{A}{\otimes} M$. Since the functor C (from the category of abstract left A-modules to the category of C-complete left A-modules) preserves direct sums, therefore $C(M^{(I)}) = \underset{i\in I}{\int} (\widehat{A} \underset{A}{\otimes} M)$. By condition (6b), this latter is equal to its image in $\underset{i\in I}{\prod} (\widehat{A} \underset{A}{\otimes} M)$. But by §2, Definition 1, this image is $\underset{i\in I}{\mathcal{Z}} (\widehat{A} \underset{A}{\otimes} M)$. By §2, Theorem 5.2.3 (with $\widehat{A}$ replacing A and $M_i = \widehat{A} \underset{A}{\otimes} M$ for all $i \in I$), this latter is equal to

$$\{(\chi_i)_{i\in I} \in \prod_{i\in I}(\widehat{A} \underset{A}{\otimes} M) : \ (\chi_i)_{i\in I} \text{ converges to zero in } \widehat{A} \underset{A}{\otimes} M$$
$$\text{for the } \widehat{A}\text{-adic topology}\}.$$

And since by Corollary 5.3.24 we have $\widehat{A} \underset{A}{\otimes} M = \widehat{M}$, this proves equation (1).

Next we show that

$$(2): \ \{(\chi_i)_{i\in I} \in (\widehat{M})^I : (\chi_i)_{i\in I} \text{ converges to zero in } \widehat{M} \text{ for the } \widehat{A}\text{-adic topology}\}$$

$$= \widehat{M^{(I)}}.$$

By hypotheses (2) and (3) of Theorem 5.3.23, $\widehat{A}$ is admissible and has a denumerable neighborhood base. Therefore (Chapter 1, §3, Remark 1) $\widehat{A}$ is strongly admissible. Therefore by conclusion (2) of Lemma 5.3.27 below, the uniform and $\widehat{A}$-adic topologies coincide in $\widehat{M}$. Therefore equation (2) follows from Corollary 5.3.29 below.   □

Remark 14. Let A be a standard topological ring such that hypotheses (2) and (3) of Theorem 5.3.23 hold. Then condition (6b′) of Corollary 5.3.25 holds iff

(a) $C(M) = \widehat{M}$, for all finitely presented abstract left A-modules M; and
(b) Condition (6b) of Theorem 5.3.23 holds.

Proof. The proof of Corollary 5.3.25 shows sufficiency. To prove necessity. Assume (6b′) of Corollary 5.3.25. Then letting I be a set of cardinality one, we obtain (a) above. By Corollary 5.3.29 below, we have that

$$\widehat{M^{(I)}} \subset (\widehat{M})^I.$$

Also, by (a) above and Corollary 5.3.2, $\widehat{M} = C(M) = \widehat{A} \underset{A}{\otimes} M$. Therefore

$$\widehat{M^{(I)}} \subset (\widehat{A} \underset{A}{\otimes} M)^I.$$

Combining with (6b'), we obtain (6b) of Theorem 5.3.23.   □

Corollary 5.3.26. *Under the hypotheses of Theorem 5.3.23, the six equivalent conditions of Theorem 5.3.23 are also equivalent to:*
(6') Both (6a') and (6b') below hold.
(6a') $\widehat{A}$ is right flat as an abstract right A-module.
(6b') Condition (6b') of Corollary 5.3.25 holds.

Proof. Corollary 5.3.25 proves necessity of (6'). Conversely, if (6') holds, then by Remark 14, condition (6b) of Theorem 5.3.23 holds. And since (6a') and (6a) are identical, therefore condition 6 of Theorem 5.3.27 holds.   □

Remark 15. Hypothesis (2) of Theorem 5.3.23 is:
(2) The standard topological ring $\widehat{A}$ is admissible.
By Chapter 1, §3, Proposition 1.3.1, a sufficient condition for this is that:
(2') The standard topological ring A is admissible. In fact, by Chapter 1, §3, Proposition 1.3.1, condition (2') above is equivalent to hypothesis (2) of Theorem 5.3.23 holding, and *in addition* that
(2a) The uniform and A-adic topologies coincide in $\widehat{A}$. Clearly, if (2a) is the case, then for every abstract left $\widehat{A}$-module N, we have that the A-adic and $\widehat{A}$-adic topologies of N coincide (since then an open neighborhood base in $\widehat{A}$ is $\{I \cdot \widehat{A} : I$ an open right ideal in $A\}$). Therefore, if the stronger condition (2') above should hold, then: In the conclusion (6b') of Corollary 5.3.25; and in conditions (3) and (4) of Lemma 5.3.19 (and of Theorem 5.3.23) one can *replace* the phrase "$\widehat{A}$-adic topology" with the phrase "A-adic topology."

Lemma 5.3.27. *Let A be a standard topological ring such that $\widehat{A}$ is admissible (respectively: strongly admissible). Then for every free abstract left A-module M (respectively: for every abstract left A-module M) we have that*
(1) The A-adic completion $\widehat{M}$ of M is complete for its $\widehat{A}$-adic topology.
(2) The $\widehat{A}$-adic and uniform topologies coincide in $\widehat{M}$.

Note: In fact if A is a standard topological ring then $\widehat{A}$ is admissible iff condition (1) holds for all free abstract left A-modules M iff condition (2) holds for all free abstract left A-modules M.

Proof. By Corollary 5.3.21, conclusion (2), we have that $(\widehat{M}, unif)$ is the completion of $(\widehat{A} \underset{A}{\otimes} M, \widehat{A}$-adic). Since $\widehat{A}$ is admissible (respectively: stongly admissible), the completion $(\widehat{M}, unif)$ of the topological $\widehat{A}$-module $(\widehat{A} \underset{A}{\otimes} M, \widehat{A}$-adic) is such that the topology is the $\widehat{A}$-adic. Therefore the $\widehat{A}$-adic and uniform topologies coincide in $\widehat{M}$, proving (2). Since $(\widehat{M}, unif)$ is complete, and by

(2) the uniform and $\widehat{A}$-adic topologies of $\widehat{M}$ coincide, therefore $\widehat{M}$ is $\widehat{A}$-adically complete.   □

Proof of the Note. If $M = A^{(I)}$, then as an abstract $\widehat{A}$-module $\widehat{M}$ is also the $\widehat{A}$-adic completion of $(\widehat{A})^{(I)}$. (This follows from Corollary 5.3.29 below, applied to the A-module $M = A$, and also to the $\widehat{A}$-module $M = (\widehat{A})$.) Therefore condition (1) asserts that the $\widehat{A}$-adic completion of a free $\widehat{A}$-module is $\widehat{A}$-adically complete, and therefore (Chapter 1, §3, Definition 1) is equivalent to the assertion that $\widehat{A}$ is admissible. Since as we have seen condition (2) implies condition (1), likewise condition (2) implies that $\widehat{A}$ is admissible.   □

Lemma 5.3.28. *Let A be a standard topological ring, let I be a set and let M_i be an abstract left A-module for all $i \in I$. Then*

$$\widehat{\underset{i \in I}{\oplus} M_i} = \{(\chi_i)_{i \in I} \in \prod_{i \in I} \widehat{M_i} : \text{ For every right open ideal } J \text{ in } A$$

$$\text{we have that } \chi_i \in \overline{J \cdot M_i} \text{ for all but finitely many } i \in I\}$$

where in this formula $\overline{J \cdot M_i}$ denotes the closure of the image of $J \cdot M_i$ in $\widehat{M_i}$ for the uniform topology.

Proof. Let $\bar{r}$ be the topology on $\prod_{i \in I} \widehat{M_i}$ such that $\prod_{i \in I} \widehat{M_i}$ becomes a topological left $\widehat{A}$-module, such that a complete system of neighborhoods of zero is $\{ \prod_{i \in I} \overline{J \cdot M_i} : J$ is an open right ideal in $A\}$. Then $(\prod_{i \in I} \widehat{M_i}, \bar{r})$ is complete.

(*Proof.* The completion of $(\prod_{i \in I} \widehat{M_i}, \bar{r})$ is

$$\varprojlim_{J \in \mathscr{R}} \left((\prod_{i \in I} \widehat{M_i}) / (\prod_{i \in I} \overline{J \cdot M_i}) \right) = \varprojlim_{J \in \mathscr{R}} \prod_{i \in I} (\widehat{M_i} / \overline{J \cdot M_i}) = \varprojlim_{J \in \mathscr{R}} \prod_{i \in I} M_i / (J \cdot M_i)$$

$$= \prod_{i \in I} \varprojlim_{J \in \mathscr{R}} M_i / J M_i = \prod_{i \in I} \widehat{M_i},$$

where $\mathscr{R}$ is the collection of open right ideals in A.)

Moreover, for the natural homomorphism: $h : \underset{i \in I}{\oplus} M_i \to (\prod_{i \in I} \widehat{M_i}, \bar{r})$, the topology induced in $\underset{i \in I}{\oplus} M_i$ is the A-adic topology.

(*Proof.* For every open right ideal J, the pre-image of the basic open neighborhood of zero $\prod_{i \in I} \overline{J \cdot M_i}$ under h is $\{(\chi_i)_{i \in I} \in \prod_{i \in I} M_i : \chi_i = 0$ for all but finitely many $i \in I$, and such that the image of χ_i in $\widehat{M_i}$ is in $\overline{J \cdot M_i}\}$. Since the map: $M_i / (J \cdot M_i) \to \widehat{M_i} / \overline{J \cdot M_i}$ is a bijection, the condition "the image of χ_i in $\widehat{M_i}$ is in $J \cdot M_i$" is equivalent to "$\chi_i \in J \cdot M_i$". Therefore the pre-image of $\prod_{i \in I} \overline{J \cdot M_i}$

under h is $\{(\chi_i)_{i \in I} \in \prod_{i \in I} M_i : \chi_i = 0$ for all but finitely many $i \in I$, and such that $\chi_i \in J \cdot M_i$ for all $i \in I\}$. And this latter clearly coincides with $J \cdot \left(\bigoplus_{i \in I} M_i \right)$.)

Therefore, by the axiomatic characterization of the completion of a topological group the A-adic completion of $\bigoplus_{i \in I} M_i$ is the closure of the image of h in $(\prod_{i \in I} \widehat{M_i}, \bar{r})$. □

Corollary 5.3.29. *Let A be a standard topological ring, let I be an infinite set and let M be an abstract left A-module. Then $\widehat{M^{(I)}} = \{(\chi_i)_{i \in I} \in (\widehat{M})^I : (\chi_i)_{i \in I}$ converges to zero in $\widehat{M}$ for the uniform topology of $\widehat{M}\}$.*

Proof. By Lemma 5.3.28, $\widehat{M^{(I)}} = \{(\chi_i)_{i \in I} \in (\widehat{M})^I$ such that, for every open right ideal J in A, we have that $\chi_i \in \overline{J \cdot M}$ for all but finitely many $i \in I\}$ (where $\overline{J \cdot M}$ is as in the statement of Lemma 5.3.28). But $\{\overline{J \cdot M} : J$ is an open right ideal in $A\}$ is a complete system of neighborhoods of zero in $(\widehat{M}, unif)$. Therefore $(\chi_i)_{i \in I}$ is such that, for all J, $\chi_i \in \overline{J \cdot M}$ for all but finitely many $i \in I$, iff $(\chi_i)_{i \in I}$ converges to zero in $(\widehat{M}, unif)$. □

5.4 The Special Properties of Noetherian Rings

Theorem 5.4.1. *(Krull) Let A be a standard topological ring such that the abstract ring A is a Noetherian commutative ring. Suppose also that*

(1) I is an open ideal in A implies that the ideal I^2 is open in A. Then, if M is a finitely generated abstract left A-module, and if N is an A-submodule of M, then we have that:

The A-adic topology of N is induced from the A-adic topology of M.

Note: Another way of stating the conclusion of the Theorem is: "then condition (2) of §3, Theorem 5.3.23, holds."

Proof. Case 1. There exists an ideal I in A, such that the topology of A is given by the positive powers of the ideal I. Then the Theorem is a very well-known theorem due originally to Krull. E.g., see [NB], Chapter 3, §3.2, Theorem 2, pp. 199 and 200.

Note: Another way of writing the conclusion of Case 1 is that, if A is any abstract commutative Noetherian ring, if M is any finitely generated abstract left A-module, and if N is any submodule of N, then for every ideal I in the ring A, and every integer $n \geq 0$, there exists an integer $m \geq 0$ such that

$$(2): \ I^n \cdot N \supset (I^m \cdot M) \cap N.$$

□

Proof. In fact, equation (2) is another way of writing that "the I-adic topology of N is the topology induced from the I-adic topology of M." So that equation

(2) is indeed a restatement of Case 1.

Case 2. General Case: Let I be an open ideal in A. Then we must show that there exists an open ideal J in A such that

$$(3): \ I \cdot N \supset (J \cdot M) \cap N.$$

By the Note (applied to the ring A, the ideal I, the integer $n = 1$, the A-module M and the A-submodule N) there exists an integer $m \geq 0$ such that

$$(4): \ I \cdot N \supset (I^m \cdot M) \cap N.$$

By condition (1) of the Theorem, we have that, if I is any open ideal in A, then I^m is open, for all integers $m \geq 0$; this is easily proved by induction on m. Hence $J = I^m$ is an open ideal such that (3) holds. $\square$

Theorem 5.4.2. *Let A be a standard topological ring such that the abstract ring A is Noetherian and commutative. Suppose also that*

(1) I is an open ideal in A implies that the ideal I^2 is open in A, and that

(2) A admits a denumerable neighborhood base at zero.

Then the hypotheses of §3, Theorem 5.3.23, are satisfied. And the six equivalent conditions of §3, Theorem 5.3.23, and therefore also the four equivalent conditions of §3, Theorem 5.3.17 and the conclusions of §3, Corollaries 5.3.24 and 5.3.25, hold.

In particular, we have that

(3) [§3, Theorem 5.3.23, (5a)] $\widehat{A}$ is flat as an A-module.

(4) [§3, Corollary 5.3.24] For every finitely generated A-module M, we have the canonical isomorphisms of C-complete left A-modules

$$\widehat{A} \underset{A}{\otimes} M \approx C(M) \approx \widehat{M}.$$

(5) [§3, Theorem 5.3.17, (1)] For every set I, $\widehat{A^{(I)}}$ is flat as an A-module.

(6) [§3, Theorem 5.3.17, (3)] For every set I, and every finitely generated A-module M, we have that $C_i(M^{(I)}) = 0$ for $i \geq 1$.

(7) [§3, Corollary 5.3.25]. For every set I, and every finitely generated A-module M, we have that $C(M^{(I)}) = \widehat{M^{(I)}}$.

(8) [§3, Theorem 5.3.17, (2)] For every set I, the functor $M \rightsquigarrow C(M^{(I)})$ is an exact functor from the category of finitely generated A-modules into the category of C-complete left A-modules.

Proof. By Chapter 1, §3, Example 1 we have that the topological ring A is admissible. Therefore hypothesis (2) of §3, Theorem 5.3.23, holds. Since A is Noetherian A is left coherent; so hypothesis (1) of §3, Theorem 5.3.23 holds. And hypothesis (2) of this Theorem is hypothesis (3) of §3, Theorem 5.3.23. Therefore all of the hypotheses of §3, Theorem 5.3.23 hold. And Theorem 5.4.1 tells us that condition (2) of the six equivalent conditions of §3, Theorem 5.3.23 holds. $\square$

Remark 1. Notice that (if one replaces "finitely generated" by "finitely presented," then) *all* of the conclusions of Theorem 5.4.2 are enjoyed by any topological ring A, such that $\widehat{A}$ is admissible, such that the topology of A admits a denumerable neighborhood base at zero, and such that the abstract ring A is left coherent — *if also condition* (2) of §3, Theorem 5.3.23 holds. That is, if also Krull's Theorem (Theorem 5.4.1 above) is true for A. The only wide class of standard topological rings of which this author is aware of at the present time that obey this latter condition (i.e., Krull's Theorem) are the ones discussed in Theorem 5.4.2 above. (There are some exceptions. E.g., it is not difficult to put conditions on a not necessarily commutative t-adic ring so that it obeys all of the hypotheses of §3, Theorem 5.3.23, of §3, Remark 14, and also §3, Theorem 5.3.23, condition (2) — but this may not be an extension of the set of standard topological rings discussed in Theorem 5.4.2 above.)

Thus, the problem of generalizing Theorem 5.4.2 above — perhaps to appropriate classes of non-commutative left coherent rings — boils down to proving e.g., condition (2) (Krull's Theorem of §3, Theorem 5.3.23 (or any of the other six equivalent conditions of (1) - (6) of §3, Theorem 5.3.23)) for those rings. It is perhaps a worthwhile problem to attempt to find such new interesting, and wide classes of rings.

Remark 2. In Theorem 5.4.2 above, conclusion (5) implies that

(5′) For every projective A-module P, we have that $\widehat{P}$ is a flat A-module.

Proof. Since P is projective, P is a direct summand of a free module $A^{(I)}$ for some set I. Therefore $\widehat{P}$ is a direct summand of $\widehat{A^{(I)}}$. And by (5) $\widehat{A^{(I)}}$ is flat. □

Remark 3. Remark 2 above applies to any commutative standard topological ring that obeys the equivalent conditions of, e.g., §3, Theorem 5.3.17. The hypothesis of commutative is, however, needed. The reason is that §3, Theorem 5.3.17 or §3, Theorem 5.3.23, merely imply that $\widehat{A^{(I)}}$ is *right flat*, and therefore gives no information about left projective modules.

As for right projective modules — notice that $\widehat{A^{(I)}}$ means the completion of $A^{(I)}$ with respect to its A-adic topology as an abstract *left* module (see the discussion at the beginning of §3, through Remark 3). This in general is different from the A-adic topology of $A^{(I)}$ as a *right* module.

If the set I is non-empty, then necessary and sufficient conditions for the right and left A-adic topologies on $A^{(I)}$ to coincide is that:

(5″) The standard topological ring A admits a neighborhood base at zero consisting of *two-sided ideals*.

Thus, when a standard topological ring obeys condition (5″) above, and also obeys the four equivalent conditions of §3, Theorem 5.3.17, then:

(5‴) P a right projective right A-module implies that the *right* A-adic completion of P is right flat as an abstract right A-modue.

The following Theorem gives a very mild extension of the class of standard topological rings such that the six equivalent conditions of §3, Theorem 5.3.23 hold.

Theorem 5.4.3. *Let A be a standard topological ring, such that*

(1) A is left coherent as an abstract ring.

(2) The completion $\widehat{A}$ of A is Noetherian and commutative.

(3) A admits a denumerable neighborhood base at zero.

(4) I an open ideal in $\widehat{A}$ implies that the ideal I^2 is open in $\widehat{A}$, and

(5) $\widehat{A}$ is right flat as a right A-module. Then all of the hypotheses of §3, Theorem 5.3.23 are satisfied. And the six equivalent conditions of §3, Theorem 5.3.23 hold, and therefore also the four equivalent conditions of §3, Theorem 5.3.17, and also the conclusions of §3, Corollaries 5.3.24 and 5.3.25.

Proof. As in the proof of Theorem 5.4.2, we see that the hypotheses of §3, Theorem 5.3.23 hold. Therefore to complete the proof, it suffices to show that condition (4) of §3, Theorem 5.3.23, holds.

In fact, let M be a finitely presented abstract left A-module and let N be a finitely generated abstract A-submodule. Then, by hypothesis (5) of this Theorem, $\widehat{A}$ is right flat over A. Therefore the induced map: $\widehat{A} \underset{A}{\otimes} N \to \widehat{A} \underset{A}{\otimes} M$ is injective. And then, since by hypothesis (2) $\widehat{A}$ is Noetherian commutative, and by hypothesis (4) the square of an open ideal in $\widehat{A}$ is open in $\widehat{A}$, we have that the $\widehat{A}$-adic topology of $\widehat{A} \underset{A}{\otimes} N$ is the topology induced from the $\widehat{A}$-adic topology of $\widehat{A} \underset{A}{\otimes} M$. And this verifies condition (6) of Theorem 5.3.23.   □

Remark 4. Hypothesis (5) of Theorem 5.4.3 cannot be deleted.

Example 1. Let A be a Noetherian commutative ring and let I be an ideal in A such that A/I is not flat as an abstract A-module. (E.g., A any commutative Noetherian local ring of dimension > 0 and I the maximal ideal.) (There are indeed many other such examples.) Regard A as a topological ring, by giving it the topology such that the smallest neighborhood of zero is I. Then $\widehat{A} = A/I$ with the discrete topology. And A obeys all of the hypotheses of Theorem 5.4.3, *except hypothesis* (5). And the hypotheses of §3, Theorem 5.3.23 are therefore satisfied. However, the six equivalent conditions of §3, Theorem 5.3.23 in this case all fail, since $\widehat{A} = A/I$ is (by hypothesis) not flat over A.

Remark 5. At first glance, it might appear that Theorem 5.4.3 is more general (but less useful) than Theorem 5.4.2. But this is in fact not so — the reason is that if A is a topological ring obeying the hypotheses of Theorem 5.4.2 — then it is not immediately clear whether or not the completion $\widehat{A}$ is Noetherian! (I have a counterexample that I won't explicate here.)

The reader might do a "double take" at the assertion that there are examples in which A is commutative Noetherian and yet $\widehat{A}$ is not Noetherian. But, the reader should note that the hypotheses of Theorem 5.4.2 above are significantly more general than the ones usually assumed for commutative Noetherian topological rings — namely, one usually deals only with the *I-adic topology* on a Noetherian ring A, when I is an ideal. (I.e., only with Case 1 of Theorem 5.4.1.) The kind of topological rings discussed in Theorem 5.4.2 above can all be constructed as follows:

Let A be a Noetherian commutative ring. Let $I_1, I_2, I_3, \ldots$ be an arbitrary sequence of ideals in A. Then let $\mathscr{I}$ be the set of all finite products of positive powers of the ideals $I_1, I_2, I_3, \ldots$. Then let $\bar{r}$ be the unique topology on A such that A is a topological ring and such that $\mathscr{I}$ is a complete system of neighborhoods of zero. Then the topological ring $(A, \bar{r})$ is an example of a topological ring obeying the hypotheses of Theorem 5.4.2 — and every topological ring obeying the hypotheses of Theorem 5.4.2 can be constructed in that way.

This is clearly considerably more general than the I-adic topology for a single ideal I. And that is why it is possible to have — and in fact there are examples — in which A is commutative Noetherian obeying the hypotheses of Theorem 5.4.2 such that $\widehat{A}$ is not Noetherian.

We conclude this section by recording a well-known result.

Proposition 5.4.4. *Let A be a commutative ring with identity and let I be a finitely generated ideal in the ring A. Let $[\widehat{A}, I\text{-adic}]$ be the completion of A for the I-adic topology. Then*

(1) $[\widehat{A}, I\text{-adic}]$ is Noetherian iff A/I is Noetherian. In particular,

(2) If A is Noetherian then $[\widehat{A}, I\text{-adic}]$ is Noetherian.

Proof. This follows from the following more general result. □

Proposition 5.4.5. *([COC], Chapter 8, p. 776, Remark 3, Lemma) Let A be a left Noetherian ring, and let I be a two-sided ideal in the ring A, such that*

(1) There exists a set $\{t_1, \ldots, t_n\}$ of generators for I as a right ideal, and such that the right ideal: $t_1 A + \cdots + t_i A$ in A is a two-sided ideal, $1 \leq i \leq n-1$. Then the I-adic completion $[\widehat{A}, I\text{-adic}]$ is left-Noetherian iff the ring A/I is left Noetherian.

Corollary 5.4.6. *([COC], Chapter 8, p. 777, Corollary) Let A be a left Noetherian ring and let I be a two-sided ideal in A, that obeys condition (1) of Proposition 5.4.5. If A is left Noetherian then $[\widehat{A}, I\text{-adic}]$ is left Noetherian.*

5.5 Topological Rings that are Strongly C-O.K

Theorem 5.5.1. *Let A be a standard topological ring. Let N be a non-negative integer. Then the following two conditions are equivalent:*

$$(1): \quad C_i(\widehat{F}) = \begin{cases} \widehat{F}, & \text{for } i = 0 \\ 0, & \text{for all integers } 1 \leq i \leq N \end{cases}$$

for all free abstract left A-modules F.

$$(2): \quad C_i(M) = \begin{cases} M, & \text{for } i = 0 \\ 0, & \text{for } 1 \leq i \leq N \end{cases}$$

for all C-complete left A-modules M.

Proof. The case $i = 0$ of condition (1) (respectively condition (2)) is identical to Chapter 3, §1, Theorem 3.1.5, condition (2) (respectively: condition (3)). Therefore by that Theorem, the case $i = 0$ of condition (1) is equivalent to the case $i = 0$ of condition (2). Clearly (2) implies (1). Assume (1). Then we have seen that (2) hlds for $i = 0$. We prove conclusion (2) for $i, 1 \leq i \leq N$, by induction on i.

Let

$$0 \to K \to \widehat{F} \to M \to 0$$

be a short exact sequence of C-complete left A-modules where $\widehat{F}$ is the completion of a free left A-module F. Then we have the exact sequence:

$$(3): \ C_1(\widehat{F}) \to C_1(M) \overset{d_1}{\to} C(K) \to C(\widehat{F}).$$

By the case $i = 0$ of condition (2), $C(K) = K$, $C(\widehat{F}) = \widehat{F}$, and therefore $C(K) \to C(\widehat{F})$ is the monomorphism $K \to \widehat{F}$. By condition (1), if $N \geq 1$ then $C_1(\widehat{F}) = 0$. Therefore from the exact sequence (3), we have that $C_1(M) = 0$.

Now assume that condition (2) has been established for $1 \leq i \leq n$, where n is a positive integer and $n < N$. To prove (2) for $i = n + 1$.

We have the exact sequence

$$(4): \ C_{n+1}(\widehat{F}) \to C_{n+1}(M) \overset{d_{n+1}}{\to} C_n(K).$$

By condition (1), $C_{n+1}(\widehat{F}) = 0$. By the inductive assumption $C_n(K) = 0$. Therefore $C_{n+1}(M) = 0$. □

Definition 1. Let A be a standard topological ring and let N be a non-negative integer. Then the topological ring A is N-C-O.K. iff the two equivalent conditions of Theorem 5.5.1 hold. The standard topological ring A is *strongly* C-O.K., iff A is N-C-O.K. for all integers $N \geq 0$.

Example 1. A standard topological ring A is 0-C-O.K. iff it is C-O.K. (By Chapter 3, §1, Definition 3).

Example 2. Clearly strongly C-O.K. implies n-C-O.K. for all integers $n \geq 0$. And $(n + 1)$-C-O.K. implies n-C-O.K. for all integers $n \geq 0$.

Proposition 5.5.2. *Let A be a topological ring such that (1) the topology is given by denumerably many right ideals $I_1 \supset I_2 \supset I_3 \supset \ldots$ and let N be a non-negative integer. Then sufficient conditions for A to be strongly C-O.K., (resp.: N-C-O.K), is that*

(2) A is C-O.K., and that

(3) $\mathrm{Tor}_i^A(A/I_j, \widehat{F}) = 0$, for all positive integers i, j (resp.: all positive integers i, j with $i \leq N + 1$) and all the free abstract left A-modules F.

Proof. It suffices to prove the parenthetical assertion. Since A is C-O.K., by Example 1 above, we have that condition (1) of Theorem 5.5.1 is satisfied for $i = 0$. By Chapter 4, §3, Theorem 4.3.1 we have the short exact sequence

$$(4): \ 0 \to \varprojlim_{j \geq 1}{}^1 \mathrm{Tor}_{i+1}^A(A/I_j, \widehat{F}) \to C_i(\widehat{F}) \to \varprojlim_{j \geq 1} \mathrm{Tor}_i^A(A/I_j, \widehat{F}) \to 0$$

for $i \geq 0$. For $N \geq i \geq 1$, equation (3) implies that the groups on both the left and right in equation (4) are zero. Therefore by equation (4) $C_i(\widehat{F}) = 0$, $N \geq i \geq 1$. □

Remark 1. By Example 1 above, condition (2) in Proposition 5.5.2, that A be C-O.K., is of course necessary (for A to be N-C-O.K. for any integer $N \geq 0$, or to be strongly C-O.K.).

Corollary 5.5.3. *Let A be a standard topological ring such that*
(1) A admits a denumerable neighborhood base at zero.
(2) A is C-O.K.
(3) For every free abstract left A-module F, we have that $\widehat{F}$ is left flat as an abstract left A-module.
Then A is strongly C-O.K.

Proof. Follows immediately from Proposition 5.5.2. □

Remark 2. Of course, under the hypothesis (1) of Corollary 5.5.3, we have that hypothesis (2) of Corollary 5.5.3 holds iff A is admissible. (See Chapter 3, §1, Theorem 3.1.6.)

Example 3. Let A be a Noetherian commutative ring. Suppose also that A is a topological ring, such that the topology is given by denumerably many ideals, and such that the square of any open right ideal is open. Then A is strongly C-O.K.

Proof. By §4, Theorem 5.4.2, we have that $\widehat{F}$ is a flat A-module, for all free A-modules F. By Chapter 1, §3, Example 1, A is admissible, and therefore (by Remark 2 above) C-O.K. But then by Corollary 5.5.3, A is strongly C-O.K. □

Remark 3. The reader should note, of course, that the hypotheses of Example 3 are satisfied for all practical problems in contemporary commutative algebra and algebraic geometry. Therefore, the condition of "strongly C-O.K." — although stronger than "C-O.K." or admissible — should be regarded as being a very common condition, that essentially "always holds in practice."

Example 4. Let A be a t-adic ring such that the t-torsion is bounded below. Then A is strongly C-O.K.

Proof. By Chapter 4, §3, Corollary 4.3.8, it suffices to prove the Example in the case $A = \mathbf{Z}[T]$. Hence we can assume that A is a proper t-adic ring. Since A is a *proper* t-adic ring, we know that (left multiplication by t): $A \to A$ is injective. Hence if F is a free left A-module, then (left multiplication by t): $F \to F$ is injective. Therefore, by Lemma 4.3.20, (left multiplication by t): $\widehat{F} \to \widehat{F}$ is injective. Therefore $\widehat{F}$ has no non-zero t-torsion. Since F is a free module, and since C is the zeroth derived functor of "$\widehat{}$", we have that $C(F) = \widehat{F}$. Therefore Chapter 4, §3, Theorem 4.3.9 with $M = \widehat{F}$, says that

$$(1): \quad div(C(\widehat{F})) = \varprojlim_{i \geq 1}{}^1 \text{ (precise } t^i\text{-torsion in } \widehat{F}).$$

$$(2): \ C_1(\widehat{F}) = \varprojlim_{i \geq 1} (\text{precise } t^i\text{-torsion in } \widehat{F}).$$

And

$$(3): \ C_n(\widehat{F}) = 0, n \geq 2.$$

Since $\widehat{F}$ has no non-zero t-torsion, equation (1) implies that the C-complete left A-module $C(\widehat{F})$ has no t-divisible elements. By Chapter 3, §2, Corollary 3.2.18, conclusion (2), we have in general that $C(N)/$ (t-divisible elements) $= \widehat{N}$, for every abstract left A-module N. It follows that $C(\widehat{F}) = \widehat{\widehat{F}}$; and, since A is admissible, $\widehat{\widehat{F}} = \widehat{F}$. Therefore $C(\widehat{F}) = \widehat{F}$. And equations (2) and (3) tell us that $C_i(\widehat{F}) = 0$ for $i \geq 2$. This being true for all free left A-modules F, we have that A is strongly C-O.K.   □

One might wonder about the relationship between a topological ring being strongly C-O.K (or N-C-O.K.) and its completion being strongly C-O.K (or N-C-O.K.). The next theorem answers this, in the fullest generality.

Theorem 5.5.4. *Let A be a standard topological ring and let N be a non-negative integer. Then the following two conditions are equivalent.*

(1) A is strongly C-O.K. (respectively N-C-O.K.)

(2) Both conditions (2a) and (2b) below hold.

(2a)

$$C_n^A(\widehat{A}) = \begin{cases} \widehat{A}, & n = 0 \\ 0, & n \geq 1 \ (\text{respectively: } N \geq n \geq 1) \end{cases}$$

and

(2b) $\widehat{A}$ is strongly C-O.K. (respectively: N-C-O.K.)

Proof. The proof is delayed until the next section.   □

We need to give a (perhaps somewhat silly) example.

Example 5. Let A be a standard topological ring, such that

(1) A admits a denumerable neighborhood base at zero.

(2) The uniform and A-adic topologies coincide in $\widehat{A}$.

(3) $\widehat{A}$ is Noetherian and commutative as an abstract ring.

(4) I is an open ideal in $\widehat{A}$ implies that I^2 is open in $\widehat{A}$.

And

(5) $\widehat{A}$ is left flat as an abstract left A-module.

Then the topological ring A is strongly C-O.K.

Proof. By (1), (3) and (4), $\widehat{A}$ obeys the hypotheses of Example 3. Therefore

(6) $\widehat{A}$ is strongly C-O.K.

Also, by equation (5) of Chapter 4, §3, Theorem 4.3.18, we have that

$$(7): \ C_n^A(\widehat{A}) = \begin{cases} \widehat{\widehat{A}}, & n = 0 \\ 0, & n > 0. \end{cases}$$

By hypothesis (2)

$$(8): \quad \widehat{\widehat{A}} = \widehat{A}.$$

(8): (7) and (8) imply condition (2a) of Theorem 5.5.4. Combining with equations (6) above, by Theorem 5.5.4, A is strongly C-O.K. $\qquad\square$

Remark 4. In Example 5 above, hypothesis (2) is necessary for the conclusion. The reason is that by Examples 1 and 2, and by Chapter 3, §1, Theorem 3.1.6, strongly C-O.K implies C-O.K. implies admissible. And by Chapter 1, §3, Proposition 1.3.1, hypothesis (2) above it is necessary for A to be admissible. However, it may be possible to deduce hypothesis (5) from the other hypotheses. Also, by Chapter 3, §1, Corollary 3.1.12, hypothesis (2) of Example 5 is equivalent to:

$(2')$ $C^A(\widehat{A}) = \widehat{A}$.

In Example 5 above, hypothesis (5) is also needed for the conclusion — see Example 6 below for a counterexample.

Remark 5. Considering the many examples (e.g., Examples 3, 4 and 5 above) of C-O.K. rings that are strongly C-O.K., one might wonder whether every C-O.K. standard topological ring is strongly C-O.K. This is not so. Example 6 below exhibits some C-O.K. topological rings that are not even 1-C-O.K.

Example 6. Let A be an abstract ring and let I be a two-sided ideal in the ring A. Regard A as a topological ring that has for smallest open neighborhood of zero, the two-sided ideal I. Then, unless $I = 0$, A is not Hausdorff. We have that

(1) $\widehat{A} = A/I$ with the discrete topology. For any abstract left A-module M, $\widehat{M} = M/(I \cdot M)$.

Thus the functor "A-adic completion" is in this case right exact. Therefore it is its own zeroth left derived functor. That is,

$$(2): \quad C(M) = \widehat{M} = M/(I \cdot M),$$

for all abstract left A-modules M. Since C_i, $i \geq 0$, are the left derived functors of C, this implies that

$$(3): \quad C_i(M) = \mathrm{Tor}_i^A(A/I, M), \text{ for all } i \geq 0.$$

For M an abstract left A-module, $C(C(M)) = (A/I) \underset{A}{\otimes} (A/I) \underset{A}{\otimes} M = (A/I) \underset{A}{\otimes} M = C(M)$. Therefore, by Chapter 3, §1, Definition 3, and Chapter 3, §1, Theorem 3.1.5, condition (1), we have that the standard topological ring A is C-O.K. (Notice that the category of C-complete left A-modules coincides with the category of abstract left (A/I)-modules.)

The ring $\widehat{A} = A/I$ has the discrete topology, and therefore is strongly C-O.K. Therefore, for any integer $N \geq 0$, we have by Theorem 5.5.4 that A is strongly C-O.K. (resp.: N-C-O.K.) iff

$$(4): \quad C_n^A(\widehat{A}) = \begin{cases} \widehat{A}, & n = 0 \\ 0, & n \geq 1 \ (\text{resp.: } 1 \leq n \leq N). \end{cases}$$

Using equations (1) and (2),

$$(5): \quad C^A(\widehat{A}) = (A/I) \underset{A}{\otimes} (A/I) = A/I = \widehat{A},$$

so condition (4) holds for $n = 0$. Therefore, by equations (1) and (3), condition (4) is equivalent to

$$(6): \quad \mathrm{Tor}_n^A(A/I, A/I) = 0,$$

for $n \geq 1$. (resp.: for $1 \leq n \leq N$). E.g., for the standard topological ring A to be 1-C-O.K., it is necessary and sufficient that

$$(7): \quad \mathrm{Tor}_1^A(A/I, A/I) = 0.$$

Recall that

$$\mathrm{Tor}_1^A(A/I, M) = \mathrm{Ker}(I \underset{A}{\otimes} M \to I \cdot M)$$

for any abstract left A-module M. Since $I \cdot (A/I) = \{0\}$, applying this formula to $M = A/I$ gives

$$\mathrm{Tor}_1^A(A/I, A/I) = I \underset{A}{\otimes} (A/I) = I/I^2.$$

Thus, A is 1-C-O.K. iff $I = I^2$.

Thus, if we take any two-sided ideal I in a ring A such that $I \neq I^2$, then the standard topological ring that we have just constructed is C-O.K., but not 1-C-O.K.

Remark 6.   In Example 6 above, suppose that the ring A is commutative Noetherian. If $I = I^2$, then the square of every open ideal in A is open. Therefore, the topological ring A obeys the hypotheses of Example 3 above. And therefore the topological ring A is strongly C-O.K. Thus, in Example 6 above, if the abstract ring A is commutative Noetherian, then A is strongly C-O.K. iff A is 1-C-O.K. iff $I = I^2$. And of course this latter condition rarely holds. (E.g., it fails if A is a commutative Noetherian local ring of dimension > 0 and I is the maximal ideal — or, more generally if A is a commutative Noetherian ring — and I is any non-zero ideal contained in the Jacobsen radical).

Remark 7.   In Example 6 above we can of course insist that A be a commutative Noetherian (e.g., take A to be any commutative Noetherian local ring of dimension > 0, and let I be the maximal ideal). Then the standard topological ring of Example 6 obeys all of the hypotheses of Example 5, except for hypothesis (5). And (as we have seen in Example 6), A does not obey the conclusion of Example 5.

Exercise 1.   (a) Let A be a ring and let $e \in A$ be an element of the center of the ring A such that $e^2 = e$. Then show that the subsets: Ae, $A(1 - e)$, of A, together with the induced multiplication and addition from A, are rings (but not usually *subrings* of A since they usually do not have the same unit elements as does A).

(b) Show that the function $a \to (ae, a(1 - e))$ is an isomorphism of abstract rings, from A onto $(Ae) \times (A(1 - e))$.

(c) Let M be an abstract left A-module. Show that eM, resp.: $(1-e)M$, is an A-module that is annihilated by $1-e$, resp!.: by e. Show that the function: $m \to (em, (1-e)m)$ is an isomorphism of A-modules, $M \approx (eM) \times ((1-e)M)$.

(d) Using what you have shown in (c), establish a canonical equivalence of categories, between Abs_A and $(Abs_{Ae}) \times (Abs_{A(1-e)})$, where Abs_A, resp.: Abs_{Ae}, $Abs_{A(1-e)}$, denotes the category of abstract left A, resp.: Ae, resp.: $A(1-e)$ modules.

(e) Using (d), show that $M \in Abs_A$ is injective (resp.: projective) as abstract left A-modle iff both eM is injective (resp.: projective) as abstract left (Ae)-module and also $(1-e)M$ is injective (resp.: projective) as abstract left $(A(1-e))$-module.

(f) Using (d), show that the functor $M \rightsquigarrow eM$ (resp.: $M \rightsquigarrow (1-e)M$), from Abs_A into Abs_{Ae} (resp.: into $Abs_{A(1-e)}$) preserves arbitrary direct sums and direct products (and even arbitrary direct, and inverse, limits, over arbitrary directed sets, or over arbitrary categories that are sets).

(g) Using (f), show that if M is a free abstract left A-module of rank K, then eM, resp.: $(1-e)M$, is a free $(A-e)$-module, resp.: $(A(1-e))$-module, of rank K (for all cardinal numbers K).

Exercise 2. Let I be a set and let A_i be an abstract ring, for all $i \in I$. Let $A = \prod_{i \in I} A_i$. For every $i \in I$, let $e_i = (\delta_{ij})_{j \in I}$. For every abstract left A-module M, let $M_i = e_i \cdot M$. Then show that for every $i \in I$, the assignment: $M \rightsquigarrow M_i$ is an exact functor that preserves both projectives and injectives, from the category of abstract left A-modules into the category of abstract left A_i-modules.

Hint. Use Exercise 1, parts (d) and (e), with $e = e_i$.

Exercise 3. (a) Let I be a set and let A_i, $i \in I$, be a family of standard topological rings. Then show that the ring $A = \prod_{i \in I} A_i$, together with the product topology, is a standard topological ring.

(b) If M is an abstract left A-module, and if $M_i = M \underset{A}{\otimes} A_i$ (where A_i is regarded as A-algebra by means of the projection: $A \to A_i$) then prove that

(1) $\widehat{M}^A \approx \prod_{i \in I} \widehat{M_i}^{A_i}$ as topological left A-modules canonically, (where $\widehat{M}^A$ denotes the A-adic completion of M, with the uniform topology and $\widehat{M_i}^{A_i}$ denotes the A_i-adic completion of the abstract left A_i-module M_i with the uniform topology, for all $i \in I$).

(c) If M is an abstract left A-module, then prove that

(2) $C_n^A(M) \approx \prod_{i \in I} C_n^{A_i}(M_i)$, canonically, for all integers $n \geq 0$.

Hint: Use Exercise 2 to show that the right side of equation (2) is a homological exact connected sequence of functors, that vanishes on projective left A-modules M in dimension $n \geq 1$. Then use Exercise 2, (and the fact that $C_n^{A_i}$, $n \geq 0$, are a system of left derived functors on Abs_{A_i}, for all $i \in I$), to compute the right side of (2) when $n = 0$ and M is a projective left A-module. Finish off by using equation (1) of part (b).

(d) Show that an abstract left A-module M is A-adically complete iff M_i is A_i-adically complete for all $i \in I$. (Hint: use part (b))

(e) Show that the topological ring A is admissible iff the topological ring A_i is admissible for all $i \in I$. (Hint: Use Exercise 1, part (g), and parts (b) and (d) of this Exercise.)

(f) Show that the topological ring A is strongly admissible iff the topological ring A_i is admissible, for all $i \in I$. (Hint: Use parts (b) and (d) of this Exercise.)

(g) Use part (c) to show that: If n is any non-negative integer, then the topological ring A is C-O.K., resp.: n-C-O.K., resp.: strongly C-O.K. iff the topological rings A_i are C-O.K., resp.: n-C-O.K., resp.: strongly C-O.K., for all $i \in I$.

Exercise 4. (a) Using Exercise 3, parts (f) and (g), construct a standard topological ring A, such that it does not admit a denumerable neighborhood base at zero, and such that A is both strongly admissible and strongly C-O.K.

(b) Show that the standard topological rings constructed in Example 6 above (that are C-O.K., but not 1-C-O.K.) are strongly admissible, and have a denumerable (in fact, finite) neighborhood base at zero.

(c) Using (b) above, and Exercise 3, parts (f) and (g), construct a standard topological ring A, such that A does not admit a denumerable neighborhood base at zero; such that A is strongly admissible, such that A is C-O.K, and such that A is not 1-C-O.K.

5.6   Exactness of Direct Sum

Definition 1. Let A be a standard topological ring and let I be a set. Then we have the category $\mathscr{C}_A$ of C-complete left A-modules, an abelian category with enough projectives. We have the functor, "I-fold direct sum,"

$$(M_i)_{i \in I} \rightsquigarrow \int_{i \in I} M_i.$$

This is a right exact functor from the category $\mathscr{C}_A^I$, an abelian category with enough projectives, into the category $\mathscr{C}_A$. Therefore we have the derived functors:

$$(M_i)_{i \in I} \rightsquigarrow \int_{\substack{n \\ i \in I}} M_i, \quad n \geq 0.$$

We call these the *higher direct sums of C-complete left A modules M_i, $i \in I$*.

Theorem 5.6.1. *Let A be a standard topological ring and let I be a set. Then for every family M_i, $i \in I$, of left A-modules, there is induced a first quadrant homological spectral sequence starting with $E_{p,q}^2$ in the category of C-complete left A-modules,*

$$E_{p,q}^2 = \int_{\substack{p \\ i \in I}} C_q(M_i) \Rightarrow C_n(\underset{i \in I}{\oplus} M_i).$$

Proof. The functor $(M_i)_{i \in I} \rightsquigarrow (C(M_i))_{i \in I}$ is a right exact functor from the category of I-tuples of abstract left A-modules into $\mathscr{C}_A^I$. It maps "enough projectives" (namely, tuples of the form $(F_i)_{i \in I}$, where the F_i are projective abstract left A-modules) into projectives. The left derived functors of this functor are $(M_i)_{i \in I} \rightsquigarrow (C_n(M_i))_{i \in I}$, $n \geq 0$. The composite with the right-exact functor:

$$\int_{i \in I} \text{ from } \mathscr{C}_n^I \text{ into } \mathscr{C}_A \text{ is the functor: } (M_i)_{i \in I} \rightsquigarrow \int_{i \in I} C(M_i).$$

But, by §1, Theorem 5.1.1, this latter functor is the functor $(M_i)_{i \in I} \rightsquigarrow C\left(\bigoplus_{i \in I} M_i\right)$.

And the left derived functors are the functors: $(M_i)_{i \in I} \rightsquigarrow C_n\left(\bigoplus_{i \in I} M_i\right)$ (since these are obviously an exact connected sequence of functors — and they vanish in dimension > 1 when M_i are all projective). Therefore the indicated spectral sequence is the spectral sequence of this composite functor. $\qquad\square$

Corollary 5.6.2. *Let A be a standard topological ring and let D be a directed set. Then for every direct system $(M_i, \alpha_{i,j})_{i,j \in D}$ of abstract left A-modules indexed by the directed set D, we have a first quadrant homological spectral sequence such that*

$$(1): \quad E_{p,q}^2 = \varinjlim_{i \in D}{}^C {}_p(C_q(M_i)) \Rightarrow C_n\left(\varinjlim_{i \in D} M_i\right).$$

<u>Note</u>: In the above equation, of course, "$\varinjlim_{i \in D}{}^C {}_p$" denotes the pth left derived functor of the right exact functor "$\varinjlim_{i \in D}{}^C$" from the category $\mathscr{C}_A{}^D$ (of direct systems of C-complete left A-modules indexed by the directed set D) into the category $\mathscr{C}_A$ (of C-complete left A-modules).

Proof. The proof is very similar to that of Theorem 5.6.1, with "$\varinjlim_{i \in D}{}^C$" replacing "$\int_{i \in I}$", and "$\varinjlim_{i \in D}{}^C {}_p$" replacing "$\int_{i \in I} {}_p$". $\qquad\square$

<u>Exercise</u>: (Requires some advanced knowledge of homological algebra) This exercise generalizes the construction in the last Corollary.

Let A be a standard topological ring and let $\mathscr{D}$ be a category that is a set. Let F be a covariant functor from the category $\mathscr{D}$ into the category of abstract left A-modules. Then we have two first quadrant, homological spectral sequences in the category of C-complete left A-modules, such that

$$(1): \quad {}^I E_{p,q}^2 = \varinjlim_{d \in \mathscr{D}}{}^C {}_p(C_q(F(d)))$$

and

$$(2): \ {}^{II}E^2_{p,q} = C_p \left(\varinjlim_{d \in \mathscr{D}} {}_q \ (F(d)) \right).$$

These spectral sequences have the same abutments K_n, $n \geq 0$ (but with in general different filtrations).

Note: The covariant functors from $\mathscr{D}$ into the category of abstract left A-modules form an abelian category $\mathscr{A}$, with natural transformations of functors for maps. For every object F in $\mathscr{A}$, we have the abutment $K_n(F)$, $n \geq 0$, of the spectral sequences (1) and (2). The assignments $F \rightsquigarrow K_n(F)$, $n \geq 0$, are the left derived functors of the functor: $F \rightsquigarrow C \left(\varinjlim_{d \in \mathscr{D}} F(d) \right)$ from A into the category of C-complete left A-modules.

If the standard topological ring A is strongly C-O.K., then the spectral sequences of Theorem 5.6.1 and Corollary 5.6.2 allow us, among other things, to write formulae for the higher direct sums $\int_p$, and for the higher direct limits $\underset{i \in I}{}$ over directed sets, $\varinjlim^C_{i \in D}$, in the category of C-completed left A-modules:

Theorem 5.6.3. *[Computation of Higher Direct Sums]*
Let A be a standard topological ring and let N be a non-negative integer. Then the following two conditions are equivalent:

(1) The topological ring A is strongly C-O.K. (respectively: N-C-O.K.)

(2) For every set I, and every family M_i, $i \in I$, of C-complete left A-modules, we have that the natural infinitely linear function is an isomorphism of C-complete left A-modules

$$C_p \left(\bigoplus_{i \in I} M_i \right) \overset{\approx}{\to} \int_p \underset{i \in I}{} M_i,$$

for all non-negative integers p (respectively: for all non-negative integers p such that $0 \leq p \leq N$).

Note: The natural transformations of functors: $C_p \left(\bigoplus_{i \in I} M_i \right) \to \int_p \underset{i \in I}{} M_i, p \geq 0$, are deduced by virtue of $\int_p$, $p \geq 0$, being the left derived functors of $\int$, and from the natural map: $C \left(\bigoplus_{i \in I} M_i \right) \to \int_{i \in I} M_i$ (which in turn is deduced from the universal mapping property definition of $C \left(\bigoplus_{i \in I} M_i \right)$ — or equivalently is the natural epimorphism into the quotient for the explicit construction of $\int_{i \in I} M_i$ in §1, Remark 3. Alternatively, (and equivalently), the maps: $C_p \left(\bigoplus_{i \in I} M_i \right)$

$\to \int_p M_i$, $p \geq 0$, are edge homomorphisms in the spectral sequence of Theo-
$i \in I$

rem 5.6.1).

Proof. (1) implies (2): Consider the spectral sequence

$$(1): \quad E^2_{p,q} = \int_{\substack{p \\ i \in I}} C_q(M_i) \Rightarrow C_n \left(\underset{i \in I}{\oplus} M_i \right)$$

of Theorem 5.6.1. M_i is by hypothesis C-complete, $i \in I$, and A is by hypothesis
N-C-O.K. Therefore

$$(2): \quad C_q(M_i) = \begin{cases} M_i, & q = 0 \\ 0, & 1 \leq q \leq N. \end{cases}$$

Substituting into (1), we see that the first through Nth rows of $E^2_{p,q}$ are identically zero. Therefore zeroth through Nth objects of the abutment must be
isomorphic, via the edge homomorphism, to the zeroth through Nth objects of
the bottom (zeroth) row. That is,

$$(3): \quad \int_{\substack{p \\ i \in I}} C_0(M_i) = C_p \left(\underset{i \in I}{\oplus} M_i \right), \; 0 \leq p \leq N.$$

Substituting the equation $C_0(M_i) = M_i$ from (2) above into (3) gives the result.
(2) implies (1): Take I to be the set consisting of the single element 0. Then

$$\int_{i \in I} M_i = M_0,$$

the identity functor from $\mathscr{C}_A$ into itself. Therefore $\int_{i \in I}$ is an exact functor and

therefore the derived functors

$$\int_{\substack{p \\ i \in I}} \equiv 0 \text{ for } p \geq 1.$$

Hence the condition (2) for this set $I = \{0\}$ reads: If M is a C-complete left
A-module then

$$C_p(M) = \begin{cases} M, & p = 0 \\ 0, & 1 \leq p \leq N. \end{cases}$$

And this is §5, Theorem 5.5.1, condition (2). Therefore by §5, Definition 1, the
topological ring A is N-C-O.K. $\qquad \square$

Corollary 5.6.4. *Let A be a topological ring that is strongly C-O.K. Then the
spectral sequence of Theorem 5.6.1 simplifies as follows: For every set I and
every family M_i, $i \in I$, of abstract left A-modules, we have the first quadrant
homological spectral sequence*

$$E^2_{p,q} = C_p \left(\underset{i \in I}{\oplus} C_q(M_i) \right) \Rightarrow C_n \left(\underset{i \in I}{\oplus} M_i \right).$$

Proof. By Theorem 5.6.3,

$$\int_p^{} C_q(M_i) = C_p\left(\bigoplus_{i \in I} C_q(M_i)\right).$$
$$\scriptstyle i \in I$$

Substituting into $E^2_{p,q}$ of Theorem 5.6.1 gives the result.   □

Corollary 5.6.5. *Let A be a topological ring such that the topology is given by denumerably many right ideals $J_1 \supset J_2 \supset J_3 \supset \ldots$, and let N be an integer ≥ 0. Suppose that A is strongly C-O.K. (respectively: N-C-O.K.). Let I be a set and let M_i be a C-complete left A-module for all $i \in I$. Then there are induced canonical isomorphisms of abstract abelian groups*

$$\int_p^{} M_i \approx C_p\left(\bigoplus_{i \in I} M_i\right) \approx \varprojlim_{j \geq 1}{}^1 \left(\bigoplus_{i \in I}(\mathrm{Tor}^A_{p+1}(A/J_j, M_i))\right)$$
$$\scriptstyle i \in i$$

for all positive integers p (respectively: for all positive integers p such that $1 \leq p \leq N$).

<u>Note 1</u>: The leftmost of these isomorphisms is the isomorphism of C-complete left A-modules of Theorem 5.6.3, condition (2).
<u>Note 2</u>: If the right ideals J_j are two-sided for all integers $j \geq 1$, then the rightmost isomorphism is an isomorphism of C-complete left A-modules.
<u>Note 3</u>: The hypotheses that A is strongly C-O.K. (resp.: N-C-O.K.) can be replaced by the weaker hypothesis:

$$C_p(M_i) = 0 \text{ for all } i \in I,$$

and all positive integers p (resp.: and all positive integers p such that $1 \leq p \leq N$).

Proof. It suffices to prove the parenthetical assertion. By Chapter 4, §3, Theorem 4.3.1 we have the short exact sequences

$$(1): \ 0 \to \varprojlim_{j \geq 1}{}^1 \mathrm{Tor}^A_{p+1}(A/A_j, M_i) \to C_p(M_i) \to \varprojlim_{j \geq 1} \mathrm{Tor}^A_p(A/J_j, M_i) \to 0$$

for all integers $p \geq 0$ and all $i \in I$. Since the M_i are C-complete and A is N-C-O.K.,
    $(1')$: $C_p(M_i) = 0$ for $1 \leq p \leq N$, for all $i \in I$. Substituting into equation (1) we see in particular that

$$(2): \ \varprojlim_{j \geq 1} \mathrm{Tor}^A_p(A/A_j, M_i) = 0, \text{ for } 1 \leq p \leq N, \text{ and all } i \in I.$$

Also, by Chapter 4, §3, Theorem 4.3.1 applied to the abstract left A-module $M = \bigoplus_{i \in I} M_i$, we have the short exact sequences

$$(3): \ 0 \to \varprojlim_{j \geq 1}{}^1 \mathrm{Tor}^A_{p+1}(A/J_j, M) \to C_p(M) \to \varprojlim_{j \geq 1} \mathrm{Tor}^A_p(A/J_j, M) \to 0$$

for all integers $p \geq 0$. The functor: $N \rightsquigarrow \mathrm{Tor}_k^A(H, N)$ preserves direct sums (from the category of abstract left A-modules into the category of abelian groups) for any abstract left A-module H. (To see this, choose a projective resolution of H.) Therefore

$$(4): \quad \mathrm{Tor}_k^A(A/J_j, M) = \bigoplus_{i \in I} \mathrm{Tor}_k^A(A/J_j, M_i),$$

for all integers $k \geq 0$ and all $j \geq 1$. In general:

Let I be a set, and for all $i \in I$ let $(G_{ij})_{j \in D}$ be an inverse system of abelian groups indexed by a fixed directed set D, such that

$$\varprojlim_{j \in D} G_{ij} = 0.$$

Then

$$\varprojlim_{j \in D} \left[\bigoplus_{i \in I} G_{ij} \right] = 0.$$

(*Proof.* Since the functor $\varprojlim\limits_{j \in D}$ preserves monomorphisms, it suffices to show that

$$\varprojlim_{j \in D} \left[\prod_{i \in I} G_{ij} \right] = 0.$$

But

$$\varprojlim_{j \in D} \left[\prod_{i \in I} G_{ij} \right] = \prod_{i \in I} \left[\varprojlim_{j \in D} G_{ij} \right] = \prod_{i \in I} 0 = 0.)$$

Applying this to the case $D = \{\text{positive integers}\}$, and

$$G_{ij} = \mathrm{Tor}_p^A(A/J_j, M_i),$$

it follows that equations (2) and (4) imply that

$$(5): \quad \varprojlim_{j \geq 1} \mathrm{Tor}_p^A(A/J_j, M) = 0, \text{ for } 1 \leq p \leq N.$$

Substituting equation (5) into the short exact sequences (3) we obtain

$$(6): \quad C_p(M) \approx \varprojlim_{j \geq 1}^1 \mathrm{Tor}_{p+1}(A/J_j, M) \text{ for } 1 \leq p \leq N.$$

Equation (4) for $k = p + 1$ and equation (6) imply the rightmost isomorphism in the conclusion of the Corollary, for $1 \leq p \leq N$. $\qquad \square$

Remark 1. Let A be a topological ring such that the topology of A is given by denumerably many right ideals: $J_1 \supset J_2 \supset J_3 \supset \dots$. Then we have seen, in Chapter 4, §3, Theorem 4.3.1, that for every integer $p \geq 0$, we have the short exact sequence

$$(1): 0 \to \varprojlim_{j \geq 1}^1 \mathrm{Tor}_{p+1}^A(A/J_j, M) \to C_p(M) \to \varprojlim_{j \geq 1} \mathrm{Tor}_p^A(A/J_j, M) \to 0.$$

If the standard topological ring A is C-O.K. ($=$ 0-C-O.K.), then we have seen, in Chapter 4, §3, Theorem 4.3.4, that for $p = 0$, the divisible part of $C_p(M)$ is the group on the left of equation (1). This suggests the possibility that — e.g., if the topological ring A is, say, strongly C-O.K., — then one might hope that the group on the left of equation (1) might always be the divisible part of $C_p(M)$ for all $p \geq 0$ (it is equivalent to say: that the group on the right of equation (1) would be the A-adic completion of $C_p(M)$, for $p \geq 0$). This is trivially true for A a proper t-adic ring with parameter t and $J_j = t^j A$, $j \geq 1$.

Proof. By Chapter 4, §3, Proposition 4.3.7, conclusion (3), $C_i \equiv 0$ for $i \geq 2$. And by Chapter 4, §3, Remark 6, $C_1(M)$ is A-adically complete. Therefore

$$(2) : div \ C_p(M) = 0, \ p \geq 1.$$

On the other hand, by Chapter 4, §3, Lemma 4.3.6, $\text{Tor}_{p+1}^A(A/t^j A, M) = 0$ for $p \geq 1$ and $j \geq 1$. Therefore the leftmost group of equation (1) in Remark 1 above is zero for $p \geq 1$. Combining with equation (2) gives the result.   □

However, the next example shows that this is definitely *false* in almost any ring even slightly more complicated than a t-adic ring — for example, — even if $p = 1$ and A is a regular local ring of dimension two with the topology given by the positive powers of the maximal ideal.

Example 1. Let A be a standard topological ring, such that the topology is given by denumerably many ideals, such that A is N-C-O.K. ($N \geq 1$) and such that the functor, "denumerable direct sum," is not exact on the category $\mathscr{C}_A$ of C-complete left A-modules. (We will see in Example 2 below that, e.g., every regular local ring of dimesion two, with topology given by powers of the maximal ideal has these properties, and is also strongly C-O.K. (the latter by §5, Example 5).)   Then the right-exact functor $\int_{i \geq 1}$ from $\mathscr{C}_A^\omega$ into $\mathscr{C}_A$ is not exact. Therefore the first left derived functor $\int_{1 \atop i \geq 1}$ of $\int_{i \geq 1}$ is not identically zero. Therefore there exists a sequence $M_1, M_2, M_3, \ldots$ of C-complete left A-modules such that

$$(2) : \int_{1 \atop i \geq 1} M_i \neq 0.$$

Then by Corollary 5.6.5, with $p = 1$, if $M = \underset{i \geq 1}{\oplus} M_i$, we have

$$(3) : \int_{1 \atop i \geq 1} M_i = C_1(M), \text{ and}$$

$$(4) : C_1(M) \approx \varprojlim_{j \geq 1}{}^1 \text{Tor}_2^A(A/J_j, M),$$

where $J_1 \supset J_2 \supset J_3 \supset \ldots$ is any sequence of open right ideals in the ring A that give the topology of A. Therefore, in equation (1) of Remark 1, with

$p = 1$, the leftmost group in that equation is the whole group $C_1(M)$. If that leftmost group were the divisible part of $C_1(M)$, it would therefore follow that the divisible part of $C_1(M)$ would be all of $C_1(M)$ — i.e., that $C_1(M)$ would be infinitely divisible. Therefore $C_1(M)$ would be an A-divisible, C-complete left A-module. By Chapter 3, §2, Proposition 3.2.1, since the topological ring A admits a denumerable neighborhood base, that would imply that $C_1(M) = 0$. But then, by equations (2) and (3), we have a contradiction.

The contradiction implies that, for the integer $p = 1$ and the abstract left A-module $M = \underset{i \geq 1}{\oplus} M_i$, that the group on the left of equation (1), for $p = 1$, is strictly larger than the divisible part of $C_1(M)$.

<u>Remark 2</u>. If we study Example 2 below, we see that the explicit counterexample is as follows: Let A, t_1, t_2 be as in Example 2 below (E.g., if A is any regular local ring of dimension ≥ 2 and t_1, t_2 are among a set of parameters). Then whether or not A is 1-C-O.K., we obtain a counterexample as in Example 1, above. Regard A as a topological ring with the topology given by powers of the ideal $(t_1 A + t_2 A)$. Then take

$$M_i = A/(t_1^i A + t_2^i A),$$

$i \geq 1$. Then M_i is A-adically discrete, and therefore A-adically complete and a fortori C-complete. Then, the argument of Example 1 above applies, showing that

$$M = \underset{i \geq 1}{\oplus} M_i$$

is a counterexample as above. In fact, as we note in Example 2 below, (whether or not A is 1-C-O.K.), the hypotheses of Note 3 of Corollary 5.6.5 are satisfied with $N = 1$ — that is $C_1(M_i) = 0$, $i \geq 1$. And

$$C_1(M) = A^\omega / \widehat{A^\omega}$$

which is definitely non-zero and, which in fact, it is easy to see, is A-adically complete. Yet, by Corollary 5.6.5 and the proof of Example 1 above, in the short exact sequence of (1) of Chapter 4, §3, Theorem 4.3.4, we have that, for $p = 1$, the leftmost group is all of $C_1(M)$, and therefore is far larger than the divisible part of $C_1(M)$, which is zero.

<u>Remark 3</u>. The reader might find one aspect of Corollary 5.6.5 "unnerving." Namely, the conclusion implies for $p \geq 1$ and for M_i C-complete, all $i \in I$, that $C_p\left(\underset{i \in I}{\oplus} M_i\right)$, (which is often non-zero — see Example 1 above), is represented as $\underset{j \geq 1}{\varprojlim^1}$ of a direct sum of inverse systems, each of which has $\underset{j \geq 1}{\varprojlim^1}$ zero! (This latter by equation (1) and (1′) in the Proof of Corollary 5.6.5.)

However, that is not unusual, if one is familiar with some of the properties of $\underset{j \geq 1}{\varprojlim^1}$, see [FLI]. For example, if $(G_j)_{j \geq 1}$ is an inverse system of abelian groups, then regard G_j as a topological group with system of neighborhoods of zero given by $\{\text{Im}(G_k \to G_j) : k \geq j\}$. Then, see [FLI], a necessary (but in general not sufficient) condition for $\varprojlim_{j \geq 1}^1 G_j$ to be zero, is that G_j is "complete but not Hausdorff" — meaning that $G_j/$(closure of zero) is complete. And always there is a natural epimorphism:

$$(1): \ \varprojlim_{j \geq 1}^1 G_j \to \varprojlim_{j \geq 1}(\widehat{G}_j/G_j),$$

where by "$\widehat{G}_j/G_j$" we mean $\text{Cok}(G_j \to \widehat{G}_j)$. Thus, suppose $(G_{ij})_{j \geq 1}$ are inverse systems, for all $i \in I$, such that $\varprojlim_{j \geq 1}^1 G_{ij} = 0$ for all $i \in I$, where I is an infinite set. Then we must have, by the above, that G_{ij} is complete but not Hausdorff. However, in the inverse system

$$\left(\bigoplus_{i \in I} G_{ij} \right)_{j \geq 1},$$

this property is not preserved (the abstract direct sum of complete topological groups, for the kind of topology involved, is usually not complete). A moment's reflection about the epimorphism (1) above, should lead one to expect that $\varprojlim_{j \geq 1}^1 \left(\bigoplus_{i \in I} G_{ij} \right)$ should be something vaguely approximating

$$\varprojlim_{j \geq 1} \left[\left(\prod_{i \in I} G_{ij} \right) \Big/ \left(\bigoplus_{i \in I} G_{ij} \right) \right].$$

In fact

Corollary 5.6.6. *Under the hypotheses of Corollary 5.6.5, for every positive integer p, (resp.: for every positive integer p with $1 \leq p \leq N-1$) we have a canonical isomorphism*

$$C_p \left(\bigoplus_{i \in I} M_i \right) \approx \varprojlim_{j \geq 1} \left[\left(\prod_{i \in I} \text{Tor}_{p+1}^A(A/J_j, M_i) \right) \Big/ \left(\bigoplus_{i \in I} \text{Tor}_{p+1}^A(A/J_j, M_i) \right) \right]$$

of abelian groups.

Note: If the ideals J_j in the ring A are two-sided, then the indicated isomorphism is an isomorphism of C-complete left A-modules.

Proof. Let $G_{ij} = \text{Tor}_{p+1}^A(A/J_j, M_i)$, for all $i \in I$ and all $j \geq 1$. Then by equations (1) and (1′) of the Proof of Corollary 5.6.5, we have that

$$(1): \ \varprojlim_{j \geq 1}^1 G_{ij} = \varprojlim_{j \geq 1} G_{ij} = 0.$$

From the explicit construction of $\varprojlim_{j \geq 1}$ and $\varprojlim^1_{j \geq 1}$ from cochains, see [COC], Introduction, Chapter 1, §7, pp. 81 and 82, we have that both $\varprojlim_{j \geq 1}$ and $\varprojlim^1_{j \geq 1}$ in the category of abelian groups commute with taking direct products. Therefore equation (1) implies

$$(2): \quad \varprojlim_{j \geq 1}^1 \left(\prod_{i \in I} G_i \right) = \varprojlim_{j \geq 1} \left(\prod_{i \in I} G_{ij} \right) = 0.$$

Consider the short exact sequence

$$(3): \quad 0 \to \left(\bigoplus_{i \in I} G_{ij} \right)_{j \geq 1} \to \left(\prod_{i \in I} G_{ij} \right)_{j \geq 1} \to \left(\left(\prod_{i \in I} G_{ij} \right) \Big/ \left(\bigoplus_{i \in I} G_{ij} \right) \right) \to 0$$

of inverse systems of abelian groups. Applying the system of derived functors $\varprojlim_{j \geq 1}, \varprojlim^1_{j \geq 1}$ to the short exact sequence, a portion of the exact sequence of six terms obtained is:

$$(4): \quad \cdots \to \varprojlim_{j \geq 1} \left(\prod_{i \in I} G_{ij} \right) \to \varprojlim_{j \geq 1} \left(\left(\prod_{i \in I} G_{ij} \right) \Big/ \left(\bigoplus_{i \in I} G_{ij} \right) \right) \overset{d_1}{\to} \varprojlim_{j \geq 1}^1 \left(\bigoplus_{i \in I} G_{ij} \right)$$

$$\to \varprojlim_{j \geq 1}^1 \left(\prod_{i \in I} G_{ij} \right) \to \cdots$$

By equation (2) the leftmost and rightmost groups in (4) are zero. Therefore

$$(5): \quad \varprojlim_{j \geq 1}^1 \left(\bigoplus_{i \in I} G_{ij} \right) \approx \varprojlim_{j \geq 1} \left[\left(\prod_{i \in I} G_{ij} \right) \Big/ \left(\bigoplus_{i \in I} G_{ij} \right) \right].$$

And the left side of equation (5) is $C_p \left(\bigoplus_{i \in I} M_i \right)$ by Corollary 5.6.5. $\qquad \square$

Remark 4. Notice in the construction of Corollary 5.6.6 that in the parenthetical case, the restriction on p is "$1 \leq p \leq N - 1$," while in Corollary 5.6.5, the corresponding restriction is only "$1 \leq p \leq N$." Of course, if the ring A is *strongly* C-O.K., this makes no difference. And of course, most topological rings that one comes across are strongly C-O.K.

The reason for the stronger condition in Corollary 5.6.6 is that one needs equation (1) of the Proof of Corollary 5.6.6. And in particular that

$$\varprojlim_{i \geq 1} \operatorname{Tor}_{p+1}^A (A/J_j, M_i) = 0.$$

And, by equations (1) and (1') of the Proof of Corollary 5.6.5, we know that this latter is so, only in the range $0 \leq p \leq N - 1$.

Note: The next Theorem makes use of direct limit, and higher direct limits, over categories that are sets – rather than just directed sets. We will make no use of it in other parts of this book; so those uncomfortable with these concepts can simply skip this Theorem.

Theorem 5.6.7. *[Spectral Sequence of the Higher Direct Limits]*
Let A be a standard topological ring. Then the following two conditions are equivalent:

(1) The topological ring A is strongly C-O.K.

(2) For every category $\mathscr{D}$ that is a set, and every covariant functor F from $\mathscr{D}$ into the category of C-complete left A-modules we have a first quadrant homological spectral sequence

$$(1): \quad E_{p,q}^2 = C_p \left(\varinjlim_{d \in \mathscr{D}} {}_qF(d) \right) \Rightarrow \varinjlim_{d \in \mathscr{D}}{}^C{}_n F(d).$$

Proof. (1) implies (2): We have the two spectral sequences (1) and (2) constructed in the Exercise following Corollary 5.6.2. Since for all $d \in \mathscr{D}$, $F(d)$ is a C-complete left A-module, and since A is strongly C-O.K.,

$$(3): \quad C_q(F(d)) = \begin{cases} F(d), & q = 0 \\ 0, & q \geq 1. \end{cases}$$

Therefore in the first spectral sequence in the Exercise, ${}^I E_{p,q}^2 = 0$ for $q \neq 0$. Therefore the nth group of the abutment is ${}^I E_{n,0}^2 = \varinjlim_{d \in D}{}^C{}_n C_0(F(d))$ — which, by equation (3) for $q = 0$, is

$$(4): \quad \varinjlim_{d \in D}{}^C{}_n F(d).$$

And the second spectral sequence (2) ${}^{II} E_{p,q}^2$ in the Exercise has the same abutment (4) as the first spectral sequence. Therefore the second spectral sequence (2) has the desired properties.

(2) implies (1): Let D be the directed set with one element 0. Then the same argument "(2) implies (1)" in the proof of Theorem 5.6.3 shows that A is strongly C-O.K.

$\square$

Remark 5. Notice in the hypotheses of Theorem 5.6.7 that the direct system $(M_d)_{d \in D}$ is a direct system of *C-complete* left A-modules. If the direct system $(M_d)_{d \in D}$ is a direct system of *abstract* left A-modules, then (even if the topological ring A is strongly C-O.K. — in fact even in the case in which A is a discrete valuation ring not a field) we will still in general have *two* spectral sequences as in the Exercise following Corollary 5.6.2. (The first spectral sequence of the Exercise following Corollary 5.6.2 then does *not* always degenerate.)

Similarly, in Theorem 5.6.3, note that it is a hypothesis that the M_i are all *C-complete*. (Indeed, the functor " $\int_{i \in I}$ " only makes sense on families of C-completes.) In the case that the M_i are only *abstract left A-modules* for $i \in I$, then we get — e.g., in the case that A is strongly C-O.K., a spectral sequence — namely, the one of Corollary 5.6.4; and if A is not strongly C-O.K., then the spectral sequence of Theorem 5.6.1.

Corollary 5.6.8. *[Computation of the Higher Direct Limits Over a Directed Set] Let A be a standard topological ring and let N be a non-negative integer. Then the following two conditions are equivalent:*

(1) The topological ring A is strongly C-O.K. (respectively: N-C-O.K.).

(2) For every directed set D and every direct system $(M_i)_{i \in D}$ of C-complete left A-modules, we have that the natural infinitely linear functions are isomorphisms of C-complete left A-modules:

$$(1): \varinjlim_{i \in D}{}^C{}_p(M_i) \overset{\approx}{\leftarrow} C_p\left(\varinjlim_{i \in D} M_i\right),$$

for all non-negative integers p (respectively: all non-negative integers p such that $0 \le p \le N$).

<u>Note</u>: The natural infinitely linear functions: $C_p\left(\varinjlim_{i \in D} M_i\right) \to \varinjlim_{i \in D}{}^C{}_p(M_i)$ are deduced in the same fashion as in the Note to Theorem 5.6.3.

Proof. The proof is very similar to that of Theorem 5.6.3 and need not be repeated. $\qquad\square$

Theorem 5.6.9. *Let A be a standard topological ring and let N be either ∞ or a positive integer. Then the following two conditions are equivalent.*

$$(1): C_n^A(\widehat{A}) = \begin{cases} \widehat{A}, & n = 0 \\ 0, & 0 \le n < N. \end{cases}$$

(2): For all abstract left $\widehat{A}$-modules M, $C_n^A(M) = C_n^{\widehat{A}}(M)$, $0 \le n < N$.

<u>Note</u>: Of course the "equals" sign, in both conditions (1) and (2), mean that "the natural infinitely linear function is an isomorphism of C-complete left A-modules."

Proof. (2) implies (1): Take $M = \widehat{A}$.
(1) implies (2): Let I be a set and let $M_i = \widehat{A}$ for all $i \in I$. Then the spectral sequence of Theorem 5.6.1 is such that

$$(1): E_{p,q}^2 = \int_p{}_{i \in I} C_q^A(\widehat{A}) \Rightarrow C_n^A\left(\underset{i \in I}{\oplus} \widehat{A}\right).$$

By condition (1) of the Theorem

$$(2): \quad C_q^A(\widehat{A}) = \begin{cases} \widehat{A}, & q = 0 \\ 0, & 1 \leq q < N. \end{cases}$$

Therefore in the spectral sequence (1) the first row, the row with $q = 1$, up to the $(N-1)$st row are identically zero. Therefore the edge homomorphism induces an isomorphism, from the zeroth up to the $(N-1)$st groups of the abutment onto the zeroth up to the $(N-1)$st groups of the zeroth row. I.e.,

$$(3): \quad C_n^A(\widehat{A}^{(I)}) = E_{n,0}^2 = \int_{\substack{n \\ i \in I}} C^A(\widehat{A})$$

for $0 \leq n < N$. Substituting equation (2) for $q = 0$ into equation (3),

$$(4): \quad C_n^A(\widehat{A}^{(I)}) = \int_{\substack{n \\ i \in I}} \widehat{A}, \ 0 \leq n < N.$$

But, Chapter 2, §4, Theorem 2.4.8, conclusion (2), $\widehat{A}$ is a projective object in $\mathscr{C}_A$. Therefore the I-tuple $(\widehat{A})_{i \in I}$ is projective in $\mathscr{C}_{\mathscr{A}}{}^I$. Since $\int_{\substack{n \\ i \in I}}$ is the nth derived functor of $\int_{i \in I}$, $\int_{\substack{n \\ i \in I}}$ vanishes on projectives for $n \geq 1$. Therefore

$$(5): \quad \int_{\substack{n \\ i \in I}} \widehat{A} = 0 \text{ for } n \geq 1.$$

Also, since $C^A(A) = \widehat{A}$ and $C^A(A^{(I)}) = \widehat{A^{(I)}}$ therefore by §1, Proposition 5.1.3 we have that

$$(6): \quad \int_{i \in I} \widehat{A} = \int_{i \in I} C^A(A) = C^A\left(\bigoplus_{i \in I} A\right) = \widehat{A^{(I)}}.$$

Substituting equations (5) and (6) into equation (4) gives

$$(7): C_n^A(\widehat{A}^{(I)}) = \begin{cases} \widehat{A^{(I)}}, & n = 0 \\ 0, & 1 \leq n < N. \end{cases}$$

Condition (1) of the Theorem for $n = 0$ is also condition (1) of Chapter 3, §1, Proposition 3.1.8. Therefore by condition (2) of Chapter 3, §1, Proposition 3.1.8, for all abstract left A-modules M,

$$(8): \quad C^A(M) = C^{\widehat{A}}(M).$$

The functors C_n^A, $0 \leq n < N$, restricted to the category of abstract left $\widehat{A}$-modules, form a homological exact connected sequence of functors. By equation

(7), the functors C_n^A for $1 \leq n < N$ vanish on free abstract left $\widehat{A}$-modules. Therefore, the functors C_n^A, $0 \leq n < N$, are such that each is the left satellite of its predecessor. But the functors $C_n^{\widehat{A}}$, $0 \leq n < \infty$, being a system of left derived functors, also have this property. Therefore equation (8) (and the uniqueness axioms for the left satellite of a half exact functor — see [CEHA] —) imply that C_n^A and $C_n^{\widehat{A}}$ coincide on the category of abstract left $\widehat{A}$-modules, $0 \leq n < N$. □

Proof of Theorem 5.5.4. It suffices to prove the parenthetical assertion. If condition (1) holds, then condition (2a) holds (since $\widehat{A}$ is complete). Therefore we can assume that condition (2a) holds. But then condition 1 of Theorem 5.6.9 holds for $N + 1$, and therefore also condition (2) of Theorem 5.6.9 for $N + 1$: That is, for every C-complete left A-module M,

$$C_n^A(M) = C_n^{\widehat{A}}(M) \text{ for } 0 \leq n \leq N.$$

But then clearly

$$C_n^A(M) = \left\{ \begin{array}{ll} M, & n = 0 \\ 0, & 1 \leq n \leq N \end{array} \right.$$

for all C-complete left A-modules M iff

$$C_n^{\widehat{A}}(M) = \left\{ \begin{array}{ll} M, & n = 0 \\ 0, & 1 \leq n \leq N \end{array} \right.$$

for all C-complete left A-modules M — i.e., A is N-C-O.K. iff $\widehat{A}$ is N-C-O.K. □

Proposition 5.6.10. *Let A be a standard topological ring and let I be a set. Then, condition (1) below implies condition (2) implies condition (3). Moreover, if the topological ring A is C-O.K. = 0-C-O.K., then these three conditions are equivalent.*

(1) The functor "I-fold direct sum" $(M_i)_{i \in I} \rightsquigarrow \int_{i \in I} M_i$, on the category of C-complete left A-modules, is exact.

(2) For every $j \geq 0$, the functor C_j, from the category of abstract left A-modules into the category of C-complete left A-modules, preserves I-fold direct sums. That is, $C_j \left(\underset{i \in I}{\oplus} M_i \right) = \int_{i \in I} C_j(M_i)$, for all families of abstract left A-modules $(M_i)_{i \in i}$.

(3) For all families of abstract left A-modules M_i, $i \in I$, we have that the natural map:

$$\int_{i \in I} C_1(M_i) \rightarrow C_1 \left(\underset{i \in I}{\oplus} M_i \right)$$

is an epimorphism.

Proof. (1) implies (2): $\int_{i \in I}$ is an exact functor iff $\int_p{}_{i \in i} \equiv 0$ for $p \geq 1$. But then in the spectral sequence of Theorem 5.6.1 we have that $E_{p,q}^2 \equiv 0$ for $p \neq 0$.

Therefore the spectral sequence degenerates, and $E_{0,n}^2 = (n$th group of the abutment), $n \geq 0$. That is,

$$\int_{i \in I} C_n(M_i) = C_n \left(\bigoplus_{i \in I} M_i \right), \ n \geq 0.$$

(2) implies (3): In fact, the condition (2) for $j = 1$ implies that the map in condition (3) is an isomorphism.

If A is C-O.K., then (3) implies (1): Condition (3) is equivalent to asserting that the edge homomorphism in the spectral sequence of Theorem 5.6.1: $E_{0,1}^2 \to (1$st group of the abutment) is an epimorphism. Since $E_{0,1}^\infty$ is "the smallest" filtered piece of (1st group of the abutment), this implies that $E_{0,1}^\infty = $ (first group of the abutment), and therefore that the cokernel $E_{1,0}^\infty$ of the inclusion: $E_{0,1}^\infty \to$ (first group of the abutment) is zero. That is, $E_{1,0}^\infty = 0$. But $E_{1,0}^2 = E_{1,0}^\infty$. Therefore $E_{1,0}^2 = 0$. That is,

$$(4): \int_{\substack{1 \\ i \in I}} C(M_i) = E_{1,0}^2 = 0.$$

This is true for any family $(M_i)_{i \in I}$ of abstract left A-modules. In particular, if M_i are C-complete left A-modules, then this equation likewise holds. But then in this case since C-O.K., we have that $C(M_i) = M_i$, for all $i \in I$. Therefore, equation (4) implies that, for all families M_i, $i \in I$, of C-complete left A-modules, that

$$\int_{\substack{1 \\ i \in I}} M_i = 0.$$

I.e., that the functor $\int_{\substack{1 \\ i \in I}} \equiv 0$. It is equivalent to saying that the right-exact

functor: $\int_{i \in I}$ is exact.   □

Corollary 5.6.11. *Let A be a standard topological ring and let N be a positive integer. Suppose that the topological ring A is strongly C-O.K. (respectively: N-C-O.K.). Then each of the three equivalent conditions of Proposition 5.6.10 are also equivalent to each of the conditions (4) and (5) below:*

(4) For every family M_i, $i \in I$, of C-complete left A-modules, we have that

$$C_j \left(\bigoplus_{i \in I} M_i \right) = 0, \ for \ 1 \leq j, (respectively: \ for \ 1 \leq j \leq N).$$

(5) Condition (4) holds for $j = 1$.

Proof. (2) implies (4): By (2), we have that, if M_i are C-complete, for all $i \in I$,

$$C_j \left(\bigoplus_{i \in I} M_i \right) = \int_{i \in I} C_j(M_i).$$

But since A is N-C-O.K., we have that $C_j(M) = 0$ for $1 \leq j \leq N$.

(4) implies (5): Obvious.

(5) implies (1): The topological ring A is by hypothesis 1-C-O.K. Therefore by Theorem 5.6.3

$$C_1 \left(\underset{i \in I}{\oplus} M_i \right) = \int_{\substack{1 \\ i \in I}} M_i,$$

for all families of C-complete left A-modules M_i, $i \in I$. Therefore condition (5) is equivalent to:

$$\int_{\substack{1 \\ i \in I}} \equiv 0.$$

That is, that $\int_{i \in I}$ is an exact functor. $\square$

Corollary 5.6.12. *Let A be an arbitrary standard topological ring. Then condition (4) of Corollary 5.6.11 is always equivalent to:*

(4′) For every C-complete left A-module M we have that $C_j(M^{(I)}) = 0$, for $1 \leq j$ (respectively: $1 \leq j \leq N$).

And Condition (5) of Corollary 5.6.11 is always equivalent to

(5′) Condition (4′) above holds for $j = 1$.

Proof. Suppose that the parenthetical part of condition (4′) holds. Let $M = \int_{i \in I} M_i$. Then M is C-complete. So by (4′) we have that

$$(6): \quad C_j(M^{(I)}) = 0, \ 1 \leq j \leq N.$$

For all $i \in I$, we have that M_i is a direct summand of M (since $M = M_i \oplus \int_{j \in I \setminus \{i\}} M_j$). Therefore $\underset{i \in I}{\oplus} M_i$ is a direct summand of $M^{(I)}$. Therefore from equation (6), if $1 \leq j \leq N$, since the functor C_j is additive (and therefore preserves the property of being a direct summand), we have that

$$C_j \left(\underset{i \in I}{\oplus} M_i \right) = 0,$$

proving (4). (4) implies (4′) is obvious. Since (5) iff [(4) in the case $N = 1$], and (5′) iff [(4′) in the case $N = 1$], therefore (5) iff (5′). $\square$

Remark 6. Let A be a topological ring that is C-O.K. and let I be a non-empty set. Then condition (4) of Corollary 5.6.11 implies that the topological ring A is strongly C-O.K. (respectively: N-C-O.K.)

(*Proof.* Let F be a free abstract left A-module. Let $i_0 \in I$. Define $M_{i_0} = \widehat{F}$, $M_i = 0$ for $i \in I - \{i_0\}$. Then condition (4) reads: $C_j(\widehat{F}) = 0$, $1 \leq j$, (respectively: $1 \leq j \leq N$).)

Remark 7. It should be noted that by §5, Examples 3 and 4, the hypothesis on a standard topological ring, "A is strongly C-O.K." holds in any "reasonable case". Therefore, in any "reasonable case", Proposition 5.6.10 and Corollary 5.6.11 really assert that, the five conditions (1), (2), ..., (5) of the Proposition and Corollary are all equivalent to each other.

Theorem 5.6.1 and Proposition 5.6.10 immediately raise the question whether or not, (in "reasonable cases") the infinite direct sum is exact in the category of C-complete left A-modules?

The next theorem shows that this is the case for proper t-adic rings. However, the following Example shows that such is *not* the case in almost every other even slightly more complicated ring; e.g., even denumerable direct sums of C-completes is not exact, e.g., for regular local rings of dimension ≥ 2 with topology given by powers of the maximal ideal. (Our proof of this fact can also be extended to, e.g., Cohen-Macauley local rings of dimension ≥ 2.)

Theorem 5.6.13. *Let A be a t-adic ring such that the t-torsion is bounded below. Let I be a set. Then the functor "I-fold direct sum" is exact on the category of C-complete left A-modules.*

Note: That is, the functor $\int_{i \in I}$ from $\mathscr{C}_A^I$ into $\mathscr{C}_A$ is exact.

Proof. Let M_i, $i \in I$ be a family of C-complete left A-modules. Then by Chapter 3, §2, Proposition 3.2.1, M_i has no non-zero infinitely A-divisible elements, and in particular has no infinitely t-divisible, t-torsion elements, (where $t \in A$ is any element such that the topology of A is the t-adic; see Chapter 4, §3, Remark 9). (An alternative proof of this: By §5, Example 4, A is strongly C-O.K. Therefore $C_1(M_i) = 0$, $i \in I$. By Chapter 4, §3, Theorem 4.3.9, Note 1 (about Remark 5), this is equivalent to saying "M_i has no infinitley t-divisible t-torsion", $i \in I$.) Therefore $\underset{i \in I}{\oplus} M_i$ has no infinitely t-divisible, t-torsion. By Chapter 4, §3, Theorem 4.3.9, Note 1 (about Remark 5),

$$(1): \quad C_1\left(\underset{i \in I}{\oplus} M_i\right) = 0.$$

We give three different ways to finish the proof.
Proof 1. By §5, Example 4, A is strongly C-O.K. Therefore by Theorem 5.6.3

$$\int_{\substack{1 \\ i \in I}} M_i = C_1\left(\underset{i \in I}{\oplus} M_i\right).$$

Therefore by equation (1) above,

$$\int_{\substack{1 \\ i \in I}} \equiv 0.$$

That is, $\int_{i \in I}$ is an exact functor.

Proof 2. By §5, Example 4, A is strongly C-O.K. Equation (1) above is condition (5) of Corollary 5.6.11. Therefore by Corollary 5.6.11, condition (1) of Proposition 5.6.10 holds.

Proof 3. Let
$$0 \to K_i \to H_i \to V_i \to 0$$
be a short exact sequence in $\mathscr{C}_A$, for all $i \in I$. Then
$$0 \to \bigoplus_{i \in I} K_i \to \bigoplus_{i \in I} H_i \to \bigoplus_{i \in I} V_i \to 0$$
is a short exact sequence of abstract left A-modules. Throwing through the system of derived functors C_n, $n \geq 0$, gives the long exact sequence in $\mathscr{C}_A$

$$(2): \quad \cdots \to C_1 \left(\bigoplus_{i \in I} V_i \right) \overset{d_1}{\to} C \left(\bigoplus_{i \in I} K_i \right) \to C \left(\bigoplus_{i \in I} H_i \right) \to C \left(\bigoplus_{i \in I} V_i \right) \to 0.$$

By equation (1) with $M_i = V_i$, $i \in I$, we have the short exact sequence

$$(3): \quad 0 \to C \left(\bigoplus_{i \in I} K_i \right) \to C \left(\bigoplus_{i \in I} H_i \right) \to C \left(\bigoplus_{i \in I} V_i \right) \to 0.$$

By Chapter 1, §3, Example 3, the t-adic ring A is C-O.K. Therefore equation (3) above and §1, Theorem 5.1.1 completes the proof. □

Remark 8. If one studies the proof of Proposition 5.6.10 and Corollary 5.6.11, one sees that Proof 3 above is really essentially the same as Proof 2 (and not very different from Proof 1).

Example 2. Let A be a ring and let t_1 and t_2 be elements in the center of the ring A. Regard A as a topological group with the topology given by powers of the right ideal $(t_1, t_2) = t_1 A + t_2 A$ generated by t_1 and t_2. Then by Chapter 1, §3, Example 2, A is a topological ring, and is admissible. Therefore, by Chapter 3, §1, Theorem 3.1.6, A is C-O.K.

Suppose that the elements t_1 and t_2 are also such that their images in $\widehat{A}$ are

(a) non-zero divisors, and are also such that

(b) $x, y \in \widehat{A}$, $t_1 x = t_2 y$, implies that t_2 divides x in $\widehat{A}$ and t_1 divides y in $\widehat{A}$.

Examples of such rings A and elements t_1, t_2 include:

1) A a unique factorization domain and t_1, t_2 non-associate irreducible elements;

2) A subexample of 1) is: If A is a regular local ring, and t_1, t_2 are among a set of uniformizing parameters;

3) If B is any (not necessarily commutative) ring, and let $A = B[t_1, t_2]$, the ring of polynomials in two central variables t_1, t_2;

4) The completion of examples 1), 2), or 3) with respect to the (t_1, t_2)-adic topology. Then I claim that the denumerable direct sum is not exact in the category of C-complete left A-modules.

Proof. By Chapter 2, §5, Proposition 2.5.8, the category of C^A-complete left A-modules coincides with the category of $C^{\widehat{A}}$-complete left $\widehat{A}$-modules. Therefore replacing A with $\widehat{A}$ if necessary, we can and do assume that A is complete.

Let $M_i = A/(At_1^i + At_2^i)$, $i \geq 1$. Then M_i is an abstract left A-module, and is discrete for the A-adic topology, since the open two-sided ideal $At_1^i + At_2^i$ annihilates $M_i, i \geq 0$. (The ideal $At_1^i + At_2^i$ is open in A since it contains $(At_1 + At_2)^{2i}$, $i \geq 0$.) M_i is A-adically discrete, therefore M_i is A-adically complete, and therefore is also C-complete, $i \geq 1$. A free acyclic homological length two of M_i is:

$$(1): \quad \cdots \longrightarrow 0 \longrightarrow 0 \longrightarrow A \xrightarrow{\phi_i} A \oplus A \xrightarrow{\psi_i} A \xrightarrow{\gamma} M_i \to 0$$
$$\qquad\qquad\quad \uparrow \qquad\quad \uparrow \qquad\quad \uparrow \qquad\qquad \uparrow \qquad\quad \uparrow$$
$$\qquad\quad dim\ 4 \qquad dim\ 3 \qquad dim\ 2 \qquad dim\ 1 \qquad dim\ 0$$

where $\phi_i : a \to (t_2^i a, -t_1^i a)$, $\psi_i : (x,y) \to t_1^i x + t_2^i y$, and $\gamma : A \to M_i$ is the natural map. Since the free modules in this resolution are of finite rank over a complete ring A, each term in the resolution is complete. Therefore the completion of (1) is an isomorphic copy of the sequence (1)

$$(\widehat{1}): \quad 0 \to \widehat{A} \to \widehat{A} \oplus \widehat{A} \to \widehat{A} \to M_i \to 0,$$

and therefore is exact. Thus

$$(2): \quad C_j(M_i) = 0, \ j \geq 1, \ i \geq 1.$$

The direct sum of the sequence (1), for $i \geq 1$, is a free homological acyclic resolution of $\underset{i \geq 1}{\oplus} M_i$

$$(3): \quad 0 \to A^{(\omega)} \to (A \oplus A)^{(\omega)} \to A^{(\omega)} \to \underset{i \geq 1}{\oplus} M_i \to 0.$$

Taking the A-adic completion, we get the cochain complex

$$(4): \quad \cdots \longrightarrow 0 \longrightarrow 0 \longrightarrow \widehat{A^{(\omega)}} \to \overbrace{\left[(A \oplus A)^{(\omega)}\right]}^{\phi} \xrightarrow{\psi} \widehat{A^{(\omega)}}$$
$$\qquad\qquad\quad \uparrow \qquad\quad \uparrow \qquad\quad \uparrow \qquad\qquad\qquad \uparrow \qquad\qquad\quad \uparrow$$
$$\qquad\quad dim\ 4 \qquad dim\ 3 \qquad dim\ 2 \qquad\qquad dim\ 1 \qquad\qquad dim\ 0$$

such that the jth homology of (4) is $C_j\left(\underset{i \geq 1}{\oplus} M_i\right)$, $j \geq 0$.

Consider the sequence $(t_2^i, -t_1^i)$, $i \geq 1$, of elements of $A \oplus A$. Since this sequence converges (t_1, t_2)-adically to zero, it defines an element

$$(5): \quad e = (t_2^i, -t_1^i)_{i \geq 1} \text{ of } \overbrace{\left[(A \oplus A)^{(\omega)}\right]}$$

The element e maps into zero in $\widehat{A^{(\omega)}}$ under the map ψ in (4). We have $\phi = \widehat{\underset{i \geq 1}{\oplus} \phi_i}$ and $\psi = \widehat{\underset{i \geq 1}{\oplus} \psi_i}$. We have that $\left(\widehat{\underset{i \geq 1}{\oplus} \phi_i}\right)((a_i)_{i \geq 1}) = (t_2^i a_i, -t_1^i a_i)_{i \geq 1}$. Therefore the only element $(a_i)_{i \geq 1}$ in A^ω such that $\left(\underset{i \geq 1}{\prod} \phi_i\right)((a_i)_{i \geq 1}) = e$ is the element

$(1)_{i\geq 1}$. Since $(1)_{i\geq 1}$ is not an element of the submodule $\widehat{A^{(\omega)}}$ of A^{ω}, it follows that $e \in \left[(A \oplus A)^{(\omega)}\right]\widehat{}$ is not in the image of ϕ. Therefore the element e defines a non-zero element in $C_1 \left(\underset{i \geq 1}{\oplus} M_i \right)$ Therefore

$$(6) : C_1 \left(\underset{i \geq 1}{\oplus} M_i \right) \neq 0.$$

If the functor, "denumerable direct sum," were exact in the category $\mathscr{C}_A$, then condition (1) of Proposition 5.6.10 would hold. Therefore by Proposition 5.6.10, condition (2) of Proposition 5.6.10 would hold. In particular,

$$C_1 \left(\underset{i \geq 1}{\oplus} M_i \right) = \int_{i \geq 1} C_1(M_i).$$

But then by equation (2), the right side is zero, so

$$C_1 \left(\underset{i \geq 1}{\oplus} M_i \right) = 0,$$

contradicting equation (6). Therefore the denumerable direct sum is not exact in the category of C-complete left A-modules. $\qquad \square$

Remark 9. In the proof of Example 2 above, it is easy to see that

$$(7) : C_1 \left(\underset{i \geq 1}{\oplus} M_i \right) \approx A^{\omega}/\widehat{A^{(\omega)}},$$

and that

$$(8) : C_j \left(\underset{i \geq 1}{\oplus} M_i \right) = 0 \text{ for } j \geq 2.$$

Proof. Since the map ϕ is injective, and since the jth homology of (4) is $C_j \left(\underset{i \geq 1}{\oplus} M_i \right)$, equation (8) follows. For each integer $i \geq 1$, using properties (a) and (b) (in the second paragraph of Example 2), it follows that the element $e_i = (t_2^i, -t_1^i)$ is a basis for $\text{Ker}(\psi_i)$. Since $\psi((x_i)_{i\geq 1}) = (\psi_i(x_i))_{i\geq 1}$, we have that $\text{Ker}\psi = \{\underset{i \geq 1}{\sum} a_i e_i : a_i \in A, i \geq 1\}$. (Notice that these are all elements of the subset $\left[(A \oplus A)^{(w)}\right]\widehat{}$ of $\underset{i \geq 1}{\prod}(A \oplus A)$.) Hence $\text{Ker}\psi \approx A^{\omega}$. And under this identification, the map ϕ is the inclusion: $\widehat{A^{(\omega)}} \to A^{\omega}$. $\qquad \square$

Remark 10. The topological ring A being as in Example 2, we have shown (in the first paragraph of Example 2), that the topological ring A is C-O.K. It can in fact be shown that $\hat{A}$ is strongly C-O.K. (We will not make any use of this; the reader can prove it as an exercise.)

Remark 11. In Example 2, if the ring A is *Noetherian* and *commutative*, then by §5, Example 3, A is strongly C-O.K.

Remark 12. The proof in Example 2 above uses Proposition 5.6.10 to show that denumerable direct sum is not exact in the category of C-complete left A-modules. In fact, the proof of this theorem is constructive and yields an explicit denumerable set of short exact sequences such that the direct sum (in the category of C-complete left A-modules) is not exact, as follows:

First, for simplicity, replacing A by $\hat{A}$ is necessary (since by Chapter 2, §5, Proposition 2.5.8, the category of C^A-complete left A-modules coincides with the category of $C^{\hat{A}}$-complete left $\hat{A}$-modules) we can and do assume that the topological ring A is complete.

Let I_i be the kernel of the natural map: $A \to M_i$; so that $I_i = At_1^i + At_2^i$. Then we have the short exact sequences of C-complete left A-modules:

$$(9): \ 0 \to I_i \to A \to M_i \to 0, \ i \geq 1.$$

Taking the direct sum in the category of abstract left A-modules we obtain the short exact sequence

$$0 \to \underset{i \geq 1}{\oplus} I_i \to A^{(\omega)} \to \underset{i \geq 1}{\oplus} M_i \to 0.$$

Throwing this short exact sequence through the system of left derived functors C_i, $i \geq 0$, we obtain the exact sequence of C-complete left A-modules:

$$(10): \ \cdots \to C_1(A^{(\omega)}) \to C_1\left(\underset{i \geq 1}{\oplus} M_i\right) \to C\left(\underset{i \geq 1}{\oplus} I_i\right) \to C(A^{(\omega)}) \to$$

$$C\left(\underset{i \geq 1}{\oplus} M_i\right) \to 0.$$

Since $A^{(\omega)}$ is a free abstract left A-module $C(A^{(\omega)}) = \widehat{A^{(\omega)}}$ and $C_1(A^{(\omega)}) = 0$. Also, in the first paragraph of Example 2, we have observed that A is C-O.K., so that by §1, Theorem 5.1.1, for any family N_i, $i \geq 1$, of C-complete left A-modules, we have that $C\left(\underset{i \geq 1}{\oplus} N_i\right) = \underset{i \geq 1}{\int} N_i$, the direct sum in the category of C-complete left A-modules. Therefore a portion of the exact sequence (10) is the exact sequence

$$(11): \ 0 \to C_1\left(\underset{i \geq 1}{\oplus} M_i\right) \to \underset{i \geq 1}{\int} I_i \to \underset{i \geq 1}{\int} A \to \underset{i \geq 1}{\int} M_i \to 0.$$

But, by equation (6) $C_1\left(\underset{i \geq 1}{\oplus} M_i\right) \neq 0$. Therefore equations (9) and (11) are an explicit example in which denumerable direct sum is not exact in the category of C-complete left A-modules.

Remark 13. In Example 2 above, the natural map:

$$\theta : \int_{i \geq 1} M_i \to \prod_{i \geq 1} M_i$$

is not injective. In fact we show that

$$(12): \mathtt{Ker}\left(\theta : \int_{i \geq 1} M_i \to \prod_{i \geq 1} M_i\right) \approx \frac{\left(\prod_{i \geq 1} I_i\right)}{\left(\widehat{\bigoplus_{i \geq 1} I_i}\right)}$$

canonically, which is clearly non-zero.

Proof. We can apply §2, Corollary 5.2.7 to show that the natural mapping θ is not injective. Better, we apply the §2, Remark following Theorem 5.2.6, to compute the kernel of θ.

In fact, consider the short exact sequence (9): $0 \to I_i \to A \to M_i \to 0$, where $I_i = At_1^i + At_2^i$, $i \geq 1$. Clearly, any sequence $x_i \in I_i = At_1^i + At_2^i$, $i \geq 1$, converges to zero in A. Therefore every element of $\prod_{i \geq 1} I_i$, is zero-bound in $\prod_{i \in I} A$, i.e., $\prod_{i \geq 1} I_i \subset \underset{i \in I}{\mathcal{Z}} A$. Therefore, the short exact sequence (5) of the Remark following Theorem 5.2.6, becomes:

$$(13): \ 0 \to \underset{i \geq 1}{\mathcal{Z}} I_i \to \prod_{i \geq 1} I_i \to \mathtt{Ker}(\theta) \to 0.$$

By definition, $\underset{i \geq 1}{\mathcal{Z}} I_i$ is the image of the natural map:

$$C\left(\bigoplus_{i \geq 1} I_i\right) = \int_{i \geq 1} I_i \to \prod_{i \geq 1} I_i.$$

Since $\prod_{i \geq 1} I_i$ is A-adically Hausdorff, it follows by Chapter 3, §2, Corollary 3.2.18, conclusion (2) (with $N = \bigoplus_{i \geq 1} I_i$) that this map factors through the epimorphism:

$$C\left(\bigoplus_{i \geq 1} I_i\right) \to \widehat{\bigoplus_{i \geq 1} I_i}.$$

Therefore $\underset{i \geq 1}{\mathcal{Z}} I_i$ is also the image of

$$\widehat{\bigoplus_{i \geq 1} I_i} \to \prod_{i \geq 1} I_i$$

— and this latter is a monomorphism since $\bigoplus_{i \geq 1} I_i$ always has the induced topology from $\prod_{i \geq 1} I_i$.

Therefore

$$(14): \underset{i \in I}{\mathcal{Z}}\, I_i = \widehat{\underset{i \geq 1}{\oplus} I_i}.$$

Equations (13) and (14) imply equation (12).   □

Remark 14. In Example 2 above, the natural map:

$$\int_{i \geq 1} I_i \to \prod_{i \geq 1} I_i$$

is not injective. The kernel of this map is isomorphic to $C_1 \left(\underset{i \geq 1}{\oplus} M_i \right)$ — and this latter is determined by equation (7) above.

Proof. We can and do assume that A is complete. We have $\int_{i \geq 1} A = C(A^{(\omega)}) = \widehat{A^{(\omega)}} \subset A^{\omega}$. Since also $\prod_{i \geq 1} I_i \subset A^{\omega}$, therefore

$$\mathbf{Ker} \left(\int_{i \geq 1} I_i \to \prod_{i \geq 1} I_i \right) = \mathbf{Ker} \left(\int_{i \geq 1} I_i \to A^{\omega} \right) = \mathbf{Ker} \left(\int_{i \geq 1} I_i \to \int_{i \geq 1} A \right),$$

and this latter is computed by equation (11).   □

Remark 15. In Remark 13, (respectively: Remark 14) above, the indicated kernel is the divisible part of the C-complete left A-module $\int_{i \geq 1} M_i = C \left(\underset{i \geq 1}{\oplus} M_i \right)$. (respectively: $\int_{i \geq 1} I_i = C \left(\underset{i \geq 1}{\oplus} I_i \right)$). This is so by §2, Corollary 5.2.10, since the C-complete left A-modules M_i (respectively: I_i) are Hausdorff, for $i \geq 1$. Therefore, we have yet another two examples (namely, M_i, $i \geq 1$; and I_i, $i \geq 1$) of denumerable families of A-adically complete left A-modules, such that the direct sum (in the category of C-complete left A-modules) is not Hausdorff.

Remark 16. In Example 2 above, we have shown that, for essentially any standard topological ring A that is "more complicated than" a t-adic ring such that the t-torsion is bounded below, that then the denumerable direct sum in the category of C-complete left A-modules is not exact. On the other hand, Theorem 5.6.13 shows that, for a t-adic ring A such that the t-torsion is bounded below, we have that for every cardinal number K, the K-fold direct sum is exact in the category of C-complete left A-modules. In the next section, we will show, however, that for any such t-adic ring A if e.g. t is in the center of A (and if the topology of A is not the discrete topology) that nevertheless the Eilenberg–Moore Axiom (S2) fails. This may be the first such known example (of an abelian category such that denumerable direct sums exist and are exact, and such that Axiom (S2) fails). One might paraphrase the result, that "Under the hypotheses of Theorem 5.6.13, if the topology of A is not discrete, then, although the denumerable direct sum (and even the K-fold direct sum, for all cardinals K) is exact in $\mathscr{C}_A$, that nevertheless the denumerable direct sum is very nearly not exact."

Thus, the philosophy is: "For a standard topological ring A that does not have the discrete topology, the direct sum in the category $\mathscr{C}_A$ of C-complete left A-modules usually is not exact. In the unusually simple cases (other than discrete rings) in which it is exact — e.g., for t-adic rings such that the t-torsion is bounded below — it "barely" is. (E.g., in the sense that, for t-adic rings with t in the center of A such that the t-torsion is bounded below, the Eilenberg–Moore Axiom (S2) fails.)

5.7 Consequences Related to the Direct Sum Not Being Exact

Lemma 5.7.1. *Let $\mathscr{A}$ be an abelian category and let I be a set. Suppose that I-fold direct sums exist in the abelian category $\mathscr{A}$, and that the I-fold direct sum is not exact. Then there exists a short exact sequence in the abelian category $\mathscr{A}$, such that the I-fold direct sum of this short exact sequence with itself is not exact.*

Proof. Let

$$(1): \ 0 \to A_i \to B_i \to C_i \to 0, \ i \in I,$$

be short exact sequences in $\mathscr{A}$ such that the sequence:

$$0 \to \bigoplus_{i \in I} A_i \to \bigoplus_{i \in I} B_i \to \bigoplus_{i \in I} C_i \to 0$$

is not exact. Let $A = \text{Im}\left(\bigoplus_{i \in I} A_i \to \bigoplus_{i \in I} B_i\right)$, $B = \bigoplus_{i \in I} B_i$ and $C = \bigoplus_{i \in I} C_i$. Then the sequence:

$$(2): \ 0 \to A \to B \to C \to 0$$

is exact. But I claim that the I-fold direct sum of this short exact sequence with itself:

$$(3): \ 0 \to A^{(I)} \to B^{(I)} \to C^{(I)} \to 0$$

is not exact.

In fact, for every $i \in I$, let

$$A^i = \text{Im}\left(\bigoplus_{j \in I - \{i\}} A_j \to \bigoplus_{j \in I - \{i\}} B_j\right), \ B^i = \bigoplus_{j \in I - \{i\}} B_j, \ C^i = \bigoplus_{j \in I - \{i\}} C_j.$$

Then for any fixed $i \in I$, the short exact sequence (2) is the direct sum of the two short exact sequences

$$(4): \ 0 \to A^i \to B^i \to C^i \to 0$$

and

$$(5): \ 0 \to A_i \to B_i \to C_i \to 0.$$

Therefore, taking the direct sum of the sequence (2) for all $i \in I$, we see that the short sequence (3) is the direct sum of the short sequence

$$(6): \ 0 \to \bigoplus_{i \in I} A^i \to \bigoplus_{i \in I} B^i \to \bigoplus_{i \in I} C^i \to 0$$

and the short sequence

$$(7): \ 0 \to \bigoplus_{i \in I} A_i \to \bigoplus_{i \in I} B_i \to \bigoplus_{i \in I} C_i \to 0.$$

Since the sequence (7) is not exact, it follows that the sequence (3) is not exact, as asserted.   □

Remark 1. Under the hypotheses of Lemma 5.7.1, if (1) are short exact sequences, for all $i \in I$, such that the direct sum, for all $i \in I$, is not exact, then we have constructed an *explicit* short exact sequence:

$$0 \to A \to B \to C \to 0$$

in the abelian category $\mathscr{A}$ such that the I-fold direct sum of this sequence with itself is not exact. Namely, by taking

$$A = \mathrm{Im}\left(\bigoplus_{i \in I} A_i \to \bigoplus_{i \in I} B_i\right), \ B = \bigoplus_{i \in I} B_i, \ \text{and} \ C = \bigoplus_{i \in I} C_i.$$

Remark 2. Suppose that $\mathscr{A}$ is an abelian category such that the I-fold direct *product* exists in the abelian category $\mathscr{A}$, and such that the I-fold direct product is not exact. Then given short exact sequences (1), for all $i \in I$ (as in Lemma 5.7.1), such that the direct product over all $i \in I$ is not exact, then one can make the dual construction using the I-fold direct product instead of the I-fold direct sum to construct another explicit short exact sequence

$$0 \to A \to B \to C \to 0$$

in $\mathscr{A}$ such that the I-fold product with itself is not exact.

Namely, by taking

$$A = \prod_{i \in I} A_i, \ B = \prod_{i \in I} B_i, \ C = \mathrm{Im}\left(\prod_{i \in I} B_i \to \prod_{i \in I} C_i\right).$$

The proof is then the same as Lemma 5.7.1—except that one defines

$$A^i = \prod_{j \in I - \{i\}} A_j, \ B^i = \prod_{j \in I - \{i\}} B_j, \ C^i = \mathrm{Im}\left(\prod_{j \in I - \{i\}} B_j \to \prod_{j \in I - \{i\}} C_j\right).$$

Note: Of course, this last Remark can be deduced from the preceding Remark, by passing to the dual category.

Remark 3. Suppose the hypotheses and notations are as in Lemma 5.7.1. Suppose in addition that the I-fold direct product is exact in the abelian category $\mathscr{A}$. Then, given short exact sequences:

$$(1): \ 0 \to A_i \to B_i \to C_i \to 0, \text{ for all } i \in I,$$

such that their I-fold direct sum is not short exact; then the sequence

$$(2): \ 0 \to \prod_{i \in I} A_i \to \prod_{i \in I} B_i \to \prod_{i \in I} C_i \to 0$$

is a short exact sequence in $\mathscr{A}$, such that the I-fold direct sum of (2) with itself is not exact in the abelian category $\mathscr{A}$.

Proof. For each $i \in I$, the short exact sequence

$$0 \to A_i \to B_i \to C_i \to 0$$

is a direct summand of the short exact sequence (2). Therefore, if the I-fold direct sum of (2) with itself were exact, then the direct sum of the short exact sequences (1), for all $i \in I$, would be exact. Since by hypothesis this is not the case, it follows that the I-fold direct sum of the short exact sequence (2) is not exact. $\qquad \square$

Example 1. Let the hypotheses and notations be as in §6, Example 2, and let $I_i = At_1^i + At_2^i$, $i \geq 1$. Then the sequence:

$$(2): \ 0 \to \prod_{i \geq 1} I_i \to \widehat{A}^\omega \to \prod_{i \geq 1} M_i \to 0$$

is a short exact sequence of C-complete left A-modules, such that the denumerable direct sum of the sequence (2) with itself, in the category of C-complete left A-modules, is not exact.

Proof. Follows from Remark 3 above and §6, Remark 12, equations (9), (11), and (6). $\qquad \square$

Remark 4. In Example 1 above, notice that I_i, $\widehat{A}$, and M_i are all A-adically complete. Therefore, the short exact sequence (2) is actually a short exact sequence of A-adically *complete* left A-modules, such that the denumerable direct sum with itself, in the category of C-complete left A-modules, is not exact.

Remark 5. Notice also that, in the sequence (2) of Example 1 above, the denumerable direct sum of the first group in the sequence (2) (respectively: third group in the sequence (2) with itself) in $\mathscr{C}_A$ is not Hausdorff. This follows from the fact that $\int_{i \geq 1} I_i$ (respectively: $\int_{i \geq 1} M_i$) is not Hausdorff, as has already been established in §6, Remark 15 (since clearly, a direct summand of a Hausdorff C-complete left A-module is Hausdorff). On the other hand, the denumerable direct sum of the middle group with itself in $\mathscr{C}_A$ *is* Hausdorff. (Since — e.g., by universal mapping properties — or by §1, Proposition 5.1.3, conclusion (1), — the direct sum in $\mathscr{C}_A$ of completions of abstract free left A-modules is the completion of an abstract free left A-module.) (There is no contradiction here — since the denumerable direct sum of the first group with itself is *not* contained in the denumerable direct sum of the second group with itself — since the denumerable direct sum of (2) with itself is *not* exact.)

Definition 1. Let $\mathscr{A}$ be an abelian category. Then $\mathscr{A}$ *obeys the Eilenberg–Moore Axiom* (P1) iff denumerable direct products of objects exist in $\mathscr{A}$, *and* the functor "denumerable direct product" from $\mathscr{A}^\omega$ into $\mathscr{A}$, is exact.

$\mathscr{A}$ *obeys the Eilenberg–Moore Axiom* (P2) iff denumerable direct products of objects exist in $\mathscr{A}$, and if, whenever $(A_i, \psi_{ij})_{i,j \geq 1}$ is an inverse system indexed by the positive integers in $\mathscr{A}$, such that the maps ψ_{ij} are all epimorphisms, then the induced map $\psi_i : \varprojlim_{i \geq 1} A_i \to A_1$ is an epimorphism.

An abelian category $\mathscr{A}$ *obeys the Eilenberg–Moore Axiom* (S1) (respectively: (S2)) iff the dual category $\mathscr{A}^o$ obeys the Eilenberg–Moore Axiom (P1) (respectively: (P2)).

Lemma 5.7.2. *Let $\mathscr{A}$ be an abelian category that obeys the Eilenberg–Moore Axiom (P2) (respectively: (S2)). Then $\mathscr{A}$ obeys the Eilenberg–Moore Axiom (P1) (respectively: (S1)).*

Proof. See [COC], Introduction, Chapter 1, §7, Remark, p. 86. $\square$

Lemma 5.7.3. *Let $\mathscr{A}$ be an abelian category such that denumerable direct sums exist. If $\mathscr{A}$ has enough injectives, then $\mathscr{A}$ obeys the Eilenberg–Moore Axiom (S2).*

Proof. Let $(A_i, \psi_{ij})_{i,j \geq 1}$ be a direct system in $\mathscr{A}$ such that ψ_{ij} is a monomorphism for all $i, j \geq 1$ with $i \leq j$. Then let Q be an injective and $f_i : A_1 \to Q$ be a monomorphism. By induction of $i \geq 1$ we define maps

$$f_i : A_i \to Q.$$

Having defined f_i for $1 \leq i \leq n$, since Q is injective, and since $\psi_{n,n+1} : A_n \to A_{n+1}$ is a monomorphism, we can choose

$$f_{n+1} : A_{n+1} \to Q$$

a map such that

$$f_{n+1} \circ \psi_{n,n+1} = f_n.$$

Then, by the universal mapping property definition of direct limit, there exists a unique map

$$f : \left(\varinjlim_{i \geq 1} A_i \right) \to Q$$

such that the composite

$$A_n \to \left(\varinjlim_{i \geq 1} A_i \right) \xrightarrow{f} Q$$

is f_n, for all $n \geq 1$. In particular, the diagram

is commutative. Since f_1 is a monomorphism, it follows that the natural map:

$$A_1 \to \varinjlim_{i \geq 1} A_i$$

is a monomorphism. □

Corollary 5.7.4. *The hypotheses being as in Lemma 5.7.3, we have that the abelian category $\mathscr{A}$ obeys Axiom (S1).*

Proof. Lemmas 5.7.2 and 5.7.3. □

Remark 6. Let $\mathscr{A}$ be an abelian category such that denumerable direct sums exist, and such that Axiom (S2) fails. Say, suppose that we have (A_i, ψ_{ij}) a direct system indexed by the positive integers such that ψ_{ij} is a monomorphism for all $i, j \geq 1$ with $i \leq j$, and such that the map $A_1 \to \varinjlim_{i \geq 1} A_i$ is not a monomorphism. Then the proof of Lemma 5.7.3 shows that A_1 does not admit a monomorphism into any injective object of $\mathscr{A}$.

Remark 7. Let $\mathscr{A}$ be an abelian category such that denumerable direct sums exist, and such that Axiom (S1) fails. Then by Corollary 5.7.4, we have that $\mathscr{A}$ does not have enough injectives. The proofs of Lemma 5.7.2 and 5.7.3 yield a specific object that cannot admit a monomorphism into any injective object of $\mathscr{A}$.

Another such construction is as follows. Suppose that

$$f_i : A_i \to B_i$$

are monomorphisms in the abelian category $\mathscr{A}$, for all $i \geq 1$, and consider

$$\bigoplus_{i \geq 1} f_i : \bigoplus_{i \geq 1} A_i \to \bigoplus_{i \geq 1} B_i.$$

Let $K = \mathrm{Ker} \left(\bigoplus_{i \geq 1} f_i \right)$. Then I claim that K does not admit a non-zero map into any injective object of $\mathscr{A}$.

Proof. Let Q be an injective object in $\mathscr{A}$ and let $h : K \to Q$ be a map. Then since Q is injective and $K \subset \underset{i \geq 1}{\oplus} A_i$, there exists a map $\alpha : \underset{i \geq 1}{\oplus} A_i \to Q$ such that $\alpha | K = h$. Let $\alpha_i : A_i \to Q$ be the ith component of α, $i \geq 1$. Then since $f_i : A_i \to B_i$ is a monomorphism, there exists a map $\beta_i : B_i \to Q$ such that $\beta_i \circ f_i = \alpha_i$, for all $i \geq 1$. Let $\beta : \underset{i \geq 1}{\oplus} B_i \to Q$ be the map whose ith coordinate is β_i, $i \geq 1$. Then $\beta \circ \left(\underset{i \geq 1}{\oplus} f_i \right) = \alpha$. Therefore map h is the composite:

$$ K \hookrightarrow \underset{i \geq 1}{\oplus} A_i \xrightarrow{\gamma} \underset{i \geq 1}{\oplus} B_i \xrightarrow{\beta} Q, $$

where $\gamma = \underset{i \geq 1}{\oplus} f_i$. Since the composite

$$ K \hookrightarrow \underset{i \geq 1}{\oplus} A_i \xrightarrow{\gamma} \underset{i \geq 1}{\oplus} B_i $$

is zero, it follows that $h = 0$. $\qquad\square$

<u>Remark 8</u>. Let the notations and hypotheses be as in §6, Example 2. Then the category of C-complete left A-modules does not have enough injectives.

In fact, the non-zero C-complete left A-module

$$ \widehat{A}^\omega / \widehat{A^{(\omega)}} $$

does not admit a non-zero map into any injective object in the category of C-complete left A-modules.

Proof. We have seen in §6, Example 2, that the abelian category $\mathscr{A}$ of C-complete left A-modules does not obey the Eilenberg–Moore Axiom (S1). Therefore by Corollary 5.7.4, $\mathscr{A}$ does not have enough injectives.

By §6, Remark 12, following Example 2, we have the short exact sequence:

$$ (11): \ 0 \to C_1 \left(\underset{i \geq 1}{\oplus} M_i \right) \to \int_{i \geq 1} I_i \to \int_{i \geq 1} A \to \int_{i \geq 1} M_i \to 0; $$

and by Remark 7 of this section, we have that

$$ C_1 \left(\underset{i \geq 1}{\oplus} M_i \right) $$

does not admit a non-zero map into any injective object of $\mathscr{A}$. But by §6, Remark 9, equation (7), we have that

$$ C_1 \left(\underset{i \geq 1}{\oplus} M_i \right) = \widehat{A}^\omega / \widehat{A^{(\omega)}}. $$

$\qquad\square$

Theorem 5.7.5. *Let A be a t-adic ring such that*

(1) There exists an element t such that the topology of A is the t-adic, and such that t is in the center of the ring A, and

(2) The topology of A is not the discrete topology.

Then the abelian category $\mathscr{C}_A$ of all C-complete left A-modules does not obey the Eilenberg–Moore Axiom (S2).

Note 1. Hypothesis (2) above is equivalent to:
$(2')$ t as in hypothesis (1) is not nilpotent.

Note 2. Hypotheses (1) and (2) can be replaced by the weaker hypothesis:

There exists a non-zero abstract left A-module M such that $atx = tax$ for all $a \in A$ and all $x \in M$, and such that (left multiplication by t): $M \to M$ is injective; where t is some element of the ring A such that the topology of A is the t-adic.

Proof. Let us first prove Note 1. The topology of A is discrete iff there exists an integer $n \geq 1$ such that $t^n A = 0$ — i.e., iff $t^n = 0$. This proves the Note.

Next, suppose that we have the hypotheses of the Theorem. Then let $T =$ (the t-torsion submodule of A), that is $T = \{a \in A : t^n a = 0 \text{ for some } n \geq 1\}$. Then T is a submodule of A. Let

$$M = A/T.$$

Then (left multiplication by t): $M \to M$ is injective.

Since t is not nilpotent, $1 \notin T$. Therefore $M = A/T$ is not the zero module. Therefore M obeys the conditions of Note 2.

Finally, let us show that if there exists an abstract left A-module M as in Note 2, then the category $\mathscr{C}_A$ does not obey the Eilenberg–Moore Axiom (S2).

Since M has non-zero t-torsion, by Chaper 4, §3, Lemma 4.3.20, it follows that $\widehat{M}$ has no non-zero t-torsion. That is

(3) The function "left multiplication by t": $\widehat{M} \to \widehat{M}$ is injective.

$(\widehat{M}, unif)$ is a topological module over the topological ring A. For every $a \in A$, the function "left multiplication by $at - ta$," from $(\widehat{M}, unif)$ into $(\widehat{M}, unif)$ is therefore continuous. Hence the kernel K of this function is closed in $(\widehat{M}, unif)$. Since $atx = tax$ for all $x \in M$, this kernel K contains the image of M. Since the image of M in $(\widehat{M}, unif)$ is dense, it follows that K is all of $\widehat{M}$. That is, that $(at - ta) \cdot \widehat{M} = 0$. Equivalently, that $atx = tax$, for all $x \in \widehat{M}$ and all $a \in A$. Equivalently,

(4) "left multiplication by t": $\widehat{M} \to \widehat{M}$ is an endomorphism of $\widehat{M}$ as abstract left A-module.

By Chapter 2, §2, Example 1, $\widehat{M}$ has a natural structure as a C-complete left A-module.

By Chapter 1, §3, Example 3, and Chapter 3, §1, Theorem 3.1.6, the t-adic ring A (and in fact every t-adic ring) is C-O.K. Therefore, by Chapter 3, §1, Definition 3, we have that condition (5) of Chapter 3, §1, Theorem 3.1.5 holds.

(That is, that an abstract left A-homomorphism of C-complete left A-modules is infinitely linear.) In particular, since by equation (4) above the homomorphism "left multiplication by t": $\widehat{M} \to \widehat{M}$ is A-linear, therefore

(5) "left multiplication by t": $\widehat{M} \to \widehat{M}$ is a map in the category $\mathscr{C}_A$ of C-complete left A-modules.

Consider the direct system in the category $\mathscr{C}_A$:

$$(6): \ \widehat{M} \xrightarrow{t} \widehat{M} \xrightarrow{t} \widehat{M} \xrightarrow{t} \cdots$$

where "t" denotes "left multiplication by t." By equations (3) and (5) above, this is a direct system indexed by the positive integers in the category $\mathscr{C}_A$, such that all the maps are monomorphisms. The direct limit of the direct system (1) in the category of abstract left A-modules is the localization of the A-module M at the element t; denote this $t^{-\infty}M$. It is a t-divisible left A-module. Therefore by Chapter 3, §2, Corollary 3.2.4

$$(7): \ C(t^{-\infty}M) = 0.$$

By §1, Theorem 5.1.2, the direct limit of the direct system (6) in the category of C-complete left A-modules is the C-completion of the direct limit in the category of abstract left A-modules. Therefore equation (7) implies that

(8) The direct limit of the direct system (6) in the category $\mathscr{C}_A$ is zero.

Therefore, we have a direct system indexed by the directed set of positive integers, of non-zero objects in the category $\mathscr{C}_A$, such that all of the maps are monomorphisms, and such that the direct limit in $\mathscr{C}_A$ is zero. Therefore the Eilenberg–Moore Axiom (S2) fails in the category $\mathscr{C}_A$. □

Remark 9. The hypothesis (2) in Theorem 5.7.5 is necessary for the conclusion. For, if A is a topological ring such that the topology is the discrete topology, then the category $\mathscr{C}_A$ of C-complete left A-modules coincides with the category of abstract left A-modules. (And of course the latter category obeys (S2) — and is even such that direct limits over directed sets are exact.)

Remark 10. In connection with Remark 9, notice that every abstract ring A can be made into a t-adic ring by giving A the discrete topology. (Then the element $t = 0 \in A$ is such that the t-adic topology is the topology of A.)

Corollary 5.7.6. *Let A be a t-adic ring that is commutative. Then the following two conditions are equivalent.*

(1) The topology of A is not the discrete topology.

(2) The category $\mathscr{C}_A$ of C-complete left A-modules does not obey the Eilenberg–Moore Axiom (S2).

Note: The hypothesis that "A is commutative" can be replaced by the less stringent hypothesis: "There exists an element t in the center of A such that the topology of A is the t-adic."

Proof. (1) implies (2) follows from Theorem 5.7.5. And (the negation of (1)) implies (the negation of (2)) is Remark 9 above. □

Example 2. Let A be a proper t-adic ring with parameter t, such that t is in the center of A. Then by Theorem 5.7.5 above, the category $\mathscr{C}_A$ of C-complete left A-modules does not obey the Eilenberg–Moore Axiom (S2). The proof of Theorem 5.7.5 constructs an explicit direct system in $\mathscr{C}_A$, indexed by the directed set of positive integers, in which the mappings are monomorphisms, and such that the direct limit in $\mathscr{C}_A$ is zero. Explicitly, for a proper t-adic ring A with parameter t in the center of A, this direct system is:

$$(9): \widehat{A} \overset{t}{\to} \widehat{A} \overset{t}{\to} \widehat{A} \overset{t}{\to} \cdots$$

Remark 11. Let the hypotheses and notations be as in Theorem 5.7.5. Then by Lemma 5.7.3 since Axiom (S2) fails, it follows that the category of C-complete left A-modules does not have enough injectives. In fact, by Remark 6 following Corollary 5.7.4, applied to the direct system (6) of Theorem 5.7.5, we have in fact that the C-complete left A-module $\widehat{M}$ does not admit a monomorphism into any injective in the category of C-complete left A-modules. This fact, and equation (8) of Theorem 5.7.5, is recorded in

Corollary 5.7.7. *Let A be a t-adic ring and let M be a C-complete left A-module such that*

(1) M has no non-zero t-torsion, and such that

(2) $t \cdot a \cdot x = a \cdot t \cdot x$, for all $a \in A$ and all $x \in M$ for some element $t \in A$ such that the topology of A is the t-adic. Then

$$\varinjlim_{i \geq 1}^{C}(M \overset{t}{\to} M \overset{t}{\to} M \overset{t}{\to} \cdots) = 0 \text{ in } \mathscr{C}_A.$$

And M does not admit a monomorphism into any injective object in the category $\mathscr{C}_A$.

Proof. Apply Remark 6 following Corollary 5.7.4 to the direct system $M \overset{t}{\to} M \overset{t}{\to} M \overset{t}{\to} \cdots$ in the category $\mathscr{C}_A$. $\square$

Theorem 5.7.8. *Let A be a t-adic ring such that the t-torsion in A is bounded below, and such that*

(1) The topology of A is not the discrete topology, and

(2) There exists an element t in the center of A such that the topology of A is the t-adic.

Then the abelian category $\mathscr{C}_A$ of all C-complete left A-modules obeys the Eilenberg–Moore Axiom (S1) but does not obey the Eilenberg–Moore Axiom (S2).

Note 1. In fact, $\mathscr{C}_A$ is such that, for every cardinal number K, the functor "K-fold direct sum": $\mathscr{C}_A^K \to \mathscr{C}_A$ is exact.

Proof. By Theorem 5.7.5, $\mathscr{C}_A$ does not obey (S2). And by §6, Theorem 5.6.13, the functor "K-fold direct sum": $\mathscr{C}_A^K \to \mathscr{C}_A$ is exact for every cardinal number K. $\square$

Remark 12. By Chapter 1, §3, Example 3 and Chapter 3, §1, Theorem 3.1.6, a t-adic ring is always C-O.K. Therefore (by Chapter 3, §1, Definition 3 and Chapter 3, §1, Theorem 3.1.5 condition (6) and by Chapter 2, §3, Theorem 2.3.12) we have that if A is a topological ring as in Theorem 5.7.8, then the abelian category $\mathscr{C}_A$ of all C-complete left A-modules, is a full, exact abelian subcategory of this category of abstract left A-modules. Explicitly (by Chapter 3, §1, Proposition 3.1.13, condition (8)), an abstract left A-module M is in the full subcategory $\mathscr{C}_A$ iff M is isomorphic to a cokernel of an abstract A-homomorphism of A-adically complete left A-modules. (For other similar characterizations of when an abstract left A-module over a C-O.K. topological ring is C-complete, see Chapter 3, §1, Proposition 3.1.13.)

And notice also, (by Chapter 4, §2, Corollary 4.2.3) that the inclusion functor from the abelian category $\mathscr{C}_A$ into the category Abs_A of abstract left A-modules is not only exact, but also preserves arbitrary inverse limits of arbitrary contravariant functors into $\mathscr{C}_A$ that are defined on set-theoretically legitimate categories.

Remark 13. Theorem 5.7.8 of course applies if A is a discrete valuation ring $\mathscr{O}$ that is not a field.

Therefore, if $\mathscr{O}$ is a discrete valuation ring that is not a field, regarded (as usual) as a topological ring by the topology given by powers of the maximal ideal, then the full subcategory of the category of all abstract $\mathscr{O}$-modules consisting of the C-complete $\mathscr{O}$-modules, is an exact, abelian subcategory that obeys Axiom (S1) but not Axiom (S2).

Chapter 6

Ext and Tor in the Category of C-complete Left A-modules

6.1 Ext in the Category of C-complete Left A-modules

If A is a standard topological ring, then we will let $\mathscr{C}_A$ stand for the category of all C-complete left A-modules and all infinitely linear functions. Thus, in particular, if M and N are objects in $\mathscr{C}_A$, then $\mathrm{Hom}_{\mathscr{C}_A}(M, N)$ denotes the set of all infinitely linear functions from M into N. (This is always a subgroup of $\mathrm{Hom}_A(M, N)$. If A is C-O.K., then $\mathrm{Hom}_{\mathscr{C}_A}(M, N) = \mathrm{Hom}_A(M, N)$.)

Remark 1. Let B be a topological ring such that the opposite ring B^o is a standard topological ring. Then we will say that B is a *left standard topological ring*. (Similarly, we will sometimes use the term *right standard*.) Thus, the topological ring B is left (respectively: right) standard iff the topology is given by left (respectively: right) ideals.

If A is a right standard (= standard) topological ring and B is a left standard topological ring, then in Chapter 2, §3, Remark 9 we have defined the notion of a C-complete (A, B)-bimodule M. It is, by definition: An abstract (A, B)-bimodule M (see [CEHA]), together with the structure on the abstract left A-module of $\mathscr{C}^A$-complete left A-module on M, and also the structure of $\mathscr{C}^B$-complete right B-module on the abstract right B-module M, such that, whenever I and J are sets, whenever $(a_j)_{j \in J}$ is a zero-bound family of elements of $\widehat{A}$, and whenever $(b_i)_{i \in I}$ is a zero-bound family of elements of $\widehat{B}$, and whenever $(x_{ji})_{(j,i)\in J\times I}$ is an arbitrary family of elements of M, we have that

$$(1):\quad \sum_{i\in I}\left(\sum_{j\in J}a_j\cdot x_{ji}\right)\cdot b_i = \sum_{j\in J}a_j\cdot\left(\sum_{i\in I}x_{ji}\cdot b_i\right).$$

Another way of writing this last compatibility condition is: Whenever I and J are sets, we have that the diagram

$$
\begin{array}{ccc}
\widehat{A^{(J)}}^A \underset{A}{\otimes} M^{I\times J} \underset{B}{\otimes} \widehat{B^{(I)}}^I & \xrightarrow{\ \phi\ } & \widehat{A^{(J)}}^A \times M^J \\
\Big\downarrow{\scriptstyle\rho} & & \Big\downarrow{\scriptstyle\psi} \\
M^I \underset{B}{\otimes} \widehat{B^{(I)}}^I & \xrightarrow{\ \ \ \ \tau\ \ \ \ } & M
\end{array}
$$

is commutative — where the maps ϕ and τ use the infinite sum structure of M as a C^B-complete right B-module, and the maps ρ and ψ use the infinite sum structure of M as a C^A-complete left A-module.

If N is a C-complete (A,B)-bimodule, then (by analogy with the standard notation for ordinary (A,B)-bimodules given in [CEHA]), we will write $_{\mathscr{C}_A}N_{\mathscr{C}_B}$. And we sometimes use the notation $\mathscr{C}_{(A,B)}$ for the category of all (A,B)-bimodules, with maps all functions that are both infinitely left A-linear and infinitely right B-linear.

Remark 2. If $_{\mathscr{C}_A}M$ and $_{\mathscr{C}_A}N_{\mathscr{C}_B}$, then we have seen (in Chapter 2, §3, Remark 9) that $\mathrm{Hom}_{\mathscr{C}_A}(M,N)$ has a natural structure as a C-complete *right* B-module — i.e., is an object of $\mathscr{C}_{B^\circ}$. We recall this construction briefly.

Since $_{\mathscr{C}_A}M$ and $_{\mathscr{C}_A}N_{\mathscr{C}_B}$, we have $_AM$ and $_AN_B$. Therefore $\mathrm{Hom}_A(M,N)$ is a right B-module. From the compatibility condition of Remark 1, (right scalar multiplication by b), from N into N, is infinitely A-linear for all $b\in B$. It follows that $\mathrm{Hom}_{\mathscr{C}_A}(M,N)$ is a right B-submodule of $\mathrm{Hom}_A(M,N)$. It remains to define a (right) B-infinite sum structure on the right B-module $\mathrm{Hom}_{\mathscr{C}_A}(M,N)$. If I is a set, if $(f_i)_{i\in I}$ are elements of $\mathrm{Hom}_{\mathscr{C}_A}(M,N)$ and if $(b_i)_{i\in I}$ is a zero-bound family of elements of $\widehat{B}$, then define

$$\sum_{i\in I}f_i\cdot b_i$$

to be the function from M into N, such that

$$(2):\quad \left(\sum_{i\in I}f_i\cdot b_i\right)(x) = \sum_{i\in I}f_i(x)\cdot b_i,$$

for all $x\in M$ (where the right B-infinite sum structure of N is used on the right side of this equation). Then, using the compatibility condition in Remark 1, we quickly verify (Chapter 2, §3, Remark 9) that $\sum_{i\in I}f_i\cdot b_i$ is infinitely A-linear.

Proof. If J is a set, $(a_j)_{j \in J}$ is a zero-bound family of elements of $\widehat{A}$ and if $(x_j)_{j \in J}$ is an arbitrary family of elements of M, then

$$\left(\sum_{i \in I} f_i \cdot b_i\right) \left(\sum_{j \in J} a_j \cdot x_j\right) = \sum_{i \in I} f_i \left(\sum_{j \in J} a_j x_j\right) \cdot b_i = \sum_{i \in I} \left(\sum_{j \in J} a_j f_i(x_j)\right) \cdot b_i$$

$$= \sum_{j \in J} a_j \left(\sum_{i \in I} f_i(x_j) \cdot b_i\right) = \sum_{j \in J} a_j \left(\sum_{i \in I} f_i \cdot b_i\right)(x_j).$$

The first equality follows from (2) above, the second since f_i is infinitely A-linear, the third from the compatibility condition (1) in Remark 1, and the last equality from (2) above. □

Therefore $\sum_{i \in I} f_i \cdot b_i \in \mathrm{Hom}_{\mathscr{C}_A}(M, N)$. The other axioms for a right B-infinite sum structure (i.e., of a left B^o-infinite sum structure) on $\mathrm{Hom}_{\mathscr{C}_A}(M, N)$ are then easily verified, see Chapter 2, §3, Remark 9.

Remark 3. If $_A M_B$ and $_A N$, then it is well-known that $\mathrm{Hom}_A(M, N)$ has the structure naturally of a left B-module, where the B-scalar product is such that

$$(b \cdot f)(x) = f(x \cdot b),$$

for all $f \in \mathrm{Hom}_A(M, N)$, $b \in B$, $x \in M$. This suggests the possibility that, if $\mathscr{C}_A M_{\mathscr{C}_B}$ and $\mathscr{C}_A N$, then $\mathrm{Hom}_{\mathscr{C}_A}(M, N)$ might possess a natural structure as a C-complete left B-module. This is *false* (even as C-complete left B-module), as is shown by Example 1 below. (However, it is nevertheless true in general that $\mathrm{Hom}_{\mathscr{C}_A}(M, N)$ has a natural structure as an *abstract left* B-module. Namely, it is a left B-submodule of $\mathrm{Hom}_A(M, N)$. The proof is easy. It uses the condition of Remark 1, and is similar to the proof, early in Remark 2 that, under the hypotheses of Remark 2, $\mathrm{Hom}_A(M, N)$ is an abstract right B-module.) Before proceeding to Example 1, let us demonstrate the difficulty with a naive attempt of a definition of such a B-infinite sum structure. The first difficulty is that if $M_{\mathscr{C}_B}$, then B is a *left* standard topological ring. To define a B-left infinite sum structure on $\mathrm{Hom}_{\mathscr{C}_B}(M, N)$, B should be *right* standard. But suppose that B is both left and right standard (this happens iff the topology of B is given by two-sided ideals). Then, in the obvious "naive attempt" of a definition of a left infinite sum structure on $\mathrm{Hom}_{\mathscr{C}_A}(M, N)$, the procedure would be: "If I is a set, and if $(b_i)_{i \in I}$ is zero-bound in $\widehat{B}$ and $(f_i)_{i \in I}$ are elements of $\mathrm{Hom}_{\mathscr{C}_A}(M, N)$, then let $\sum_{i \in I} b_i f_i$ be the function from M into N such that $\left(\sum_{i \in I} b_i f_i\right)(x) = \sum_{i \in I} f_i(x b_i)$."
The difficulty with this attempted definition is that the right side does not make sense. It is an infinite sum with no coefficients in N.

The next Example shows that there is no way to "patch up" this difficulty. Namely, it is an Example in which the abstract left B-module does not admit an infinite sum structure — even though the topological ring A is $\mathbf{Z}$ with the discrete topology.

Example 1. Let $A = \mathbf{Z}$, the ring of integers with the discrete topology, and let $B = \mathbf{Z} < T >$, the ring of formal power series in one variable T over the ring of integers. Then B is a commuutative, complete admissible topological ring (in fact B is Noetherian, and the topology is the T-adic; so that B is a proper t-adic ring with parameter T). Let $M = \mathbf{Z} < T >$. Then M is a (right) C-complete B-module. Since $\mathbf{Z}$ has the discrete topology, we have seen that the notions of "abstract $\mathbf{Z}$-module" and "C-complete $\mathbf{Z}$-module" coincide. Therefore M is a C-complete $(\mathbf{Z}, B)$-bimodule (i.e., $_{\mathscr{C}_{\mathbf{Z}}} M_{\mathscr{C}_B}$). Then I claim that, for every non-zero C-complete left A-module N — i.e., for every non-zero abelian group N — that the (left) B-module $\mathrm{Hom}_{\mathbf{Z}}(M, N)$ does not possess any infinite sum structure as a B-module.

Proof. If $f \in \mathrm{Hom}_{\mathbf{Z}}(M, N)$, then using the structure as an abstract (left) B-module, we have the element $T \cdot f \in \mathrm{Hom}_{\mathbf{Z}}(M, N)$. Explicitly,

$$(1): (T \cdot f)(u) = f \cdot (u \cdot T), \text{ for all } u \in M.$$

Every formal power series $x \in \mathbf{Z} < T >$ can be explicitly written uniquely in the form

(2): $x = a_0 + T \cdot y$,

where $a_0 \in \mathbf{Z}$ (is the "constant term of x"), and $y \in \mathbf{Z} < T >$. If now $g \in \mathrm{Hom}_{\mathbf{Z}}(M, N)$ is arbitrary, then define a function f from $M = \mathbf{Z} < T >$ into N, by

(3): $f(x) = g(y)$,

where y is determined from x by equation (2), for all $x \in M$. Then $f \in \mathrm{Hom}_{\mathbf{Z}}(M, N)$; and, by equations (2) and (3), $f(Ty) = g(y)$, for all $y \in M$. But then, by equation (1), $T \cdot f = g$. $g \in \mathrm{Hom}_{\mathbf{Z}}(M, N)$ being arbitrary, it follows that the abstract (left) $\mathbf{Z} < T >$-module $\mathrm{Hom}_{\mathbf{Z}}(M, N)$ is T-divisible. Therefore this module is a divisible abstract (left) module over the topological ring $\mathbf{Z} < T >$. Since the topological ring $\mathbf{Z} < T >$ admits a denumerable neighborhood base at zero, by Chapter 3, §2, Corrolary 3.2.5, we have that every divisible C-complete (left) $\mathbf{Z} < T >$-module is a zero module. Since $\mathrm{Hom}_{\mathbf{Z}}(M, N) \neq \{0\}$ (since this abelian group contains the direct summand $\mathrm{Hom}_{\mathbf{Z}}(\mathbf{Z}, N) \approx N \neq 0$), it follows that the abstract (left) $\mathbf{Z} < T >$-module $\mathrm{Hom}_{\mathbf{Z}}(M, N)$ cannot admit any structure as a C-complete (left) $\mathbf{Z} < T >$-module. $\square$

Remark 4. The detailed structure of the $\mathbf{Z} < T >$-module $\mathrm{Hom}_{\mathbf{Z}}(\mathbf{Z} < T >, \mathbf{Z})$ discussed in Example 1 is as follows: It is a fairly well-known theorem that

$$\mathrm{Hom}_{\mathbf{Z}}(\mathbf{Z}^{\omega}, \mathbf{Z}) \approx \mathbf{Z}^{(\omega)}$$

— i.e., that the only homomorphisms of abelian groups from $\mathbf{Z}^{\omega}$ into $\mathbf{Z}$ are finite integer linear combinations of the canonical projections $\pi_n : \mathbf{Z}^{\omega} \to Z$, $n \geq 0$. Since $\mathbf{Z} < T > \approx \mathbf{Z}^{\omega}$ as abelian group, it follows that $\mathrm{Hom}_{\mathbf{Z}}(\mathbf{Z} < T >, \mathbf{Z})$ is a free abelian group on the functions π_i, $i \geq 0$, where for $i \geq 0$,

$$\pi_i \left(\sum_{j \geq 0} a_j T^j \right) = a_i.$$

Then

$$(T \cdot \pi_{i+1}) \left(\sum_{j \geq 0} a_j T^j \right) = \pi_{i+1} \left(\left(\sum_{j \geq 0} a_j T^j \right) \cdot T \right) = \pi_{i+1} \left(\sum_{j \geq 0} a_j T^{j+1} \right)$$

$$= a_i, \ i \geq -1$$

where $a_{-1} = 0$, so that $T \cdot \pi_{i+1} = \pi_i$, $i \geq 0$, $T \cdot \pi_0 = 0$. It follows that, as $\mathbf{Z} < T >$-module,

$$\mathrm{Hom}_{\mathbf{Z}}(\mathbf{Z} < T >, \mathbf{Z}) \approx (T^{-1}\mathbf{Z} < T >)/(\mathbf{Z} < T >),$$

the quotient of $\mathbf{Z} < T >$-module of the ring $T^{-1}\mathbf{Z} < T >$ of formal Laurent series over $\mathbf{Z}$ in one variable T, by the ring $\mathbf{Z} < T >$ of formal power series over $\mathbf{Z}$ in one variable T. (The isomorphism is given explicitly by making π_i correspond to the coset of T^{-i-1}, $i \geq 0$.)

Definition 1. Let A be a standard topological ring and let N be an arbitrary fixed C-complete left A-module. Then the functor: $M \rightsquigarrow \mathrm{Hom}_{\mathscr{C}_A}(M, N)$ is a left exact contravariant functor from the category $\mathscr{C}_A$ into the category of abelian groups. Since $\mathscr{C}_A$ has enough projectives, we have the derived functors of this functor. We denote these $\mathrm{Ext}^n_{\mathscr{C}_A}(M, N)$, $n \geq 0$.

Remark 5. Let A be a right standard topological ring and let B be a left standard topological ring. Let N be a fixed C-complete (A, B)-bimodule. Then, in Remark 2 above, we have seen that, for every object M in $\mathscr{C}_A$, that the abelian group $\mathrm{Hom}_{\mathscr{C}_A}(M, N)$ possesses a natural structure as C-complete *right* B-module. Therefore in this case, the functor: $M \rightsquigarrow \mathrm{Hom}_{\mathscr{C}_A}(M, N)$ of Definition 1 is a contravariant left exact functor from the category $\mathscr{C}_A$ into the category $\mathscr{C}_{B^o}$ of all C-complete *right* B-modules. [1] It follows that $\mathrm{Ext}^n_{\mathscr{C}_A}(M, N)$ has a natural structure as a C-complete *right* B-module, for all $n \geq 0$ and all objects M in $\mathscr{C}_A$ — such that the assignment: $M \rightsquigarrow \mathrm{Ext}^n_{\mathscr{C}_A}(M, N)$, $n \geq 0$, is a system of right derived contravariant functors from the category $\mathscr{C}_A$ into the category $\mathscr{C}_{B^o}$.

Proof. Let E^n, $n \geq 0$ be the right derived functors of $M \rightsquigarrow \mathrm{Hom}_{\mathscr{C}_A}(M, N)$ regarded as contravariant left-exact functor from the abelian category with enough projectives $\mathscr{C}_A$ into the abelian category $\mathscr{C}_{B^o}$. Let V denote the "stripping functor" from the category $\mathscr{C}_{B^o}$ into the category of abelian groups, an exact functor. Then $V \circ E^n$, $n \geq 0$, obeys the axioms for a system of contravariant right derived functors from $\mathscr{C}_A$ into the category of abelian groups. Therefore by Definition 1

$$V(E^n(M)) \approx \mathrm{Ext}^n_{\mathscr{C}_A}(M, N), \ n \geq 0,$$

---

[1]Note: In general, if B is a ring, by the *opposite ring* B^o we mean the ring having the same underlying additive abelian group as B, and such that $(x \cdot y$ in $B^o) = (y \cdot x$ in $B)$. Thus, an abstract left (respectively: right) B^o-module is the same as an abstract right (respectively: left) B-module; and a C-complete left (respectively: right) B^o-module is the same as a C-complete right (respectively: left) B^o-module. In particular, therefore, $\mathscr{C}_{B^o}$ is the category of all C-complete right B-modules. And similarly, $\mathscr{C}_{(A,B)} = \mathscr{C}_{(B^o,A^o)}$, for all pairs A and B, where A (respectively: B) is a right (respectively: left) standard topological ring.

canonically as abelian groups. Using this canonical isomorphism, we therefore obtain a natural structure of C-complete right B-module on $\text{Ext}^n_{\mathscr{C}_A}(M,N)$, for all $n \geq 0$, and all $M \in \mathscr{C}_A$. And $\text{Ext}^n_{\mathscr{C}_A}(M,N)$, being so regarded as C-complete right B-modules for all $M \in \mathscr{C}_A$ and all $n \geq 0$, form a system of derived functors from $\mathscr{C}_A$ into $\mathscr{C}_{B^\circ}$ — the derived functors of $\text{Ext}^0_{\mathscr{C}_A}(M,N) = \text{Hom}_{\mathscr{C}_A}(M,N)$. $\square$

Remark 6. Let A be a standard topological ring that is commutative. Then of course every C-complete left A-module has a natural structure as a C-complete (A,A)-bimodule. Therefore, by Remark 2, if M and N are objects in $\mathscr{C}_A$, then the abelian group $\text{Hom}_{\mathscr{C}_A}(M,N)$, and more generally the abelian group $\text{Ext}^n_{\mathscr{C}_A}(M,N)$, for all integers $n \geq 0$, possesses a natural structure as C-complete A-module. And, again by Remark 5, for every C-complete A-module N, the assignment:

$$M \rightsquigarrow \text{Ext}^n_{\mathscr{C}_A}(M,N), \ n \geq 0,$$

is a system of contravariant right derived functors from $\mathscr{C}_A$ into $\mathscr{C}_A$.

Theorem 6.1.1. *Let A be an arbitrary standard topological ring, let N be an arbitrary C-complete left A-module and let M be an arbitrary abstract left A-module. Then there is induced a first quadrant cohomological spectral sequence in the category of abelian groups, such that*

(1): $E_2^{p,q} = \text{Ext}^p_{\mathscr{C}_A}(C_q(M),N) \Rightarrow \text{Ext}^n_A(M,N)$.

Note 1. If the C-complete left A-module N has the structure of a C-complete (A,B)-bimodule, where B is a left standard topological ring, then the spectral sequence (1) occurs in the abelian category $\mathscr{C}_{B^\circ}$ of C-complete right B-modules.

Note 2. In particular, if the abstract ring A is *commutative*, then the spectral sequence (1) occurs in the category $\mathscr{C}_A$ of all C-complete A-modules.

Remark 7. The reader should note that, in the spectral sequence (1) of Theorem 6.1.1, that the groups of the abutment, $\text{Ext}^n_A(M,N)$, are the (ordinary abstract) "Ext," using only the structure of A as an abstract ring and the structures of M and N as abstract left A-modules — that is, the derived functors (in either or both variables) of the usual $\text{Hom}_A(M,N)$.

Proof. The functor C from the category of abstract left A-modules into the category of $\mathscr{C}_A$ is right exact, and maps projectives into projectives. The functor $M \rightsquigarrow \text{Hom}_{\mathscr{C}_A}(M,N)$ is contravariant and left exact from $\mathscr{C}_A$ into the category of abelian groups (in the case of Note 1, "... into the category $\mathscr{C}_{B^\circ}$"). By the universal mapping property definition of $C(M)$, we have that $\text{Hom}_{\mathscr{C}_A}(C(M),N) = \text{Hom}_A(M,N)$. Therefore the indicated spectral sequence is the spectral sequence of the composite functor:

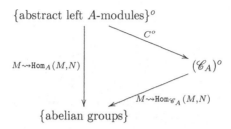

$$\{\text{abelian groups}\}$$

□

Theorem 6.1.2. *Let A be a topological ring and let N be a non-negative integer. Suppose that the topological ring A is strongly C-O.K. (respectively: N-C-O.K.). Then for every pair M, H of C-complete left A-modules, we have that*

$$\text{Ext}^n_{\mathscr{C}_A}(M, H) = \text{Ext}^n_A(M, H)$$

for all integers $n \geq 0$ (respectively: for all integers n such that $N \geq n \geq 0$).

Proof. It suffices to prove the parenthetical assertion. In the spectral sequence (1) of Theorem 6.1.1, we have that $E_2^{p,q} = 0$ for $1 \leq q \leq N$. Therefore $E_2^{n,0} = $ (nth group of the abutment) for $0 \leq n \leq N$. That is, $\text{Ext}^n_{\mathscr{C}_A}(C(M), H) = \text{Ext}^n_A(M, H)$, $0 \leq n \leq N$. Since A is N-C-O.K. for $N \geq 0$, A is C-O.K., and $C(M) = M$. □

Corollary 6.1.3. *Under the hypotheses of Theorem 6.1.2, we have that the edge homomorphism is a monomorphism of abelian groups:*

$$\text{Ext}^{N+1}_{\mathscr{C}_A}(M, H) \to \text{Ext}^{N+1}_A(M, H).$$

Remark. Of course if A is strongly C-O.K. (or even $(N + 1)$-C-O.K.), then by Theorem 6.1.2 this map is an isomorphism.

Proof. In the spectral sequence of Theorem 6.1.1, consider the $E_2^{N+1,0}$-term. For $r \geq 2$, $d_r^{N+1,0}$ maps $E_r^{N+1,0}$ into $E_r^{N+1+r,-r}$. Since $-r \leq -2$, the range is not in the first quadrant. Therefore

$$(1): \quad d_r^{N+1,0} = 0 \text{ for } r \geq 2.$$

Also, for $r \geq 2$, consider

$$d_r^{N-r+1,r-1} : E_r^{N-r+1,r-1} \to E_r^{N+1,0}.$$

If $r \geq N + 2$, then $N - r + 1 \leq -1$, and the domain is not in the first quadrant and therefore

$$(2): \quad E_r^{N-r+1,r-1} = 0.$$

If $r \leq N+1$, then $r-1 \leq N$, and therefore $E_2^{N-r+1,r-1} = \text{Ext}_{\mathscr{C}_A}^{N-r+1}(C_{r-1}(M), H)$. Since A is by hypothesis N-C-O.K. and $r-1 \leq N$, and M is C-complete, by §5, Definition 1, $C_{r-1}(M) = 0$. Therefore we again obtain equation (2) above. Therefore, for $r \geq 2$, $d_r^{N-r+1,r-1} = 0$ and $d_r^{N+1,0} = 0$. It is equivalent to say that

$$(3): \quad E_2^{N+1,0} = E_\infty^{N+1,0}.$$

But, since $E_\infty^{N+1+a,-a} = 0$ for $a > 0$ (since $E_2^{N+1+a,-a} = 0$ for $a > 0$, since the term $E_2^{N+1+a,-a}$ is not first quadrant for $a > 0$), we have that $E_\infty^{N+1,0}$ is a subobject of the $(N+1)$st group of the abutment. That is,

$$E_\infty^{N+1,0} \hookrightarrow \text{Ext}_A^{N+1}(M, H).$$

But by equation (3),

$$\text{Ext}_{\mathscr{C}_A}^{N+1}(C(M), H) = E_2^{N+1,0} = E_\infty^{N+1,0}.$$

Combining these two equations, and since also $C(M) = M$ (since M is C-O.K.) gives the proof.   □

Remark 8. Let A and N be as in Theorem 6.1.2. Then, by Chapter 5, §5, Theorem 5.5.4, we also have that

Corollary 6.1.4. *Let the hypotheses be as in Theorem 6.1.2. Then for every pair M and H of C-complete left A-modules, we have that $\text{Ext}_{\mathscr{C}_A}(M, H) = \text{Ext}_A^n(M, H) = \text{Ext}_{\widehat{A}}^n(M, H)$ for all integers $n \geq 0$ (respectively: all integers n such that $N \geq n \geq 0$).*

Proof. By Remark 8 above, Theorem 6.1.2 also applies to the topological ring $\widehat{A}$. And $\mathscr{C}_A = \mathscr{C}_{\widehat{A}}$ (by Chapter 2, §3, Corollary 2.3.2).   □

Remark 9. Corollary 6.1.4 is perhaps not too surprising, since for $n = 0$ it asserts that $\mathscr{C}_A$ is a full subcategory of the category of abstract left A, respectively: $\widehat{A}$, modules (— which we know since A and $\widehat{A}$ are C-O.K.).

Remark 10. As we have observed in Chapter 5, §5, Examples 3, 4, and 5, the hypothesis of Theorem 6.1.2 that the topological ring A be strongly C-O.K., "almost always" holds. Therefore, by Theorem 6.1.2 one does not have to be too careful with the "$\mathscr{C}_A$" subscript in "$\text{Ext}_{\mathscr{C}_A}^n(M, N)$", since this is "almost always" isomorphic to $\text{Ext}_A^n(M, N)$ for all $n \geq 0$.

Proposition 6.1.5. *Let A be a standard topological ring and n be a non-negative integer. Suppose that A is n-C-O.K. Let M be a C-complete left A-module. Suppose that the projective dimension of M as an abstract left A module (as defined in [CEHA], Chapter 6, beginning of §2, p. 109) is $\leq n$. Then if P_* is a homological projective acyclic resolution of M in the category of C-complete left A-modules, then*

   (1): $\text{Im}(P_{n+1} \to P_n)$ is a projective C-complete left A-module, and

(2): If

$$Q_i = \begin{cases} P_i, & i < n \\ \mathrm{Im}(P_{n+1} \to P_n), & i = n \\ 0, & i \geq n+1 \end{cases}$$

then Q_* is also a homological acyclic projective resolution of M in the category of C-complete left A-modules. And $Q_i = 0$ for $i \geq n+1$.

Proof 1. $\mathrm{Ext}_A^{n+1}(M, N) = 0$, for all abstract left A-modules N. Therefore by Corollary 6.1.3, if N is any C-complete left A-module, then

$$\mathrm{Ext}_{\mathscr{C}_A}^{n+1}(M, N) \hookrightarrow \mathrm{Ext}_A^{n+1}(M, N) = 0.$$

Therefore the abelian category with enough projectives $\mathscr{C}_A$ has left projective dimension $\leq n$. The result of the Proposition follows by the argument in [CEHA], Chapter 6, §2, Proposition 2.1, p. 110. □

Proof 2. By hypothesis, the functor $M \rightsquigarrow \mathrm{Ext}_A^n(M, N)$ is right-exact from the category of abstract left A-modules into the category of abelian groups. By Theorem 6.1.2,

$$\mathrm{Ext}_{\mathscr{C}_A}^n(M, N) = \mathrm{Ext}_A^n(M, N)$$

for all C-complete left A-modules N. Therefore, the functor

$$N \rightsquigarrow \mathrm{Ext}_{\mathscr{C}_A}^n(M, N)$$

is right exact, from $\mathscr{C}_A$ into the category of abelian groups. But then, by the well-known argument in [CEHA], Chapter 6, §2, Proposition 2.1, p. 110, we obtain the conclusion of the Proposition. □

Remark 11. In general, let $\mathscr{A}$ be an abelian category with enough projectives. Then, given an object M in $\mathscr{A}$, define the *projective dimension* of M to be the smallest integer n such that there exists a homological acyclic projective resoluion P_* of M, such that $P_i = 0$ for $i \geq n+1$. The projective dimension of M is by definition $+\infty$ if there is no such integer. Then, the proof of [CEHA], Chapter 6, §2, Proposition 2.1 goes through, mutatis mutandis.

[CEHA], Chapter 6, §2, Proposition 2.1
Let $\mathscr{A}$ be an abelian category with enough projectives, and let M be an object in $\mathscr{A}$ and let n be an integer ≥ 0. Then the following four conditions are equivalent:

(a) proj dim $M \leq n$.

(b) $\mathrm{Ext}_{\mathscr{A}}^{n+1}(M, H) = 0$, for all objects H in $\mathscr{A}$.

(c) The functor from the abelian category $\mathscr{A}$ into the category of abelian groups, $H \rightsquigarrow \mathrm{Ext}_A^n(M, H)$, is right exact.

(d) If $0 \to X_n \to X_{n-1} \to \cdots \to X_0 \to M \to 0$ is an exact sequence in the category $\mathscr{A}$, such that $X_0, \ldots, X_{n-1}$ are projective in $\mathscr{A}$, then X_n is projective.

(For the proof, see p. 110 of [CEHA]).

With the above terminology, another way of phrasing Proposition 6.1.5, is:

Corollary 6.1.6. *Let A be a standard topological ring such that A is strongly C-O.K. Let M be a C-complete left A-module. Then the projective dimension of M qua C-complete left A-module is less than or equal to the projective dimension of M qua abstract left A-module.*

Note: More generally, the hypothesis that "A is strongly C-O.K." can be weakened to: "A is n-C-O.K.," where $n \geq$ proj. dim. M *qua* abstract left A-module.

Proof. This is a restatement of Proposition 6.1.5.   □

Remark 12. The projective dimension of a C-complete left A-module M, over a strongly C-O.K. topological ring A, can be strictly smaller than the projective dimension of M as an abstract left A-module.

Example 2. Let A be a standard topological ring and let F be a free abstract left A-module of infinite rank. Then it is almost never true that $\widehat{F}$ is projective as an abstract left A-module. [2] When this is the case, the projective dimension of $\widehat{F}$ as an abstract left A-module is therefore ≥ 1. However, $\widehat{F}$ is a projective object in the category $\mathscr{C}_A$ of C-complete left A-modules (Chapter 2, §4, Theorem 2.4.8). Therefore the projective dimension of $\widehat{F}$ *qua* C-complete left A-module is zero.

Example 3. Let A be a standard topological ring that is not complete. Then the completion $\widehat{A}$ of A is almost never projective as an abstract left A-module.[3] Therefore as in Example 2, we have that proj. dim. A *as an abstract left A-module* ≥ 1, but that proj. dim. A *as a C-complete left A-module* equals zero.

Remark 13. In general, if $\mathscr{A}$ is an abelian category with enough projectives, then (by analogy with the definition in [CEHA], Chapter 6, beginning of §2, p. 109 — which is the special case in which the category $\mathscr{A}$ is the category of left modules over a ring) define the *projective dimension* of $\mathscr{A}$ to be the supremum of proj. dim. M, for all objects M in $\mathscr{A}$. In particular, if A is a standard topological ring, then we define the *projective dimension* of the topological ring A to be the projective dimension of the abelian category $\mathscr{C}_A$. Then

---

[2]E.g., consider the case in which A is a principal ideal domain not a field, and the topology is given by powers of some maximal ideal $\mathbf{p}$. Then every projective A-module is free. Let $K(\mathbf{p}) = A/\mathbf{p}$. If $\widehat{F}$ were projective, then $\widehat{F}$ would be free of rank equal to $dim_{K(\mathbf{p})} \widehat{F} \underset{A}{\otimes} K(\mathbf{p})$. Since $\widehat{F} \underset{A}{\otimes} K(\mathbf{p}) \approx F \underset{A}{\otimes} K(\mathbf{p})$ (since A is admissible, by Chapter 1, §3, Example 1), it would follow that $\widehat{F} \approx F$. Therefore the module F would be complete. But if $(e_i)_{i \in I}$ is a basis for F, if $(a_i)_{i \in I}$ is any family of elements of $\widehat{A}$ that converges to zero, and such that for infinitely many $i \in I$, $a_i \neq 0$, then $\sum_{i \in I} a_i e_i \in \widehat{F}$, but not in F. Therefore F is not A-adiccally complete. Therefore $\widehat{F}$ is not projective as an abstract left A-module.

[3]E.g., consider the case in which A is, say, a discrete valuation ring that is not complete. Then every projective abstract A-module is free. If $\widehat{A}$ were projective, then it would be free. The rank of $\widehat{A}$ as a free module would be the dimension of $\widehat{A} \underset{A}{\otimes} K(A)$ as a $K(A)$ vector space. Since this latter is one, $\widehat{A}$ would be free of rank one. But then $\widehat{A} \approx A$, contradicting the hypothesis that A is not complete.

Corollary 6.1.7. *Let A be a standard topological ring that is strongly C-O.K. Then the projective dimension of A as a standard topological ring is less than or equal to the projective dimension of the abstract ring A.*

Note: More generally, if A is a standard topological ring and n is a non-negative integer such that A is n-C-O.K., and such that the projective dimension of A as an abstract ring is $\leq n$, then the projective dimension of A as a topological ring is $\leq n$.

Proof. Follows immediately from Proposition 6.1.5. □

6.2   C-complete Tensor Product

Let A be a commutative standard topological ring, and let M and N be C-complete A-modules. Then in Chapter 2, §3, Remark 8, we have defined the *C-complete tensor product* $M \overset{C}{\underset{A}{\otimes}} N$. Let us briefly recall the definition.

If H is another C-complete A-module, and if $f : M \times N \to H$ is a function, then f is *infinitely bilinear* iff for every $m \in M$ (respectively: $n \in N$), the function $n \to f(m,n)$, (respectively: $m \to f(m,n)$), is infinitely linear, from N (respectively: M) into H. Then a *C-complete tensor product of M and N* is:

(1): A C-complete A-module denoted $M \overset{C}{\underset{A}{\otimes}} N$.

(2): An infinitely bilinear function: $M \times N \to M \overset{C}{\underset{A}{\otimes}} N$ (Notation: the image of (m,n) is denoted $m \overset{C}{\underset{A}{\otimes}} n$, all $m \in M$, $n \in N$) such that

(3): (1) and (2) above are universal with these properties.

Then in Chapter 2, §3, Remark 8, we proved that the C-complete tensor product $M \overset{C}{\underset{A}{\otimes}} N$ exists, and is unique up to canonical isomorphism. We give an alternate construction, especially useful in the case in which A is C-O.K.

Theorem 6.2.1. *Let A be a commutative standard topological ring that is C-O.K. Then for every pair M, N of C-complete A-modules, we have that*

$$M \overset{C}{\underset{A}{\otimes}} N = C(M \underset{A}{\otimes} N)$$

— that is, the C-complete tensor product of M and N, is the C-completion of the abstract A-module $M \underset{A}{\otimes} N$.

Proof. One immediately verifies the universal mapping property. □

Remark 1. A more general form of Theorem 6.2.1, that holds for commutative standard topological rings that are not necessarily C-O.K., is:

Corollary 6.2.2. *Let A be an arbitrary commutative standard topological ring, and let M and N be C-complete left A-modules. Then*

$$M \overset{C}{\underset{A}{\otimes}} N \approx [C(M \underset{A}{\otimes} N)]/H,$$

where H is the C-complete A-submodule of $C(M \underset{A}{\otimes} N)$ generated by $S \cup T$, where

$$S = \{ \iota \left[\left(\sum_{i \in I} a_i x_i \right) \underset{A}{\otimes} y \right] - \sum_{i \in I} a_i [\iota (x_i \underset{A}{\otimes} y)] : I \text{ is a set, } (a_i)_{i \in I} \in \widehat{A} \text{ is zero-bound,}$$

$$x_i \in M, \text{ all } i \in I, \text{ and } y \in N \},$$

$$T = \{ \iota \left[x \underset{A}{\otimes} \left(\sum_{i \in I} a_i y_i \right) \right] - \sum_{i \in I} a_i [\iota (x \underset{A}{\otimes} y_i)] : I \text{ is a set, } (a_i)_{i \in I} \in \widehat{A} \text{ is zero-bound,}$$

$$x \in M, \text{ and } y_i \in N, \text{ all } i \in I \},$$

and where $\iota : M \underset{A}{\otimes} N \to C(M \underset{A}{\otimes} N)$ is the natural map.

Proof. One immediately verifies the universal mapping property. $\square$

Corollary 6.2.3. *Let A be a commutative standard topological ring, and let N be a C-complete A-module. Then the functor, $M \rightsquigarrow M \overset{C}{\underset{A}{\otimes}} N$, from the category of C-complete A-modules into itself, preserves arbitrary direct limits of arbitrary covariant functors over arbitrary categories that are sets. In particular,*

(1): If I is any set and M_i, $i \in I$, are C-complete A-modules, then the natural map is an isomorphism of C-complete A-modules

$$\left[\int_{i \in I} (M_i \overset{C}{\underset{A}{\otimes}} N) \right] \overset{\approx}{\to} \left[\int_{i \in I} M_i \right] \overset{C}{\underset{A}{\otimes}} N.$$

(2): If D is a directed set, and if $(M_i, \alpha_{ij})_{i,j \in D}$ is a directed system of C-complete A-modules indexed by the directed set D, then the natural map is an isomorphism of C-complete A-modules

$$\varinjlim_{d \in D}{}^{C} (M_d \overset{C}{\underset{A}{\otimes}} N) \overset{\approx}{\to} \left[\varinjlim_{d \in D}{}^{C} M_d \right] \overset{C}{\underset{A}{\otimes}} N.$$

(3): If $\mathscr{D}$ is a category that is a set and if $M : \mathscr{D} \to \mathscr{C}_A$ is a covariant functor, then the natural map is an isomorphism of C-complete A-modules

$$\varinjlim_{d \in \mathscr{D}}{}^{C} (M_d \overset{C}{\underset{A}{\otimes}} N) \overset{\approx}{\to} \left[\varinjlim_{d \in \mathscr{D}}{}^{C} M_d \right] \overset{C}{\underset{A}{\otimes}} N.$$

(4): The functor $M \rightsquigarrow M \overset{C}{\underset{A}{\otimes}} N$ is right exact, from $\mathscr{C}_A$ into $\mathscr{C}_A$.

<u>Note</u>. Conclusion (4) of the Corollary, and the corresponding observation about the functor $M \rightsquigarrow N \overset{C}{\underset{A}{\otimes}} M$, in the terminology of [CEHA], is equivalent to the assertion: "The functor of two variables, $(M, N) \rightsquigarrow M \overset{C}{\underset{A}{\otimes}} N$ from the category $\mathscr{C}_A$ into itself, is right exact."

Proof. Follows immediately from the universal mapping property, as in the proof of Chapter 5, §1, Proposition 5.1.3. (The corollary also follows, alternatively, from the "adjoint functor" relationship established in Chapter 2, §3, Remark 8.) □

<u>Exercise 1</u>. Let A be a commutative standard topological ring. Let M and N be C-complete A-modules. Then in Chapter 2, §3, Remark 9, we defined $\mathscr{C}_A M \underset{A}{\otimes} N_{\mathscr{C}_A}$. Prove that if A is C-O.K., then

$$\mathscr{C}_A M \underset{A}{\otimes} N_{\mathscr{C}_A} \approx M \overset{C}{\underset{A}{\otimes}} N.$$

(Thus, at least in the case of topological rings that are C-O.K., the construction of Chapter 2, §3, Remark 9 is a generalization of the C-complete tensor product, that makes sense for non-commuative rings.)

(It would appear, however, that Exercise 1 requires that the commutative standard topological ring A be C-O.K.)

<u>Exercise 2</u>. From the universal mapping property definition, prove that

$$M \overset{C}{\underset{A}{\otimes}} N \approx N \overset{C}{\underset{A}{\otimes}} M$$

canonically as C-complete A-modules, for all C-complete A-modules M and N. (This is a very easy exercise.)

<u>Exercise 3</u>. For all C-complete A-modules M,

$$M \overset{C}{\underset{A}{\otimes}} \widehat{A} \approx M,$$

canonically as C-complete left A-modules.

<u>Hint</u>: Use the universal mapping property.

<u>Exercise 4</u>. Prove that, for all abstract A-modules M and N,

$$C(M) \overset{C}{\underset{A}{\otimes}} C(N) \approx C(M \underset{A}{\otimes} N),$$

canonically as C-complete A-modules.

<u>Hint</u>: Use the universal mapping property.

Corollary 6.2.4. *Let A be a commutative standard topological ring and let M_i, $i \in I$, be an indexed family of abstract A-modules and let N be a C-complete A-module. Then*

$$\left[C \left(\underset{i \in I}{\oplus} M_i \right) \right] \overset{C}{\underset{A}{\otimes}} N \approx \int_{i \in I} \left[C(M_i) \overset{C}{\underset{A}{\otimes}} N \right]$$

canonically as C-complete A-modules.

Proof. Chapter 5, §1, Proposition 5.1.3, conclusion (1), and the fact, Corollary 6.2.3, conclusion (1), that the functor $H \rightsquigarrow H \overset{C}{\underset{A}{\otimes}} N$ preserves infinite direct sums.   □

Corollary 6.2.5. *Let A be a commutative standard topological ring, let $\mathscr{D}$ be a category that is a set and let M be a covariant functor from $\mathscr{D}$ into the category of abstract A-modules. Then for every C-complete A-module N,*

$$\left[C \left(\varinjlim_{d \in \mathscr{D}} M_d \right) \right] \overset{C}{\underset{A}{\otimes}} N \approx \varinjlim_{d \in \mathscr{D}}^{C} \left[C(M_d) \overset{C}{\underset{A}{\otimes}} N \right].$$

Proof. Chapter 5, §1, Proposition 5.1.3, conclusion (2) and Corollary 6.2.3, conclusion (3).   □

Corollary 6.2.6. *Let A be a commutative standard topological ring, let I be a set and let M be a C-complete A-module. Then we have a canonical isomorphism of C-complete A-modules*

$$\widehat{A^{(I)}} \overset{C}{\underset{A}{\otimes}} M \approx \int_{i \in I} M.$$

Proof. Since C is the zeroth left derived functor of "A-adic completion," and $A^{(I)}$ is a free A-module, $C(A^{(I)}) = \widehat{A^{(I)}}$. Therefore taking $M_i = A$, all $i \in I$ in Corollary 6.2.4, (and noting, by taking I to be of cardinality one, that $C(A) = \widehat{A}$), and using Exercise 3 gives the result.   □

6.3 The C-complete Torsion Product

In general, let $\mathscr{A}$, $\mathscr{B}$, and $\mathscr{C}$ be abelian categories, such that $\mathscr{A}$ and $\mathscr{B}$ have enough projectives, and let $F : \mathscr{A} \times \mathscr{B} \to \mathscr{C}$ be a covariant, right-exact functor of two variables from $\mathscr{A}$ and $\mathscr{B}$ into $\mathscr{C}$. If B is a fixed object in $\mathscr{B}$, then the assignment: $A \to F(A, B)$ is a right-exact functor from $\mathscr{A}$ into $\mathscr{C}$. Then, following the conventions in [CEHA], we let $^{I}F_i$, $i \geq 0$, denote the left derived functors of this functor. In [CEHA], these are called "the left derived functors of F, with the first variable active, and the second variable passive". Similarly, for every object A in $\mathscr{A}$, we have the right exact functor $B \to F(A, B)$ from $\mathscr{B}$ into $\mathscr{C}$, the left derived functors of which are denoted $^{II}F_i$, $i \geq 0$, and called "the left derived functors of F, with the first variable passive, and the second variable active".

If A is an object in $\mathscr{A}$ and B is an object in $\mathscr{B}$, and if P_*, resp.: Q_*, is an exact acylic projective resolution of A in $\mathscr{A}$, resp.: of B in $\mathscr{B}$, then $F(P_i, Q_j)_{i,j \geq 0}$ is a double complex in the abelian category $\mathscr{C}$. In [CEHA] (Chapter V, §8, pp. 94–97), it is shown that the homology of the associated singly graded chain complex is independent, up to canonical isomorphisms, of the projective resolutions: P_* of A in $\mathscr{A}$ and Q_* of B in $\mathscr{B}$ chosen. And the resulting homology is denoted $^{I,II}F_i$, $i \geq 0$, and are functors of two variables from $\mathscr{A}$ and $\mathscr{B}$ into the category $\mathscr{C}$. They are called "the left derived functors of F, with both variables active". And there are well-known axioms for $^{I,II}F_i$, $i \geq 0$, that determine $^{I,II}F_*$ uniquely up to canonical isomorphisms; see [CEHA].

Also, in [CEHA] (Chapter V, §8), it is shown that there is induced a canonical map, of homological connected sequences of functors in the first variable,

$$(1): \ ^{I,II}F_i \to {}^{I}F_i, \ i \geq 0,$$

and also a map of homological connected sequences of functors in the second variable

$$(2): \ ^{I,II}F_i \to {}^{II}F_i, \ i \geq 0$$

from $\mathscr{A} \times \mathscr{B}$ into $\mathscr{C}$. The maps (1) and (2) are also natural transformations of functors of two variables for each integer $i \geq 0$. (See [CEHA], Chapter V §8, the section on "partial derived functors," for more details.) The functor F of two variables is called *balanced* ([CEHA]) if the maps (1) and (2) are natural equivalences of functors, for every integer $i \geq 0$.

A very well-known example of a balanced functor is the functor $(M, N) \rightsquigarrow M \underset{A}{\otimes} N$, from the category $\mathscr{A}$ of abstract (B, A)-bimodules and the category $\mathscr{B}$ of abstract (A, C)-bimodules into the category $\mathscr{C}$ of abstract (B, C)-bimodules, where A, B, and C are any rings. The functors $^{I}F_i$, $^{II}F_i$, and $^{I,II}F_i$, (which are then all canonically naturally equivalent) are unambiguously denoted $\mathrm{Tor}_i^A(M, N)$, $i \geq 0$ (see [CEHA]). This immediately raises the question, "If A is a commutative standard topological ring, then is the right-exact functor of two variables, C-complete tensor product, $(M, N) \rightsquigarrow M \overset{C}{\underset{A}{\otimes}} N$, from the abelian category $\mathscr{C}_A$ into itself, a balanced functor?" The next theorem answers this question.

Theorem 6.3.1. *Let A be a commutative standard topological ring. Then the following five conditions are equivalent.*

[1]: The right exact functor of two variables, "C-complete tensor product," $(M, N) \rightsquigarrow M \overset{C}{\underset{A}{\otimes}} N$, from $\mathscr{C}_A \times \mathscr{C}_B$ into $\mathscr{C}_A$, is balanced.

[2]: For every set I, the functor "I-fold direct sum," from the category $\mathscr{C}_A^I$ into $\mathscr{C}_A$, is exact.

[2']: Let I be a set and let $f_i : M_i \to N_i$ be monomorphisms of C-complete A-modules. Then

$$\int_{i \in I} f_i : \int_{i \in I} M_i \to \int_{i \in I} N_i$$

is a monomorphism.

[3]: For every pair M, N of C-complete A-modules, we have that the natural infinitely linear functions of C-complete A-modules,

$$^{I,II}F_i(M, N) \to {}^{I}F_i(M, N), \ i \geq 0,$$

are bijective (where $F(M,N) = M \overset{C}{\underset{A}{\otimes}} N$).

[4]: For every pair M, N of C-complete A-modules, we have that the natural infinitely linear functions of C-complete A-modules

$$^{I,II}F_i(M,N) \to {}^{II}F_i(M,N), \ ,i \geq 0$$

are bijective (where $F(M,N) = M \overset{C}{\underset{A}{\otimes}} N$).

Proof. Since $\int_{i \in I}$ is a right-exact functor, [2] iff [2'] is clear.

[3] iff [4] follows from §2, Exercise 2.

[1] iff ([3] and [4]): is [CEHA], Chapter V, §9, p. 96, last two paragraphs on that page.

[1] iff [2]: By [CEHA], Chapter V, §8, Theorem 8.1, p. 95, condition [1] holds iff for every projective P in $\mathscr{C}_A$, the functors:

(3): $M \rightsquigarrow M \overset{C}{\underset{A}{\otimes}} P,$

(4): $M \rightsquigarrow P \overset{C}{\underset{A}{\otimes}} M,$

from $\mathscr{C}_A$ into itself, are exact. By §2, Exercise 2, it is equivalent to say that the functor (4) is exact. By Chapter 2, §4, Corollary 2.4.9, an object P in $\mathscr{C}_A$ is projective iff P is isomorphic to a direct summand of $\widehat{A^{(I)}}$, for some set I. Therefore the functor (4) is exact, for all projective P in $\mathscr{C}_A$, iff for every set I, the functor

$$(5): \ M \rightsquigarrow \widehat{A^{(I)}} \overset{C}{\underset{A}{\otimes}} M$$

from $\mathscr{C}_A$ into itself, is exact. By §2, Corollary 6.2.6, this latter is equivalent to the assertion that the functor:

$$M \rightsquigarrow \int_{i \in I} M,$$

from $\mathscr{C}_A$ into itself, is exact, for all sets I. And by Chapter 5, §7, Lemma 5.7.1, this latter is equivalent to condition [2].   □

Remark 1. We have observed, in Chapter 5, §6, Example 2, that for most standard topological rings A, even the denumerable direct sum in the category of C-complete left A-modules, is not exact. Therefore, Theorem 6.3.1 implies that (for those commutative standard topological rings A), the functor "C-complete tensor product over A" is not balanced.

Definition 1. Let A be a commutative standard topological ring A, such that for every set I, the functor "I-fold direct sum," from $\mathscr{C}_A^I$ into $\mathscr{C}_A$, is exact. (As we have oberved in Remark 1 above, this is rarely the case. However, see Example 1 below.) Then, for every pair M, N of C-complete A-modules define

$$\mathsf{Tor}_i^{\mathscr{C}_A}(M,N), \ i \geq 0,$$

to be the left derived functors of the balanced functor of two variables, $(M, N) \rightsquigarrow M \overset{C}{\underset{A}{\otimes}} N$, from the category of C-complete A-modules into itself. We call these the *C-complete torsion products of M and N*.

Example 1. Let A be a commutative t-adic ring such that the t-torsion is bounded below (Chapter 4, §3, Definition 3). Then we have seen, in Chapter 5, §6, Theorem 5.6.13, that A obeys the hypotheses of Definition 1. Therefore, if M and N are C-complete A-modules, then we have the C-complete left A-modules $\text{Tor}_i^{\mathscr{C}_A}(M, N)$, for $i \geq 0$.

Subexample 1.1. Every discrete valuation ring that is not a field is a proper t-adic ring.

Subexample 1.2. $\mathbf{Z}[T]$, together with the T-adic topology is a proper t-adic ring.

Remark 2. There has been some discussion, in the literature, of "p-adic Banach spaces." These are modules M over a discrete valuaion ring $\mathscr{O}$ of mixed characteristic, that are the tensor product over $\mathscr{O}$ with its quotient field, of an $\mathscr{O}$-adically complete module (and also such that some other conditions hold). Given two such M and M', there has been some discussion in the literature of the "complete tensor product"

$$M \widehat{\underset{\mathscr{O}}{\otimes}} M'.$$

This operation is, of course, a very special case of the much more general operation, "C-complete tensor product of two C-complete $\mathscr{O}$-modules over $\mathscr{O}$" (see §2). Perhaps the reason that the concept of "complete tensor product of p-adic Banach spaces" has proven to be a bit useful in the literature is that such a ring $\mathscr{O}$, by subexample 1.1, obeys the hypotheses of Definition 1. Therefore, the functor "C-complete tensor product over $\mathscr{O}$", is balanced; so that $\text{Tor}_{\mathscr{C}_{\mathscr{O}}}^i(M, N)$, $i \geq 0$, is defined for any C-complete $\mathscr{O}$-modules M amd N, unambiguously.

Most likely, the applications in the literature, of "the complete tensor product of p-adic Banach spaces," would run more smoothly if one were to use the more general "C-complete tensor product of C-complete $\mathscr{O}$-modules," *and* its left derived functor $\text{Tor}_{\mathscr{C}_{\mathscr{O}}}^1(M, N)$, $i \geq 0$. (Then the left derived functors in dimension ≥ 2, of *any* additive functor on $\mathscr{C}_{\mathscr{O}}$, necessarily vanish identically — see §1, Corollary 6.1.6.)

To see how to compute $M \overset{C}{\underset{A}{\otimes}} N$ and $\text{Tor}_i^{\mathscr{C}_A}(M, N)$ in general, where M and N are in $\mathscr{C}_A$, and A is a commuative t-adic ring such that the torsion is bounded below (e.g., this includes the case of a discrete valuation ring not a field, just discussed), see Corollary 6.3.11 below, and also Examples 3 and 4 below (Example 4 is an example of a typical such computation).

Definition 1 makes sense in the (comparitively rare) cases in which the functor C-complete tensor product is balanced. The next definition extends Definition 1 — necessarily in three different directions — in the case that this is not so.

Definition 2. Let A be a commuative standard topological ring. Then, for every pair M, N of C-complete A-modules, we let $^I\text{Tor}_i^{\mathscr{C}_A}(M, N)$, respectively:

$^{II}\mathrm{Tor}_i^{\mathscr{C}_A}(M,N)$, respectively: $^{I,II}\mathrm{Tor}_i^{\mathscr{C}_A}(M,N)$ denote the left derived functors of the right exact functor of two variables, "C-complete tensor product over A," $(M,N) \rightsquigarrow M \overset{C}{\underset{A}{\otimes}} N$, from the category $\mathscr{C}_A$ into itself, where M, respectively: N, respectively: M and N, are the active variables, and N, respectively: M, respectively: no variables, are the passive variables (where we are using the terminology of [CEHA]).

Then, e.g. by [CEHA], the functors $^{I}\mathrm{Tor}_i^{\mathscr{C}_A}(M,N)$, $i \geq 0$, $^{II}\mathrm{Tor}_i^{\mathscr{C}_A}(M,N)$, $i \geq 0$, $^{I,II}\mathrm{Tor}_i^{\mathscr{C}_A}(M,N)$, $i \geq 0$, $^{I,II}\mathrm{Tor}_i^{\mathscr{C}_A}(M,N)$, $i \geq 0$, for fixed N, fixed M, fixed M, for fixed N; is an exact connected sequence of functors in the variable M, variable N, variable N and variable M respectively. And we have the natural transformations of functors of two variables:

(1): $^{I,II}\mathrm{Tor}_i^{\mathscr{C}_A}(M,N) \to {}^{I}\mathrm{Tor}_i^{\mathscr{C}_A}(M,N)$, $i \geq 0$, respectively,

(2): $^{I,II}\mathrm{Tor}_i^{\mathscr{C}_A}(M,N) \to {}^{II}\mathrm{Tor}_i^{\mathscr{C}_A}(M,N)$, $i \geq 0$,

discussed earlier in this section, which are maps of homological exact connected sequences of functors in the variable M for fixed N, respectively: in the variable N for fixed M.

Lemma 6.3.2. *Let A be a commutative standard topological ring. Then the five equivalent conditions of Theorem 6.3.1 are also equivalent to the following condition:*

[5]: For every object M in $\mathscr{C}_A$ (it suffices to consider only projective objects M in $\mathscr{C}_A$), there exist a choice of connecting homomorphisms, such that the sequence of functors: $N \rightsquigarrow {}^{I}\mathrm{Tor}_i^{\mathscr{C}_A}(M,N)$ becomes a homological exact connected sequence of functors from $\mathscr{C}_A$ into $\mathscr{C}_A$.

Proof. [3] implies [5]: If [3] holds, then $^{I}\mathrm{Tor}_i^{\mathscr{C}_A}(M,N) \approx {}^{I,II}\mathrm{Tor}_i^{\mathscr{C}_A}(M,N)$, $i \geq 0$. And since the latter is an exact connected sequence of functors in the variable N, indeed such connecting homomorphisms exist.

([5] for M projective) implies [1]: In fact, since M is projective, $^{I}\mathrm{Tor}_i^{\mathscr{C}_A}(M,N)$ $= 0$ for $i \geq 1$ and for all N in $\mathscr{C}_A$. If such connecting homomorphisms exist, then it follows that the functor: $N \rightsquigarrow M \overset{C}{\underset{A}{\otimes}} N$ is exact for M projective. By §2, Exercise 2, therefore also $N \rightsquigarrow N \overset{C}{\underset{A}{\otimes}} M$ is exact for M projective. But then, by [CEHA], Chapter V, §8, Theorem 8.1, p. 95, it follows that [1] holds. $\quad\square$

Lemma 6.3.3. *Let A be a commutative standard topological ring and let N be a fixed C-complete A-module. Then the following two conditions are equivalent:*

(1): $^{I,II}\mathrm{Tor}_i^{\mathscr{C}_A}(P,N) = 0$ for $i \geq 1$ and for all projective C-complete A-modules P.

(2): The natural map $^{I,II}\mathrm{Tor}_i^{\mathscr{C}_A}(M,N) \to {}^{I}\mathrm{Tor}_i^{\mathscr{C}_A}(M,N)$ is an isomorphism in $\mathscr{C}_A$, for all C-complete A-modules M and all non-negative integers i.

Note: For condition (1) to hold, it suffices that it holds for $P = \widehat{F}$, for all free A-modules F.

Corollary 6.3.4. *Let A be a commutative standard topological ring. Then the five equivalent conditions of Theorem 6.3.1, and condition [5], of Lemma 6.3.2, are also equivalent to*

[6] *P projective in $\mathscr{C}_A$, N C-complete, implies*

$$^{I,II}\text{Tor}_i^{\mathscr{C}_A}(P,N) = 0 \text{ for } i \geq 1.$$

Note: For condition [6] to hold, it suffices that it holds for all C-complete A-modules N, and all $P = \widehat{F}$, where F is a free A-module.

Proof of Lemma 6.3.3. For $i = 0$, the map in equation (2) is an isomorphism. For fixed N, the non-negative homological exact connected sequence of functors $M \rightsquigarrow {}^{I}\text{Tor}_i^{\mathscr{C}_A}(M,N)$ from $\mathscr{C}_A$ into $\mathscr{C}_A$ is characterized axiomatically by the fact that P projective in $\mathscr{C}_A$ implies ${}^{I}\text{Tor}_i^{\mathscr{C}_A}(P,N) = 0$ for $i \geq 1$. Therefore condition (1) is necessary and sufficient that the map of homological exact connected sequences of functors from $\mathscr{C}_A$ into $\mathscr{C}_A$

$$^{I,II}\text{Tor}_i^{\mathscr{C}_A}(M,N) \to {}^{I}\text{Tor}_i^{\mathscr{C}_A}(M,N), \; i \geq 0, \; M \in \mathscr{C}_A,$$

be an isomorphism of homological connected sequences of functors. Therefore (1) iff (2).

The Note follows from Chapter 2, §4, Corollary 2.4.9. $\qquad\square$

Proof of Corollary 6.3.4. [6] (respectively: condition [3] of Theorem 6.3.1) holds iff (1) (respectively (2)) of Lemma 6.3.3 holds, for all C-complete A-modules N. Therefore the Corollary and the Note follow from Lemma 6.3.3 and the Note to Lemma 6.3.3. $\qquad\square$

Remark 3. Theorem 6.3.1 implies that, if A is a commutative standard topological ring that does *not* obey the hypotheses of Definition 1, then the natural transformations of functors: $^{I,II}\text{Tor}_i^{\mathscr{C}_A}(M,N) \to {}^{I}\text{Tor}_i^{\mathscr{C}_A}(M,N)$ are not natural equivalences of functors for all $i \geq 0$. Lemma 6.3.2 implies, more strongly, that when this is the case, it is even impossible to "make ${}^{I}\text{Tor}_i^{\mathscr{C}_A}(M,N)$, $i \geq 0$, into a homological exact connected sequence of functors in the variable N. Therefore, there cannot exist natural equivalences of functors, ${}^{I}\text{Tor}_i^{\mathscr{C}_A} \approx {}^{I,II}\text{Tor}_i^{\mathscr{C}_A}$, for all $i \geq 0$. In fact, the proof of Lemma 6.3.2 shows even that, then, there cannot even exist a natural equivalence of functors between ${}^{I}\text{Tor}_1^{\mathscr{C}_A}$ and $^{I,II}\text{Tor}_1^{\mathscr{C}_A}$.

Proposition 6.3.5. *Let A be a commutative standard topological ring and let N be a C-complete A-module. Then the following two conditions are equivalent.*

[1] *For all sets I and all integers $n \geq 1$,*

$$\int_{\substack{n \\ i \in I}} N = 0.$$

[2] *For all C-complete A-modules M, the natural map is an isomorphism of C-complete A-modules*

$$^{I,II}\text{Tor}_n^{\mathscr{C}_A}(M,N) \xrightarrow{\approx} {}^{I}\text{Tor}_n^{\mathscr{C}_A}(M,N),$$

for all integers $n \geq 0$.

Proof. Let P_* be an acyclic projective resolution of N in the category $\mathscr{C}_A$. Then [1] is true, iff for all sets I, the chain complex

$$(1): \int_{i \in I} P_*$$

in the category $\mathscr{C}_A$ is acyclic.

By §2, Corollary 6.2.6, this chain complex (1) is isomorphic to

$$(2): \widehat{A^{(I)}} \overset{C}{\underset{A}{\otimes}} P_*.$$

Therefore condition [1] holds iff (2) is acyclic, for all sets I.

Since $\widehat{A^{(I)}}$ is projective in $\mathscr{C}_A$ (Chapter 2, §4, Theorem 2.4.8, conclusion (2)), a projective resolution of $\widehat{A^{(I)}}$ in $\mathscr{C}_A$ is the chain complex that is $\widehat{A^{(I)}}$ in dimension zero, and zero in dimensions $\neq 0$. Recalling the construction of $^{I,II}\mathrm{Tor}_i^{\mathscr{C}_A}$ (as we described at the beginning of this section; as in [CEHA]), using this projective resolution of $\widehat{A^{(I)}}$, and the projective resolution P_* of N in $\mathscr{C}_A$, the explicit construction of the derived functors of a functor of two variables "with both variables active" we have that the homology of the chain complex (2) in dimension i is

$$(3): {}^{I,II}\mathrm{Tor}_i^{\mathscr{C}_A}(\widehat{A^{(I)}}, N), \quad i \geq 0.$$

But we have observed that condition [1] holds iff the chain complex (2) is acyclic. Therefore condition [1] holds iff

[1'] $^{I,II}\mathrm{Tor}_i^{\mathscr{C}_A}(\widehat{A^{(I)}}, N) = 0$ for $i \geq 1$.

But by Lemma 6.3.3 and the Note to Lemma 6.3.3, this latter is equivalent to condition [2].   □

Lemma 6.3.6. *The hypotheses being as in Proposition 6.3.5, suppose also that*

$$(1): C_n(N) = \begin{cases} N, & \text{for } n = 0 \\ 0, & \text{for } n \geq 1. \end{cases}$$

Then the three equivalent conditions, [1] and [2] of Proposition 6.3.5 and (1) of Lemma 6.3.3, are also equivalent to

[3] $C_n(N^{(I)}) = 0$ for $n \geq 1$, for all sets I.

Proof. For every set I, we have the spectral sequence, Chapter 5, §6, Theorem 5.6.1,

$$(2): E_{p,q}^2 = \int_{i \in I}^{p} C_q(N) \Rightarrow C_n(N^{(I)}).$$

By hypotheses (1) of the Lemma, we have that $E_{p,q}^2 = 0$ for $q \neq 0$ and that $E_{n,0}^2 = \int_{i \in I} N$. Therefore the edge homomorphisms in the spectral sequence (2) are isomorphisms

$$\int_{\substack{n \\ i \in I}} N \approx C_n(N^{(I)})$$

for all sets I and all integers $n \geq 0$. Therefore condition [3] of the Lemma is equivalent to condition [1] of Proposition 6.3.5.   □

Corollary 6.3.7. *Let A be a commutative standard topological ring, and let N be an abstract module over $\widehat{A}$ of finite presentation as an $\widehat{A}$-module. Then, by Chapter 2, §3 Corollary 2.4.12, there exists a unique structure of C-complete left A-module on N, that induces the given abstract $\widehat{A}$-module structure. Suppose that*

(1) $\widehat{A}$ is flat as an A-module.

(2) $N \approx \widehat{A} \underset{A}{\otimes} N'$, where N' is an abstract A-module of finite presentation,

and that

(3) N admits an infinite chain of syzygies as an abstract $\widehat{A}$-module (Chapter 5, §3, Definitions 1 and 2) (e.g., by (1), this is the case if N' admits an infinite chain of syzygies as an abstract A-module).

Then the three equivalent conditions [1] and [2] of Proposition 6.3.5, and condition (1) of Lemma 6.3.3, are also equivalent to each of the following two conditions:

[4] For every free abstract A-module F, we have that $\mathrm{Tor}_i^A(\widehat{F}, N') = 0$ for $i \geq 1$.

[5] For every projective C-complete A-module P, we have that $\mathrm{Tor}_i^A(P, N') = 0$ for $i \geq 1$.

Note 1. Sufficient conditions for hypotheses (1) and (2) to hold is that A be complete.

Note 2. Sufficient conditions for hypothesis (3) to hold is that either A or $\widehat{A}$ be coherent.

Note 3. Sufficient conditions for both hypotheses (1) and (3) to hold, is that A be Noetherian, that the topology of A is given by denumerably many ideals, and that the square of an open ideal is open.

Note 4. Under the hypotheses of this corollary, N *considered as an abstract $\widehat{A}$ module* obeys the hypotheses of Lemma 6.3.6 *as an abstract $\widehat{A}$-module.* Therefore, the equivalent conditions, [1], [2], [4], and [5] are also equivalent to the condition

$[3_{\widehat{A}}]$: $C_n^{\widehat{A}}(N^{(I)}) = 0$ for $n \geq 1$.

Proof. By Chapter 2, §4, Corollary 2.4.9, every projective P in $\mathscr{C}_A$ is isomorphic, in $\mathscr{C}_A$, to a direct summand of $\widehat{F}$, for some abstract A-module F. Therefore [4] and [5] are equivalent.

Let I be a set. Then by Lemma 6.3.8 below,

(4): $\mathrm{Tor}_i^A(\widehat{A^{(I)}}, N') \approx \mathrm{Tor}_i^{\widehat{A}}(\widehat{A^{(I)}}, N)$, for all integers $i \geq 0$.

Therefore, replacing A by $\widehat{A}$ and N' by N if necessary, to show that [4] is equivalent to the equivalent conditions [1] and [2] of Proposition 6.3.5, and also to prove Note 4, it suffices to prove these in the case in which A is complete. Then by Chapter 5, §3, Corollary 5.3.14, and by hypothesis (3),

$$C_i(N) = \begin{cases} N, & i = 0 \\ 0, & i \geq 1. \end{cases}$$

Therefore the hypotheses of Lemma 6.3.6 are satisfied. This proves Note 4. By Lemma 6.3.6, the equivalent conditions of Proposition 6.3.5 hold iff $C_i(N^{(I)}) = 0$, for $i \geq 1$, and for all sets I. But by Chapter 5, §3, Theorem 5.3.12, and by hypothesis (3) of this corollary,

$$C_i(N^{(I)}) \approx \mathrm{Tor}_i^A(\widehat{A^{(I)}}, N),$$

for all integers $i \geq 0$ and for all sets I. Therefore the equivalent conditions of Proposition 6.3.5 hold iff

$\mathrm{Tor}_i^A(\widehat{A^{(I)}}, N) = 0$ for $i \geq 1$ and for all sets I — that is, iff condition [4] holds. This completes the proof of the corollary.

Next, let us return to the case, in which A is not necessarily complete, and prove the parenthetical observation made in hypothesis (3).

If N' admits an infinite chain of syzygies as an abstract A-module, then let P_* be a free acyclic resolution of N' as an abstract A-module such that P_i is free of finite rank as an A-module, for $i \geq 0$. Then, since by hypothesis (1), $\widehat{A}$ is flat as an A-module, $\widehat{A} \underset{A}{\otimes} P_*$ is a free acyclic resolution of $N = \widehat{A} \underset{A}{\otimes} N'$ as an abstract $\widehat{A}$-module. Therefore N admits an infinite chain of syzygies as an abstract $\widehat{A}$-module.

Note 1 is evident. Note 2 follows from Chapter 5, §3, Exercise 3. If we have the hypotheses of Note 3, then by Chapter 5, §4, Theorem 5.4.2, conclusion (3), we have that hypothesis (1) of the corollary holds. And since a Noetherian ring is coherent, by Note 2 we have that hypothesis (3) of the corollary holds. □

Lemma 6.3.8. *Let $\phi : A \to B$ be a homomorphism of abstract rings, such that B is flat considered as an abstract right A-module. Let M be any abstract left A-module, and let F be any abstract right B-module. Then there is induced a canonical isomorphism of abelian groups,*

$$\mathrm{Tor}_i^A(F, M) \approx \mathrm{Tor}_i^B(F, B \underset{A}{\otimes} M),$$

for all integers $i \geq 0$.

Proof. Let P_* be a free acyclic resolution of M as a left A-module. Then since B is right flat as an A-module. $B \underset{A}{\otimes} P_*$ is a free acyclic resolution of $B \underset{A}{\otimes} M$ as an abstract left B-module. Therefore, using the projective acyclic resolution P_* (respectively: $B \underset{A}{\otimes} P_*$) of M (respectively: of $B \underset{A}{\otimes} M$), we have that $\mathrm{Tor}_i^A(F, M)$ (respectively: $\mathrm{Tor}_i^B(F, B \underset{A}{\otimes} M)$) is isomorphic canonically to the ith homology group of the chain complex $F \underset{A}{\otimes} P_*$ (respectively: $F \underset{B}{\otimes} (B \underset{A}{\otimes} P_*)$). The identity $F \underset{B}{\otimes} (B \underset{A}{\otimes} P_*) \approx F \underset{A}{\otimes} P_*$ therefore completes the proof. $\qquad\square$

Proposition 6.3.9. *Let A be a commutative standard topological ring. Let N be an abstract A-module. Then the following three conditions are equivalent.*

[1] Both [1a] and [1b] hold.

[1a] $C_n(N) = 0$ for $n \geq 1$.

[1b] $\int_n \underset{i \in I}{} C(N) = 0$ for $n \geq 1$, for all sets I.

[2] $C_n(N^{(I)}) = 0$ for $n \geq 1$ and for all sets I.

[3] For every abstract A-module M, there exist two first quadrant homological spectral sequences in the category of C-complete left A-modules,

(1): $E^2_{p,q} = {}^I\mathrm{Tor}_p^{\mathscr{C}_A}(C_q(M), C(N)) \Rightarrow K_n$,

(2): ${}'E^2_{p,q} = C_p(\mathrm{Tor}_q^A(M, N)) \Rightarrow K_n$,

converging to the same C-complete A-modules K_n, $n \geq 0$, but with perhaps different filtrations.

Note 1. Whether or not the three equivalent conditions of the proposition hold, the spectral sequence (1) of condition [3] always holds. And K_n, $n \geq 0$, are the left derived functors of the right exact functor: $M \rightsquigarrow C(M \underset{A}{\otimes} N)$, from the category of abstract A-modules into the category of C-complete A-modules.

Note 2. When the three equivalent conditions of Proposition 6.3.9 hold, then in the spectral sequence (1) of condition [3], one can replace ${}^I\mathrm{Tor}_p^{\mathscr{C}_A}(C_q(M), C(N))$ with ${}^{I,II}\mathrm{Tor}_p^{\mathscr{C}_A}(C_q(M), C(N))$.

Proof. [2] iff [3]: Let Abs_A denote the category of abstract A-modules. Then we have the commutative diagram of abelian categoires and right exact functors

$$(3): \qquad \begin{array}{ccc} Abs_A & \overset{C}{\longrightarrow} & \mathscr{C}_A \\ {\scriptstyle M \rightsquigarrow M \underset{A}{\otimes} N} \downarrow & & \downarrow {\scriptstyle H \rightsquigarrow H \overset{C}{\underset{A}{\otimes}} C(N)} \\ Abs_A & \overset{C}{\longrightarrow} & \mathscr{C}_A \end{array}$$

(Commutativity of this diagram is equivalent to the identity

$$C(M) \overset{C}{\underset{A}{\otimes}} C(N) \approx C(M \underset{A}{\otimes} N),$$

which follows from §2, Exercise 4.) The functor $C : Abs_A \to \mathscr{C}_A$ maps projectives into projectives (Chapter 2, §4, Theorem 2.4.8, conclusion (2)). Therefore, the spectral sequence (1) of condition [3] is always defined, and converges to the left derived functors of the composite functor K_0 of the diagram (3). This proves Note 1. And also, it follows that condition [3] holds, iff the hypotheses for the spectral sequence of the composite functor: $C \circ (M \rightsquigarrow M \underset{A}{\otimes} N)$ hold.

That is, iff for every set I, we have that $C_n(A^{(I)} \underset{A}{\otimes} N) = 0$ for $n \geq 1$. Since $A^{(I)} \underset{A}{\otimes} N \approx N^{(I)}$, we have therefore proved the equivalence of conditions [2] and [3].

[1] iff [2]: Taking I to be a set of cardinality one, we see that [2] implies [1a]. Therefore both [1] and [2] imply [1a]. Therefore we can, and do, assume that [1a] holds. By Chapter 5, §6, Theorem 5.6.1, for every set I we have the first quadrant homological spectral sequence in $\mathscr{C}_A$:

$$E^2_{p,q} = \int_{\substack{p \\ i \in I}} C_q(N) \Rightarrow C_n(N^{(I)}).$$

By condition [1a], $E^2_{p,q} = 0$ for $q \geq 1$. Therefore the spectral sequence degenerates and $E^2_{n,0} = (n\text{th}$ group of the abutment). That is,

$$\int_{\substack{n \\ i \in I}} C(N) \approx C_n(N^{(I)}), \ n \geq 0.$$

This equation for $n \geq 1$ implies that [1b] iff [2].

If the equivalent conditions in the proposition hold, then in particular condition [1b] holds. Therefore, by Proposition 6.3.5, the natural map

$$^{I,II}\mathrm{Tor}^{\mathscr{C}_A}_p(C_q(M), C(N)) \to {}^{I}\mathrm{Tor}^{\mathscr{C}_A}_p(C_q(M), C(N))$$

is an isomorphism in $\mathscr{C}_A$. This proves Note 2. $\square$

<u>Example 1</u>. Let A be a commuative t-adic ring such that the torsion is bounded below. (Chapter 4, §3, Definition 3.) Let t be any element of A such that the topology of A is the t-adic. (Then, since A is commutative, it follows readily that the t-torsion in A is bounded below.) Let N be an abstract A-module. Then N obeys the three equivalent conditions of Proposition 6.3.9 iff N has no non-zero infinitely t-divisible, t-torsion elements.

Proof. We work with condition [2] of Proposition 6.3.9. By part (3) of Chapter 4, §3, Theorem 4.3.9, $C_i \equiv 0$ for $i \geq 2$. And by Chapter 4, §3, Remark 5, as generalized in Note 1 to Theorem 4.3.9, $C_1(N^{(I)}) = 0$ iff $N^{(I)}$ has no infinitely t-divisible, t-torsion — i.e., iff N has no infinitely t-divisible, t-torsion. $\square$

<u>Example 2</u>. Let A be a commutative, coherent ring, that is a standard topological ring (i.e., such that we have a topology on A given by ideals). Then the following two conditions are equivalent.

[1] For every finitely presented abstract A-module N, the three equivalent conditions of Proposition 6.3.9 hold.

[2] For every free abstract A-module F, we have that $\widehat{F}$ is flat as an A-module.

Proof. By Chapter 5, §3, Theorem 5.3.17, condition [2] of Proposition 6.3.9 (which is condition [3] of Chapter 5, §3, Theorem 5.3.17) is equivalent to condition [2] of this example (which is condition [1] of Chapter 5, §3, Theorem 5.3.17). □

Example 2.1. More generally, let A be a commutative standard topological ring, and let N be an A-module that admits an infinite chain of syzygies (Chapter 5, §3, Definitions 1 and 2). Then N obeys the three equivalent conditions of Proposition 6.3.9 iff $\mathrm{Tor}_i^A(\widehat{F}, N) = 0$, $i \geq 1$, for all free A-modules F.

Proof. Chapter 5, §3, Theorem 5.3.12, implies that condition [2] of Proposition 6.3.9 holds iff the indicated condition holds. □

Corollary 6.3.10. *[Spectral Sequence of the C-complete Torsion Products] Let A be a commutative standard topological ring and let M and N be abstract left A-modules. Suppose that*

(1) $C_i(M) = 0$ for $i \geq 1$ and that

(2) $C_i(N^{(I)}) = 0$ for $i \geq 1$ and for all sets I.

Then there is induced a first quadrant, homological spectral sequence in the category of C-complete A-modules:

$$E_{p,q}^2 = C_p(\mathrm{Tor}_q^A(M, N)) \Rightarrow {}^I\mathrm{Tor}_n^{\mathscr{C}_A}(C(M), C(N)).$$

Proof. Condition (2) above is identical to condition [2] of Proposition 6.3.9. Therefore, we have the two spectral sequences of condition [3] of Proposition 6.3.9:

(3): $E_{p,q}^2 = {}^I\mathrm{Tor}_p^{\mathscr{C}_A}(C_q(M), C(N)) \Rightarrow K_n$,

(4): $'E_{p,q}^2 = C_p(\mathrm{Tor}_q^A(M, N)) \Rightarrow K_n$,

with the same abutment K_n. By condition (1), for the first spectral sequence we have that $E_{p,q}^2 = 0$ for $q \geq 1$. Therefore, this spectral sequence degenerates, and $E_{n,0}^2 = K_n$, $n \geq 0$. I.e, $K_n = {}^I\mathrm{Tor}_n^{\mathscr{C}_A}(C(M), C(N))$. Substituting this into the second spectral sequence gives the desired result. □

Corollary 6.3.11. *[Computation of the C-complete Torsion Product Over t-Adic Rings] Let A be a commutative t-adic ring such that the torsion is bounded below (Chapter 4, §3, Definition 3). Let M and N be abstract A-modules, such that M and N do not have any non-zero infinitely t-divisible, t-torsion elements. Then for every non-negative integer n, there are induced short exact sequences in $\mathscr{C}_A$*

$$(1):\ 0 \to C(\mathrm{Tor}_n^A(M, N)) \to \mathrm{Tor}_n^{\mathscr{C}_A}(C(M), C(N)) \to C_1(\mathrm{Tor}_{n-1}^A(M, N)) \to 0.$$

Proof. Since N has no non-zero infinitely t-divisible, t-torsion elements, neither does $N^{(I)}$, for all sets I. Therefore by Chapter 4, §3, Remark 5, as generalized by Note 1 to Theorem 4.3.9, we have that $C_1(N^{(I)}) = 0$, $C_1(M) = 0$. Therefore the hypotheses of Corollary 6.3.10 are satisfied, so that we have the spectral sequence of Corollary 6.3.10. Now we have $C_i \equiv 0$ for $i \neq 0, 1$. Therefore the spectral sequence is concentrated in the two columns, $p = 0, 1$. And therefore the spectral sequence becomes the indicated short exact sequences (with edge homomorphisms being the maps; and the groups in the middle of the indicated short exact sequence are the abutment).   □

Example 3. Let $\mathscr{O}$ be a discrete valuation ring not a field. Let M and N be abstract $\mathscr{O}$-modules such that M and N have no non-zero infinitely divisible torsion elements. Then

(1) $C(M) \overset{C}{\underset{\mathscr{O}}{\otimes}} C(N) = C(M \underset{\mathscr{O}}{\otimes} N)$,

(2) $\mathrm{Tor}_1^{\mathscr{C}_{\mathscr{O}}}(C(M), C(N))$ fits into the short exact sequence:

$$0 \to C(\mathrm{Tor}_1^{\mathscr{O}}(M, N)) \to \mathrm{Tor}_1^{\mathscr{C}_{\mathscr{O}}}(C(M), C(N)) \to C_1(M \underset{\mathscr{O}}{\otimes} N) \to 0$$

and

(3) $\mathrm{Tor}_i^{\mathscr{C}_{\mathscr{O}}}((C(M), C(N)) = 0$ for $i \geq 2$.

Also,

(4) $\mathrm{Tor}_1^{\mathscr{O}}(M, N)$ has no non-zero infinitely divisible torsion elements.

Proof. (1) and (2) follow from the short exact sequence of Corollary 6.3.11. ((1) follows alternately from §2, Exercise 4.) Since the ring $\mathscr{O}$ is of projective dimension 1, by §1, Proposition 6.1.5, every object in $\mathscr{C}_{\mathscr{O}}$ admits a projective resolution in $\mathscr{C}_{\mathscr{O}}$ of length ≤ 1. Therefore the left derived functors of any additive functor on $\mathscr{C}_{\mathscr{O}}$ vanish in dimension ≥ 2. In particular we have conclusion (3) of the Example.

Applying the short exact sequence of Corollary 6.3.11 in the case $n = 2$ we obtain $\mathrm{Tor}_2^{\mathscr{C}_{\mathscr{O}}}(C(M), C(N)) = C_1(\mathrm{Tor}_1^A(M, N))$. Using (3) above, therefore $C_1(\mathrm{Tor}_1^A(M, N)) = 0$. By Chapter 4, §3, Theorem 4.3.9, conclusion (2), this is equivalent to conclusion (4) of the example.   □

Example 3.1. In Example 3 above, the discrete valuation ring $\mathscr{O}$ can be replaced by any commutative t-adic ring $\mathscr{O}$ such that the torsion is bounded below, and such that $\mathscr{O}$ is of projective dimension ≤ 1. (E.g., one can take $\mathscr{O}$ to be any principal ideal domain, or Dedekind domain, and t any non-zero element.)

Remark 4. Let A be a commutative t-adic ring such that the torsion is bounded below (respectively: let $\mathscr{O}$ be a discrete valuation ring not a field). Then of course, the hypotheses of Corollary 6.3.11 (respectively: of Example 3) are satisfied by any pair M, N of C-complete A-modules (respectively: $\mathscr{O}$-modules). The idea of the greater generality of the corollary (respectively: example) is as follows: If we can write $M = C(M')$ and $N = C(N')$ where M' and N' are *relatively simply constructed* abstract A-modules (respectively: $\mathscr{O}$-modules), then the exact sequence of Corollary 6.3.11 (respectively: the conclusions of Example 3) might give comparatively easy computations of the C-complete tensor product, and the C-complete torsion products, of M and N.

(Of course one can always apply the corollary (respectively: the example) directly to the C-completes M and N themselves, but this might not always give the most efficient computation.)

Example 4. Let A be a commutative proper t-adic ring with parameter t (e.g., A can be any discrete valuation ring not a field, and t any non-zero, non-unit in A). Let $M = N = C\left(\bigoplus_{n \geq 0} A/t^n A\right)$. (This is the module discussed in Chapter 5, §2, Examples 9 and 11.) Then compute $\mathrm{Tor}_i^{\mathscr{C}_A}(M, N)$, $i \geq 0$.

Solution. Take $M' = N' = \bigoplus_{n \geq 0} A/t^n A$. Then M' and N' have no non-zero divisible elements, so that the short exact sequences of Corollary 6.3.11 apply. We deduce

$$(1) M \overset{C}{\underset{A}{\otimes}} N = C(M' \underset{A}{\otimes} N') \approx C\left(\bigoplus_{n \geq 0} (A/t^n A)^{(\omega)}\right).$$

Since $M' \underset{A}{\otimes} N' \approx \bigoplus_{n \geq 0} (A/t^n A)^{(\omega)}$ and has no non-zero divisible elements, by Chapter 4, §3, Example 5, $C_1(M' \underset{A}{\otimes} N') = 0$. Therefore by Corollary 6.3.11,

$$\mathrm{Tor}_1^{\mathscr{C}_A}(M, N) \approx C(\mathrm{Tor}_1^A(M', N'))$$

$\mathrm{Tor}_1^A(M', N')$ is computed using the fact that Tor_1^A commutes with infinite direct sums [CEHA], and Chapter 4, §3, Lemma 4.3.6, to be

$$\mathrm{Tor}_1^A\left(\bigoplus_{n \geq 0} A/t^n A, \bigoplus_{m \geq 0} A/t^m A\right) = \bigoplus_{n,m \geq 0} \mathrm{Tor}_1^A(A/t^n A, A/t^m A)$$

$$\approx \bigoplus_{n,m \geq 0} (A/t^{\inf(n,m)} A) \approx \bigoplus_{n \geq 0} (A/t^n A)^{(\omega)}.$$

So that once again,

$$(2): \mathrm{Tor}_1^{\mathscr{C}_A}(M, N) \approx C\left(\bigoplus_{n \geq 0} (A/t^n A)^{(\omega)}\right).$$

Also, by equation (1) of Corollary 6.3.11, and Chapter 4, §3, Lemma 4.3.6,

$$(3): \mathrm{Tor}_i^{\mathscr{C}_A}(M, N) = 0 \text{ for } i \neq 0, 1.$$

In the case $A = \mathscr{O}$, a complete discrete valuation ring, the C-complete $\mathscr{O}$-module on the right side of equations (1) and (2) is discussed in Chapter 5, §2, Remark 8, and for an arbitrary proper t-adic ring A in Remark 9. (As noted there, it is of some interest in the study of p-adic cohomology in algebraic geometry.) To reiterate: This C-complete A-module has a huge divisible submodule. Its divisible submodule is A-adically complete, and is isomorphic to

$$\left[(\hat{A}^\omega)^{(\omega)}\right]^{\widehat{}} \Big/ \left[\widehat{A^{(\omega)}}^{(\omega)}\right]^{\widehat{}}.$$

And modulo its divisible part (Chapter 3, §2, Corollary 3.2.18, conclusion (2)), the module on the right side of equations (1) and (2) is isomorphic to

$$\widetilde{\bigoplus_{n \geq 0} B_n^{(\omega)}}, \text{ where } B_n = A/t^n A.$$

Lemma 6.3.12. *Let A be a commutative standard topological ring, and let M and N be C-complete A-modules. Then for every non-negative integer i, we have a commutative diagram in the category of abstract left A-modules*

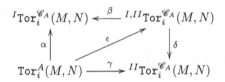

For each non-negative integer i, these maps $\alpha, \beta, \gamma, \delta, \epsilon$, for all M and N in $\mathscr{C}_A$, are (five) natural transformations of functors, of covariant functors of two variables from $\mathscr{C}_A$ and $\mathscr{C}_A$ into the category Abs_A of abstract A-modules.

Note 1. The natural transformations β and δ are the maps (1) and (2) discussed in Definition 2.

Note 2. The maps α, for all objects M and N in $\mathscr{C}_A$ and for all non-negative integers i, can be characterized uniquely by the following two properties:

(a) For every fixed C-complete A-module N, the maps α for all $i \geq 0$ and all M in $\mathscr{C}_A$, are a map of homological connected sequences of functors from $\mathscr{C}_A$ into Abs_A.

(b) For $i = 0$, and for all M and N in $\mathscr{C}_A$, the map α is the natural map: $M \underset{A}{\otimes} N \to M \overset{C}{\underset{A}{\otimes}} N$ constructed in the proof of the Lemma.

Note 3. A similar axiomatic characterization can be given for the maps γ, for all $i \geq 0$ and for all M, N in $\mathscr{C}_A$.

Note 4. The maps ϵ, for all integers $i \geq 0$ and for all M, N in $\mathscr{C}_A$, can be characterized uniquely by the following three properties:

(a) For every fixed N in $\mathscr{C}_A$, the maps ϵ for all $i \geq 0$ and for all M in $\mathscr{C}_A$, are a map of homological connected sequence of functors, from $\mathscr{C}_A$ into Abs_A.

(b) Ibid, but interchanged "M" and "N."

(c) Same as (b) of Note 2, with "ϵ" replacing "α."

Proof. The natural infinitely A-bilinear map: $M \times N \to M \overset{C}{\underset{A}{\otimes}} N$ is in particular A-bilinear. Therefore we deduce a natural homomorphism of A-modules: $M \underset{A}{\otimes} N \to M \overset{C}{\underset{A}{\otimes}} N$. Then, by the axiomatic characterization of "derived" functors in [CEHA], for all M, N in $\mathscr{C}_A$ and for all $i \geq 0$, there exist unique maps α (respectively: γ, respectively: ϵ) in the category of abstract A-modules, such that the axioms in Note 2, (respectively: Note 3, respectively: Note 4) hold. Define β (respectively: δ) to be the maps (1) (respectively: (2)) constructed in Definition 2. Then all of the results in Notes 1 - 4 hold. In particular, by Note 4, properties (a) and (c), it follows that the composite $\beta \circ \epsilon$, obeys Note 2, properties (a) and (b). Therefore by the uniqueness part of Note 2, $\beta \circ \epsilon = \alpha$. Similarly $\gamma \circ \epsilon = \gamma$. □

Definition 3. Let A be a commutative standard topological ring and let M and N be C-complete A-modules. We say that *the functor C-complete tensor product is balanced at the pair of C-complete A-modules M and N* iff the natural maps:

(1): $^{I,II}\mathsf{Tor}_i^{\mathscr{C}_A}(M, N) \to {}^I\mathsf{Tor}_i^{\mathscr{C}_A}(M, N)$ and

(2): $^{I,II}\mathsf{Tor}_i^{\mathscr{C}_A}(M, N) \to {}^{II}\mathsf{Tor}_i^{\mathscr{C}_A}(M, N)$ are isomorphisms of C-complete A-modules, for all integers $i \geq 0$. When this is the case, for each non-negative integer i, we use the symbol $\mathsf{Tor}_i^{\mathscr{C}_A}(M, N)$ to stand for any of the three canonically isomorphic C-complete A-modules

$$^I\mathsf{Tor}_i^{\mathscr{C}_A}(M, N), \ \ ^{II}\mathsf{Tor}_i^{\mathscr{C}_A}(M, N), \ or \ ^{I,II}\mathsf{Tor}_i^{\mathscr{C}_A}(M, N).$$

Note. Definition 3 of course generalizes Definition 1.

Remark 5. Of course, Definition 3 can be posed, for any right exact covariant functor $F(M, N)$ of two variables, from abelian categories $\mathscr{A}$ and $\mathscr{B}$ into an abelian category $\mathscr{C}$, such that $\mathscr{A}$ and $\mathscr{B}$ both have enough projectives. And then, the functor F is balanced at (M, N) for all M in $\mathscr{A}$ and for all N in $\mathscr{B}$ iff the functor F is balanced.

In particular, by Theorem 6.3.1, for any commutative standard topological ring A, the functor C-complete tensor product is balanced at every pair of C-complete A-modules M, N iff the functor, "I-fold direct sum," from $\mathscr{C}_A^I$ into $\mathscr{C}_A$, is exact, for all sets I. An immediate consequence of Proposition 6.3.5 is

Corollary 6.3.13. *Let A be a commutative standard topological ring, and let M and N be C-complete A-modules. Suppose that*

$$\int_{\substack{n \\ i \in I}} M = \int_{\substack{n \\ i \in I}} N = 0 \ for \ n \geq 1$$

and for all sets I. Then the functor C-complete tensor product is balanced at the pair M, N.

Proof. Follows immediately from Proposition 6.3.5. □

The rest of this section is devoted to a study of when the functor "C-complete tensor product" is balanced at pairs M, N, where both M and N are C-complete, and one of M or N is finitely presented over $\widehat{A}$. The answer is: "yes, in all important cases." (Corollary 6.3.16 below.) And, in fact, in all such important cases, when one of M or N is finitely presented over $\widehat{A}$, $\mathsf{Tor}_i^{\mathscr{C}_A}(M, N) \approx \mathsf{Tor}_i^{\widehat{A}}(M, N)$, $i \geq 0$. (See Corollary 6.3.16.)

Proposition 6.3.14. *Let A be a coherent commutative standard topological ring. Suppose,*

(1) For every free abstract A-module F, we have that $\widehat{F}$ is flat as an A-module.

Suppose also, that

(2) The topological ring A is strongly C-O.K.

[By hypothesis (1), a sufficient condition for (2) above is:

(2′) The topological ring A is C-O.K., and admits a denumerable neighborhood base at zero.]

Then for every C-complete A-module M, and for every finitely presented A-module N', if we let $N = \widehat{A} \underset{A}{\otimes} N'$, then we have that:

(3) The functor C-complete tensor product is balanced at the pair M and N.

(4) If the topological ring A is also complete, then for every such M and N', we also have that the maps $\alpha, \beta, \gamma, \delta, \epsilon$ in the commutative diagram of Lemma 6.3.12 are all isomorphisms of abstract A-modules, for all integers $i \geq 0$.

Note 1. If M and N' are as in the Proposition, then whether or not A is complete,

$$\mathrm{Tor}_i^{\widehat{A}}(M,N) \approx {}^I\mathrm{Tor}_i^{\mathscr{C}_A}(M,N) \approx {}^{II}\mathrm{Tor}_i^{\mathscr{C}_A}(M,N) \approx {}^{I,II}\mathrm{Tor}_i^{\mathscr{C}_A}(M,N)$$

$$\approx \mathrm{Tor}_i^{\mathscr{C}_A}(M,N)$$

(where this last is defined in Definition 3 above), canonically as abstract left $\widehat{A}$-modules (and all but the first of these canonical isomorphisms are isomorphisms of C-complete A-modules).

Otherwise stated, if we consider the diagram of Lemma 6.3.12, with $\widehat{A}$ replacing A (i.e., with $\mathrm{Tor}_i^{\widehat{A}}(M,N)$ in the bottom left corner in lieu of $\mathrm{Tor}_A^i(M,N)$), then the five maps $\alpha, \beta, \gamma, \delta, \epsilon$ in that diagram are all isomorphisms of abstract $\widehat{A}$-modules, for all non-negative integers i.

Note 2. The hypothesis (2) of this proposition can be replaced by the weaker hypothesis,

$(2_{\widehat{A}})$ The topological ring $\widehat{A}$ is strongly C-O.K.

Note 3. Whether or not hypothesis (2), (2′) or $(2_{\widehat{A}})$ holds, hypothesis (1) is necessary for the conclusion (3) of the proposition. In fact, if we delete hypothesis (1) and (2), and if $\widehat{A}$ is flat as an A-module, then the natural map

$$ {}^{I,II}\mathrm{Tor}_i^{\mathscr{C}_A}(M,N) \to {}^I\mathrm{Tor}_i^{\mathscr{C}_A}(M,N) $$

is an isomorphism, for all C-complete A modules M, and for all $N = \widehat{A} \underset{A}{\otimes} N'$, where N' is finitely presented as an abstract A-module, and for all integers $i \geq 0$, iff hypothesis (1) of this proposition holds.

Before proving Proposition 6.3.14, we prove a stronger result.

Corollary 6.3.15. *Let A be a commutative standard topological ring. Let M be a C-complete A-module, and let N be an abstract $\widehat{A}$-module such that*

(1) N admits an infinite chain of syzygies (Chapter 5, §3, Definitions 1 and 2) as abstract $\widehat{A}$-module.

Suppose also that

(2) The standard topological ring A is strongly C-O.K., and that

(3) $\text{Tor}_i^{\widehat{A}}(\widehat{F}, N) = 0$ *for all integers* $i \geq 1$, *and for all free modules F over A. Then*

(1) For every C-complete A-module M, the functor C-complete tensor product is balanced at the pair M, N.

(2) If also the topological ring A is complete, then for every C-complete A-module M, we have that the maps $\alpha, \beta, \gamma, \delta, \epsilon$ *in the diagram of Lemma 6.3.12, are isomorphisms of abstract A-modules, for all integers* $i \geq 0$.

Note 1. Suppose we delete the hypothesis (2), that "the topological ring A is strongly C-O.K.," and also that we delete hypothesis (3). Then the proof still shows that

(4) For the topological ring $\widehat{A}$, the map γ of Lemma 6.3.12 is an isomorphism of homological connected sequences of functors from $\mathscr{C}_{\widehat{A}}$ into $\mathscr{C}_{\widehat{A}}$. And then also

(5) The map β of Lemma 6.3.12 is an isomorphism for all M in $\mathscr{C}_A$ and for all $i \geq 0$, iff hypothesis (3) of this corollary holds.

Note 2. The hypothesis (2) of Corollary 6.3.15, that "the topological ring A is strongly C-O.K.," can be replaced by the less restrictive hypothesis

$(2_{\widehat{A}})$ The topological ring $\widehat{A}$ is strongly C-O.K.

Proof of Corollary 6.3.15. The categories $\mathscr{C}_A$ and $\mathscr{C}_{\widehat{A}}$ are canonically isomorphic (Chapter 2, §3, Corollary 2.3.2). Since A is strongly C-O.K., by Chapter 5, §5, Theorem 5.5.4, $\widehat{A}$ is strongly C-O.K. Therefore replacing A by $\widehat{A}$ if necessary, we can and do assume that A is complete. (And then Note 2 trivially holds.) By hypothesis N admits an infinite chain of syzygies. Let F_* be a free acyclic resolution of N as abstract A-module such that F_i is free of finite rank as an A-module, $i \geq 0$. Then, since A is complete, and F_i is free of finite rank, $C(F_i) = \widehat{F_i} = F_i$, $i \geq 0$, and therefore by Chapter 2, §4, Theorem 2.4.8, F_i is projective in $\mathscr{C}_A$, for $i \geq 0$. Therefore F_* is a projective resolution of N, both in Abs_A and in $\mathscr{C}_A$. Also, by §2, Exercise 3, since F_i is free of *finite* rank,

$$M \overset{C}{\underset{A}{\otimes}} F_i = M \underset{A}{\otimes} F_i,$$

for all integers $i \geq 0$. Therefore, by the usual construction of left derived functors using projective resolutions,

$$^{II}\text{Tor}_i^A(M, N) = H_i(M \overset{C}{\underset{A}{\otimes}} F_*) = H_i(M \underset{A}{\otimes} F_*) = \text{Tor}_i^A(M, N), \ i \geq 0.$$

That is, the natural map

$$(6): \ \gamma: \text{Tor}_i^A(M, N) \to {}^{II}\text{Tor}_i^{\mathscr{C}_A}(M, N)$$

in the commutative diagram of Lemma 6.3.12 is an isomorphism, for all integers $i \geq 0$. This proves equation (4) of Note 1.

By Corollary 6.3.7, in the case $A = \widehat{A}$, $N' = N$, the equivalence of condition [4] of Corollary 6.3.7 and condition [2] of Proposition 6.3.5, we have that $\operatorname{Tor}_i^{\widehat{A}}(\widehat{F}, N) = 0$ for all integers $i \geq 1$ and for all free A-modules F, iff the map β in Lemma 6.3.12 is an isomorphism, for all integers $i \geq 1$ and for all C-complete A-modules M. This proves equation (5) of Note 1, and therefore completes the proof of Note 1.

It remains to prove the remainder of the Corollary, in the case in which A is complete.

We have shown that the map

$$(7): \quad \beta: \; {}^{I,II}\operatorname{Tor}_i^{\mathscr{C}_A}(M, N) \to {}^{I}\operatorname{Tor}_i^{\mathscr{C}_A}(M, N)$$

is an isomorphism for $i \geq 0$, and for all M in $\mathscr{C}_A$.

By Example 2.1 and hypothesis (3), N obeys the three equivalent conditions of Proposition 6.3.9. In particular, hypothesis (2) of Corollary 6.3.10 (which is identical to condition [2] of Proposition 6.3.9) holds. If M is a C-complete A-module, then since by hypothesis A is strongly C-O.K.,

$$(8): \quad C_i(M) = 0 \text{ for } i \geq 1.$$

That is, we have that hypothesis (1) of Corollary 6.3.10 holds. Therefore by Corollary 6.3.10 we have a first quadrant homological spectral sequence in the category $\mathscr{C}_A$,

$$(9): \quad E_{p,q}^2 = C_p(\operatorname{Tor}_q^A(M, N)) \Rightarrow {}^{I}\operatorname{Tor}_n^{\mathscr{C}_A}(C(M), C(N)).$$

By equation (6), $\operatorname{Tor}_i^A(M, N)$ is C-complete, for all $i \geq 0$. Since by hypothesis the topological ring A is strongly C-O.K., this implies that

$$(10): \quad C_p(\operatorname{Tor}_q^A(M, N)) = \begin{cases} \operatorname{Tor}_q^A(M, N), & p = 0 \\ 0, & p \geq 1. \end{cases}$$

Since N admits an infinite chain of syzygies, N is of finite presentation as an A-module. Therefore by Chapter 5, §3, Corllary 5.3.3,

(11): $C(N) = N$.

Also, since the ring A is strongly C-O.K., A is a fortiori C-O.K. Therefore, since M is C-complete,

(12): $C(M) = M$.

Substituting equations (10), (11), and (12) into the spectral sequence (9), it follows that the spectral sequence (9) degenerates, and that the edge homomorphisms are isomorphisms

$$(13): \quad \operatorname{Tor}_n^A(M, N) \stackrel{\approx}{\to} {}^{I}\operatorname{Tor}_n^{\mathscr{C}_A}(M, N),$$

for all integers $n \geq 0$. That is, the natural map

$$\alpha: \operatorname{Tor}_i^A(M, N) \to {}^{I}\operatorname{Tor}_i^{\mathscr{C}_A}(M, N), \; i \geq 0$$

of Lemma 6.3.12 is an isomorphism of homological connected sequences of functors in the variable M from the category $\mathscr{C}_A$ into the category of abstract A-modules.

Equations (6), (7), and (13) imply that, for all integers $i \geq 0$, and all C-complete A-modules M, all of the maps $\alpha, \beta, \gamma, \delta$ and ϵ, in the commutative diagram of Lemma 6.3.12, are isomorphisms. And in particular, it follows, Definition 3, that the C-complete tensor product is balanced at the pair M, N, for all C-complete A-modules M. $\qquad\Box$

Proof of Proposition 6.3.14. We first prove Note 3. Assume that $\widehat{A}$ is flat as an A-module. Let N' be an A-module of finite presentation, and let $N = \widehat{A} \underset{A}{\otimes} N'$. Since the ring A is coherent, there exists an infinite chain of syzygies P_* of N'. Therefore the hypotheses of Corollary 6.3.7 are satisfied. Therefore, by Corollary 6.3.7 (equivalence of condition [4] of Corollary 6.3.7 and condition [2] of Proposition 6.3.5), we have that the map β of Lemma 6.3.12 is an isomorphism, for all M in $\mathscr{C}_A$, and for all finitely presented A-modules N' (where $N = \widehat{A} \underset{A}{\otimes} N'$) iff $\mathrm{Tor}_i^A(\widehat{F}, N') = 0$ for all integers $i \geq 1$, for all free abstract A-modules F, and for all abstract A-modules N' of finite presentation as abstract A-module. And this latter is equivalent to:

$\widehat{F}$ is flat over A, for all free abstract left A-modules F.

(*Proof*: By [COC], Introduction, Chapter 1, Section 6, Lemma 4, p. 72, an abstract A-module G is flat over A iff for every finitely generated ideal I in A, the natural map: $I \underset{A}{\otimes} G \to I \cdot G$ is injective. This is equivalent to $\mathrm{Tor}_1^A(G, A/I) = 0$. Since A/I is of finite presentation as an A-module, therefore G is flat as an A-module iff $\mathrm{Tor}_i^A(G, N') = 0$ for all integers $i \geq 1$, and for all finitely presented abstract A-modules N'. Taking $G = \widehat{F}$ gives the indicated equivalence.)

Therefore, we have shown that, if $\widehat{A}$ is flat as an A-module, then condition (1) of the proposition is equivalent to the map β of Lemma 6.3.12 being an isomorphism, for all M in $\mathscr{C}_A$, and for all N' that are finitely presented as A-modules. This proves Note 3 to the proposition.

If hypothesis $(2')$ of the proposition holds, then, using hypothesis (1) and Chapter 5, §5, Corollary 5.5.3, we have that hypothesis (2) of the proposition holds. And by Chapter 5, §5, Theorem 5.5.4, condition $(2_{\widehat{A}})$ of Note 2 to the proposition is implied by condition (2) of the proposition.

Assume now that condition (1) of the proposition, and condition $(2_{\widehat{A}})$ of Note 2 to the proposition, both hold. Then, since A is free as an A-module, by condition (1), we have that $\widehat{A}$ is flat as an A-module. Therefore, for all finitely presented abstract A-modules N', if $N = \widehat{A} \underset{A}{\otimes} N'$, then by Lemma 6.3.8, since $\widehat{A}$ is flat as an abstract A-module, for every such N',

$$\mathrm{Tor}_i^A(\widehat{F}, N') \approx \mathrm{Tor}_i^{\widehat{A}}(\widehat{F}, N), \ i \geq 0.$$

Therefore by hypothesis (1) of the proposition

$$\mathrm{Tor}_i^{\widehat{A}}(\widehat{F}, N) = 0, \ i \geq 1$$

for all finitely presented A-modules N' and for all free A-modules F, where $N = \widehat{A} \underset{A}{\otimes} N'$.

Therefore the topological ring A and the abstract $\widehat{A}$-module N obey the hypotheses of Corollary 6.3.15 as modified by Note 2 to Corollary 6.3.15. Therefore, by Corollary 6.3.15, we have the conclusions (3) and (4) of the proposition. And by conclusion (2) of Corollary 6.3.15 for the ring $\widehat{A}$ and the abstract $\widehat{A}$-module N, we have the conclusion of Note 1 of the proposition.   □

Corollary 6.3.16. *Let A be a Noetherian commutative ring, and suppose that A is a topological ring, such that the topology is given by denumerably many ideals, and such that the square of every open ideal is open. Then, for every C-complete A-module M, and every finitely generated A-module N', if $N = \widehat{A} \underset{A}{\otimes} N'$, then the functor C-complete tensor product over A is balanced at the pair M, N.*

If the topological ring A is also complete, then for every C-complete A-module M, and for every finitely generated A-module N, then all of the five maps $\alpha, \beta, \gamma, \delta, \epsilon$ in the commutative diagram of Lemma 6.3.12 are isomorphisms of abstract A-modules, for all integers $i \geq 0$.

Note 1. (Same as Note 1 to Proposition 6.3.14.)

Proof. By Chapter 5, §4, Theorem 5.4.2, hypothesis (1) of Proposition 6.3.14 holds. By Chapter 5, §5, Example 3, hypothesis (2) of Proposition 6.3.14 holds. Therefore the corollary follows from Proposition 6.3.14.   □

Remark 6. Corollary 6.3.16 applies to, in practice, a larger class of topological rings than one comes across in contemporary algebraic geometry and contemporary commutative algebra. However, it should be noted that there is an important, *very* nontrivial, hypothesis on one of the C-complete A-modules — namely, that $N \approx \widehat{A} \underset{A}{\otimes} N'$, where N' is finitely generated as an A-module. Of course, most C-complete A-modules N *do not* obey this extremely strong finiteness condition.

Again, one should emphasize— see Remark 1 above — that the functor "complete tensor product over A" is *rarely* a balanced functor. (E.g., C-complete tensor product is not balanced over A, if A is a Cohen-Macauley local ring of dimension ≥ 2 with the topology given by powers of the maximal ideal (see Remark 1 above, and Chapter 5, §6, Example 2).

The finiteness condition considered above on N should be thought of as being a very special condition.

6.4   Change of Topological Ring

Definition 1. Let $\phi : A \to B$ be a continuous ring homomorphism of standard topological rings. Then, for every C^B-complete left B-module M, we endow M with the structure of C^A-complete left A-module, as follows.

First, regard M as an abstract left A-module by means of ϕ in the usual fashion (e.g., see [NB]). Next, we define an infinite sum structure (Chapter 2, §2, Definition 1) on the abstract left A-module M as follows: First, let $\widehat{\phi} : \widehat{A} \to \widehat{B}$ denote the unique extension of ϕ to a continuous ring homomorphism on the cocompletions. If I is any set, and if $(a_i)_{i \in I} \in \widehat{A}^I$ is a zero-bound family of elements of $\widehat{A}$, then since $\widehat{\phi}$ is continuous, we have that the family $(\widehat{\phi}(a_i))_{i \in I} \in \widehat{B}^I$ is a zero-bound family of elements of $\widehat{B}$. If $(x_i)_{i \in I} \in M^I$, then define

$$(1): \quad \sum_{i \in I} a_i \cdot x_i = \sum_{i \in I} \widehat{\phi}(a_i) \cdot x_i,$$

where the infinite linear combination on the right side of the equation makes use of the structure of M as a C^B-complete left B-module. Then it is an easy exercise, using the fact that ϕ is a continuous ring homomorphism from A into B, to show that definition (1) obeys the Axioms (IS1) - (IS4) of Chapter 2, §2, Definition 1.

Thus, every C^B-complete left B-module M has (by means of the continuous ring homomorphism ϕ), a natural structure as a C^A-complete left A-module. We will call this the *underlying C^A-complete left A-module of M*. Otherwise stated, we have a functor from the abelian category $\mathscr{C}_B$ into $\mathscr{C}_A$. This functor is clearly exact and faithful. And, by Chapter 4, §2, Theorem 4.2.1, (respectively: Proposition 4.2.2) we have that the functor, "underlying C^A-complete left A-module," from the category $\mathscr{C}_B$ into $\mathscr{C}_A$, preserves arbitrary direct products of objects indexed by arbitrary sets (respectively: preserves arbitrary inverse limits over arbitrary set-theoretically legitimate categories).

Remark 1. The hypothesis in Definition 1, that the ring homomorphism ϕ be *continuous*, is unnecessarily restrictive for the construction of Definition 1. It suffices that: A and B be standard topological rings, and that we are given $\widehat{\phi} : \widehat{A} \to \widehat{B}$ an abstract ring homomorphism, of the completions, such that $\widehat{\phi}$ is *sequentially continuous* for the uniform topologies of $\widehat{A}$ and $\widehat{B}$ — i.e., such that for every sequence of elements $(x_i)_{i \geq 1}$ of $\widehat{A}$ indexed by the positive integers that converges to zero in $\widehat{A}$, we have that the sequence $(\widehat{\phi}(x_i))_{i \geq 1}$ converges to zero in $\widehat{B}$. (That this is sufficient for the construction follows readily from Chapter 5, §2, Exercise 1.)

However, it should be noted that, if the topological ring A admits a denumerable neighborhood base at zero, then a function $\widehat{\phi} : \widehat{A} \to \widehat{B}$ is sequentially continuous iff it is continuous. Therefore, in this case, the observation of this Remark is not really a generalization. (It is then merely just the obvious fact that we need only a continuous homomorphism $\widehat{A} \to \widehat{B}$, rather than a continuous homomorphism $A \to B$.)

All of the results of this section generalize without change, with a "sequentially continuous ring homomorphism $\widehat{\phi} : \widehat{A} \to \widehat{B}$" replacing the (slightly less general) "continuous ring homomorphism $\phi : A \to B$." (Nevertheless, for reasons of aesthetics, we prefer stating results for continuous ring homomorphisms $\phi : A \to B$. To generalize one just changes the phrase.)

Definition 2. Let $\phi : A \to B$ be a continuous homomorphism of standard topological rings. Let M be a C^A-complete left A-module. Then by *an extension of M to B* we mean

(1) A C^B-complete left B-module, which we denote $[B, M]$.

(2) A map in the category $\mathscr{C}_A$ (that is, an infinitely A-linear function) $\kappa : M \to [B, M]$. (Here, we are regarding the C^B-complete left B-module $[B, M]$, as being a C^A-complete left A-module, by means of Definition 1.) such that

(3) The pair $[B, M]$ and κ, are universal with these properties.

Clearly, if an extension $[B, M]$ of M to B exists in the sense of Definition 2, then it is unique (as an object in $\mathscr{C}_B$) up to canonical isomorphisms.

Lemma 6.4.1. *Let $\phi : A \to B$ be a continuous ring homomorphism of standard topological rings. Let M be a C^A-complete left A-module. Then there exists an extension $[B, M]$ of M to B.*

Proof. The C^B-complete left B-module, $[B, M] = C^B(B \underset{A}{\otimes} M)/H$ obeys the universal mapping property of Definition 2, where H is the C^B-complete left submodule of $C^B(B \underset{A}{\otimes} M)$ generated by the subset

$$S = \left\{ \iota \left(1_B \underset{A}{\otimes} \left(\sum_{i \in I} a_i \cdot x_i \right) \right) - \sum_{i \in I} \widehat{\phi}(a_i) \cdot [\iota(1_B \underset{A}{\otimes} x_i)] : I \text{ is a set}, (a_i)_{i \in I} \right.$$

is a family of elements of $\widehat{A}$ that is zero-bound, $x_i \in M$, all $i \in I$,

$$\left. \text{and } \iota : B \underset{A}{\otimes} M \to C^B(B \underset{A}{\otimes} M) \text{ is the natural map} \right\}.$$

$\square$

Corollary 6.4.2. *Let $\phi : A \to B$ be a continuous homomorphism of standard topological rings. Suppose that the topological ring A is C-O.K. Then for every C^A-complete left A-module, we have that*

$$[B, M] = C^B(B \underset{A}{\otimes} M).$$

Proof. Since A is C-O.K., every A-linear function of C^A-complete left A-modules is infinitely A-linear. (Chapter 3, §1, Theorem 3.1.5, condition [5], and Chapter 3, §1, Definition 3.) Therefore, in Definition 2 above, equation (2), the condition on the map κ_M, that "κ_M is a map in the category $\mathscr{C}_A$," can be replaced by the then equivalent condition, "κ_M is a homomorphism of abstract left A-modules." One then verifies that $C^B(B \underset{A}{\otimes} M)$ obeys the universal mapping property. $\square$

Theorem 6.4.3. *Let $\phi : A \to B$ be a continuous homomorphism of standard topological rings. Then the assignment, $M \rightsquigarrow [B, M]$, is a functor from the category $\mathscr{C}_A$ into the category $\mathscr{C}_B$. This functor preserves arbitrary direct limits* *of arbitrary covariant functors indexed by arbitrary set-theoretically legitimate categories. In particular,*

(1) For every set I, and for every family M_i, $i \in I$, of C^A-complete left A-modules, we have that

$$\left[B, \int_{i \in I}^{A} M_i \right] \approx \int_{i \in I}^{B} [B, M_i]$$

canonically in $\mathscr{C}_B$. (Here, $\int_{i \in I}^{A}$, respectively: $\int_{i \in I}^{B}$, denotes the I-fold direct sum in the category $\mathscr{C}_A$, respectively: in the category $\mathscr{C}_B$.)

(2) For every directed set D, and for every direct system $(M_i, \alpha_{ij})_{i,j \in D}$ indexed by the directed set D in the category $\mathscr{C}_A$, we have that the natural map is an isomorphism in $\mathscr{C}_B$

$$\left[B, \varinjlim_{d \in D}^{A} M_d \right] \overset{\approx}{\leftarrow} \varinjlim_{d \in D}^{B} [B, M_d].$$

(Here, $\varinjlim_{d \in D}^{A}$, respectively: $\varinjlim_{d \in D}^{B}$, denotes the direct limit in the category $\mathscr{C}_A$, respectively: in the category $\mathscr{C}_B$, of direct systems over the directed set D.)

(3) For every category $\mathscr{D}$ that is a set, and every covariant functor M from $\mathscr{D}$ into $\mathscr{C}_A$, we have the natural map is an isomorphism in $\mathscr{C}_B$,

$$\left[B, \varinjlim_{d \in D}^{A} M(d) \right] \overset{\approx}{\leftarrow} \varinjlim_{d \in D}^{B} [B, M(d)].$$

(4) The functor $M \rightsquigarrow [B, M]$ from the abelian category $\mathscr{C}_A$ into the abelian category $\mathscr{C}_B$ is right exact.

Proof. Entirely similar to that of §2, Corollary 6.2.3. □

Remark 2. From Definition 2, we deduce a canonical isomorphism of abelian groups

$$(1): \operatorname{Hom}_{\mathscr{C}_A}(M, N) \approx \operatorname{Hom}_{\mathscr{C}_B}([B, M], N)$$

for all N in $\mathscr{C}_B$ and all M in $\mathscr{C}_A$ (where on the left side of equation (1), N in $\mathscr{C}_B$ is regarded as an object in $\mathscr{C}_A$ by means of the "underlying C^A-complete left A-module functor" of Definition 1). Otherwise stated, the functor $M \rightsquigarrow [B, M]$ from $\mathscr{C}_A$ into $\mathscr{C}_B$, is the right adjoint of the functor "underlying left A-module from $\mathscr{C}_B$ into $\mathscr{C}_A$".

Theorem 6.4.3 also follows from this fact (any functor that admits a left adjoint necessarily commutes with arbitrary direct limits when defined). But this proof of Theorem 6.4.3 is really not different from the proof given, of verifying the universal mapping property. Also, from equation (1), and the fact that "underlying C^A-complete left A-module" is an exact functor, and from [PPWC], Chapter 1, §1, Lemma 1, p. 114, it follows that

Corollary 6.4.4. *The functor "extension to B," $M \rightsquigarrow [B, M]$, from $\mathscr{C}_A$ into $\mathscr{C}_B$, maps projectives into projectives.*

However, Example 2 below is a better theorem. First, let us note that, since A and B are both topological rings, then given an abstract left B-module M, the notation "$\widehat{M}$" is ambiguous. One must write "$\widehat{M}^B$" or "$\widehat{M}^A$", to indicate whether one means the completion for the B-adic, or for the A-adic, topology.

Example 1. $[B, \widehat{A}^A] = \widehat{B}^B$.

Proof. Follows immediately from the universal mapping property. □

Example 2. For every set I, that natural map is an isomorphism of C^B-complete left B-modules

$$[B, \widehat{A^{(I)}}^A] \overset{\approx}{\to} \widehat{B^{(I)}}^B .$$

Proof. Since $\overset{A}{\underset{i \in I}{\int}} \widehat{A}^A = \widehat{A^{(I)}}^A$, $\overset{A}{\underset{i \in I}{\int}} \widehat{B}^B = \widehat{B^{(I)}}^B$, the result follows from Example 1 and from Theorem 6.4.3, conclusion (1). (An essentially equivalent alternative proof can be given, using the universal mapping property definition of $[B, \widehat{A^{(I)}}^A]$ directly, and also using Chapter 2, §4, Corollary 2.4.10.) □

Example 3. Let M be any C^A-complete left A-module, and let

$$(1): \ \widehat{A^{(I)}}^A \to \widehat{A^{(J)}}^A \to M \to 0$$

be an exact sequence in the category $\mathscr{C}_A$. Then there is induced an exact sequence

$$(2): \ \widehat{B^{(I)}}^B \to \widehat{B^{(J)}}^B \to [B, M] \to 0.$$

Proof. Example 2, and Theorem 6.4.3, conclusion (4). □

Remark 3. Example 3 above — which, like Examples 1 and 2, can be proved directly using the universal mapping property definition, Definition 2, of $[B, M]$ — can be used to give an alternate proof of Lemma 6.4.1. In fact, Example 3 is, in practice, the "working construction" of $[B, M]$.

Remark 4. In Example 3, the map $\widehat{A^{(I)}}^A \to \widehat{A^{(J)}}^A$ in equation (1) is uniquely determined (Chapter 2, §5, Remark 2) by its matrix $(a_{ij})_{(i,j) \in I \times J}$, where $a_{ij} \in \widehat{A}$, for all $i \in I$, all $j \in J$ (so that, for all $i \in I$, the ith column $(a_{ij})_{j \in J} \in \widehat{A}^J$ is zero-bound). Then, the map $\widehat{B^{(I)}}^B \to \widehat{B^{(J)}}^B$ in equation (2) is the one given by the matrix $(\hat{\phi}(a_{ij}))_{(i,j) \in I \times J}$.

Remark 5. In terms of generators and relations (Chapter 2, §4, Definition 1), Example 3 can be stated as follows. Let M be any C^A-complete left A-module. Then choose any sets $(e_j)_{j \in J}$ of generators and $(r_i)_{i \in I}$ of relations for M as C^A-complete left A-module. (It is equivalent to giving an exact sequence in the category $\mathscr{C}_A$, as in equation (1) of Example 3.) Thus $e_j \in M$, all $j \in J$.

Every element of M is an infinite linear combination of the e_j, $j \in J$; and $r_i = (a_{ij})_{j \in J} \in \widehat{A^{(J)}}^A$, for all $i \in I$, are such that $\sum_{j \in J} a_{ij} e_j = 0$ in M, for all $i \in I$; and these are "enough relations" to generate all the relations between $(e_j)_{j \in J}$ in the C-complete left A-module M.

Then, equation (2) asserts that $[B, M]$ is the C^B-complete left B-module, given by the generators $(e_j)_{j \in J}$, and by the relations $(\tilde{r}_i)_{i \in I}$, where $\tilde{r}_i = (\widehat{\phi}(a_{ij}))_{j \in J} \in \widehat{B^{(J)}}^B$, for all $i \in I$.

Remark 6. Yet another way of stating Example 3 is as follows. Let M be any C^A-complete left A-module. Let $(a_{ij})_{(i,j) \in I \times J}$ be a matrix, the coefficients of which are elements of $\widehat{A}$, such that for every $i \in I$, the family $(a_{ij})_{j \in J} \in \widehat{A}^J$ is zero-bound in $\widehat{A}$, and such that the matrix $(a_{ij})_{(i,j) \in I \times J}$ determines the C^A-complete left A-module M in the sense of Chapter 2, §5, Remark 2. Then, the C^B-complete left B-module $[B, M]$ is the one determined by the matrix $(\widehat{\phi}(a_{ij}))_{(i,j) \in I \times J}$.

Remark 7. Let $\phi : A \to B$ be a continuous homomorphism of standard topological rings. Suppose that the topological ring A is C-O.K. Then it is not difficult to show that, in the notations of Chapter 2, §3, Remark 8,

$$\mathscr{C}_B \widehat{B}^B \underset{A}{\otimes} M_{\mathscr{C}_{\mathbf{Z}}} \approx [B, M],$$

canonically in $\mathscr{C}_B$, for all M in $\mathscr{C}_A$ — i.e., that the above is a natural equivalence of functors from $\mathscr{C}_A$ into $\mathscr{C}_B$. (In this equation, "$\mathbf{Z}$" denotes the discrete topological ring of integers. Therefore $\mathscr{C}_{\mathbf{Z}}$ is the cateogry of abstract abelian groups.) Hence, if the topological ring A is C-O.K., then Definition 2 above can be interpreted as being a special case of the very general "C-complete generalized tensor product" constructed in Chapter 2, §3, Remark 8.

Remark 8. In the special case that $\phi : A \to B$ is a homomorphism of *discrete* topological rings, then it is well-known, for every abstract left A-module M (= C^A-complete left A-module M, when A has the discrete topology), that the abstract left B-module $B \underset{A}{\otimes} M$ (= C^B-complete left B-module, when B has the discrete topology), obeys the universal mapping property definition for $[B, M]$. In fact, the same argument shows that:

Let $\phi : A \to B$ be a continuous homomorphism of standard topological rings, such that B has the discrete topology. (Note: When B has the discrete topology, the condition that an abstract ring homomorphism $\phi : A \to B$ be *continuous* is equivalent to the condition that "The two-sided ideal $\mathrm{Ker}(\phi)$ in the ring A is an open ideal".) Then

$$[B, M] \approx B \underset{A}{\otimes} M$$

canonically in $\mathscr{C}_B$. This identity *which depends strongly on B having the discrete topology*, suggests the question, "Is there in some generality an isomorphism between $[B, M]$ and $\widehat{B} \overset{C}{\underset{A}{\otimes}} M$ (the C-complete tensor product over A as defined in §2)"?

In fact, the functor "extension to B," of Definition 2, cannot, in general, be interpreted as a special case of the C-complete tensor product over a commutative ring A (§2). For one thing, we have not assumed that A is commutative. But even when the standard topological ring A is commutative, it is still not true in general that $[B, M]$ is isomorphic to $\widehat{B}^B \overset{C}{\underset{A}{\otimes}} M$. We study this question briefly in the next proposition and two corollaries, and in the next few examples.

Proposition 6.4.5. *Let* $\phi : A \to B$ *be a continuous homomorphism of standard topological rings. Suppose that the abstract ring A is commutative and that $\phi(A)$ is contained in the center of the ring B. Then the following three conditions are equivalent.*

[1] For every C-complete A-module M, the natural map in $\mathscr{C}_A$

$$\widehat{B}^B \overset{C}{\underset{A}{\otimes}} M \to [B, M]$$

is an isomorphism of C^A-complete A-modules.

[2] For every set I, and for every family N_i, $i \in I$, of C^B-complete left B-modules, the natural map in $\mathscr{C}_A$

$$\overset{A}{\underset{i \in I}{\int}} N_i \to \overset{B}{\underset{i \in I}{\int}} N_i$$

is an isomorphism of C^A-complete A-modules.

[3] Condition [2] above holds, for all sets I, in the special case $N_i = \widehat{B}^B$, for all $i \in I$.

Proof. [1] iff [3]. $\widehat{B}^B$ is an object in $\mathscr{C}_B$. Therefore by Definition 1 $\widehat{B}^B$ has a natural structure as C^A-complete A-module. Therefore the C-complete tensor product over A,

$$\widehat{B}^B \overset{C}{\underset{A}{\otimes}} M$$

is defined, and is an object in $\mathscr{C}_A$. Since $\phi(A)$ is contained in the center of the ring B, one sees that the function $(b, m) \to b \cdot \kappa_M(m)$ from $\widehat{B}^B \times M$ into $[B, M]$ is infinitely A-bilinear (where the C^B-complete left B-module $[B, M]$ is given the structure of C^A-complete A-module by Definition 1). Therefore (see the beginning of §2 — or else Chapter 2, §3, Remark 8) we have a natural map in the category $\mathscr{C}_A$

$$(1) : \quad \widehat{B}^B \overset{C}{\underset{A}{\otimes}} M \to [B, M]$$

for all C-complete A-modules M. Since the functors $M \rightsquigarrow \widehat{B}^B \overset{C}{\underset{A}{\otimes}} M$, $M \rightsquigarrow [B, M]$ are both right exact considered as functors from $\mathscr{C}_A$ into $\mathscr{C}_A$, the map (1) is an isomorphism for all M in (1) iff the map (1) is an isomorphism in the case $M = \widehat{A^{(I)}}$, for every set I. But then, by §2, Corollary 6.2.6, the left side of equation (1) is

$$\overset{A}{\underset{i \in I}{\int}} \widehat{B}^B.$$

And by Example 2 above the right side is

$$\widehat{B^{(I)}}^B = \overset{B}{\underset{i \in I}{\int}} \widehat{B}^B.$$

Therefore condition [1] is equivalent to condition [3].

[2] implies [3] is clear.

[3] implies [2]. For any fixed set I, the functors $(N_i)_{i \in I} \rightsquigarrow \overset{A}{\underset{i \in I}{\int}} N_i$, $(N_i)_{i \in I} \rightsquigarrow \overset{B}{\underset{i \in I}{\int}} N_i$, from $\mathscr{C}_B^I$ into $\mathscr{C}^A$, are right exact. Therefore the natural map in $\mathscr{C}_A$:

$$(2): \quad \overset{A}{\underset{i \in I}{\int}} N_i \to \overset{B}{\underset{i \in I}{\int}} N_i$$

is an isomorphism, for all $(N_i)_{i \in I}$ in $\mathscr{C}_B^I$, iff the map (2) is an isomorphism whenever $N_i = \widehat{B^{(J_i)}}^B$, where J_i is a set, for all $i \in I$. We can assume that the sets J_i are pairwise disjoint, for $i \in I$. But, if condition [3] holds, then

$$\overset{A}{\underset{i \in I}{\int}} \widehat{B^{(J_i)}}^B = \overset{A}{\underset{i \in I}{\int}} \overset{B}{\underset{j \in J_i}{\int}} \widehat{B}^B = \overset{A}{\underset{i \in I}{\int}} \overset{A}{\underset{j \in J_i}{\int}} \widehat{B}^B = \overset{A}{\underset{j \in \mathscr{J}}{\int}} \widehat{B}^B$$

where $\mathscr{J} = \underset{i \in I}{\cup} J_i$, and similarly (with "$B$" replacing "$A$" throughout in the last computation) $\overset{B}{\underset{i \in I}{\int}} \widehat{B^{(J_i)}}^B = \overset{B}{\underset{j \in \mathscr{J}}{\int}} \widehat{B}^B$. And therefore $\overset{A}{\underset{i \in I}{\int}} \widehat{B^{(J_i)}}^B = \overset{B}{\underset{i \in I}{\int}} \widehat{B^{(J_i)}}^B$, proving that (2) is an isomorphism in this case. $\qquad \square$

Corollary 6.4.6. *The hypotheses being as in Proposition 6.4.5, suppose that the standard topological ring A is C-O.K. Then the three equivalent conditions of Proposition 6.4.5 are also equivalent to*

[4] For every set I, the natural map

$$C^A\left((\widehat{B}^B)^{(I)}\right) \to \widehat{B^{(I)}}^B$$

is an isomorphism.

Proof. Since A is C-O.K., by Chapter 5, §1, Theorem 5.1.1

$$(1): \quad \int\limits_{i \in I}^{A} \widehat{B}^B = C^A \left((\widehat{B}^B)^{(I)} \right),$$

for all sets I. Always,

$$(2): \quad \int\limits_{i \in I}^{B} \widehat{B}^B = \widehat{B^{(I)}}^B.$$

Substituting equations (1) and (2) into condition [3] of the Proposition gives condition [4] of the Corollary.   □

Corollary 6.4.7. *Let the hypotheses be as in Proposition 6.4.5. Suppose that the topology of the topological ring B is the A-adic topology (i.e., that the topology of B is the one given by the two-sided dieals, $\{I \cdot B$ such that I is an open ideal in $A\}$ [4]). Suppose also that the topological ring A is C-O.K. Then the four equivalent conditions, conditions [1], [2], [3] of Proposition 6.4.5 and condition [4] of Corollary 6.4.6 are also equivalent to*

[5] $C^A(\widehat{B}^{(I)}) = \widehat{B^{(I)}}$, for all sets I.

Note: Since, in this corollary, we are assuming that the topology of B is the A-adic topology, therefore for every B-module M, we have that the A-adic and B-adic topologies of M coincide. Therefore in condition [5], it is not necessary to specify whether we mean the "A-adic completion" or the "B-adic completion" (and similarly we can write "C" instead of "C^A" or "C^B").

Proof. By the Note, condition [5] is just condition [4] rewritten.   □

Example 4. Let A be a commutative t-adic ring such that the t-torsion is bounded below. Let $\phi : A \to B$ be a homomorphism of abstract rings such that $\phi(A)$ is contained in the center of B. Regard B as a topological ring with the A-adic topology. Then B is a t-adic ring. Then the five equivalent conditions, [1], [2], [3], [4], [5] of Proposition 6.4.5 and of Corollaries 6.4.6 and 6.4.7 hold, iff the t-torsion in $\widehat{B}$ is bounded below.

Proof. We show that the t-torsion is bounded below in $\widehat{B}$ iff condition (5) of Corollary 6.4.7 above holds.

$\widehat{B^{(I)}}$ satisfies the universal mapping property for the A-adic completion of $B^{(I)}$; therefore

$$\widehat{\widehat{B}^{(I)}} = \widehat{B^{(I)}}.$$

Therefore, replacing B by $\widehat{B}$ if necessary, we can assume that B is complete. By Chapter 1, §3, Lemma 1.3.4, we have that

$$\widehat{B^{(I)}} \subset B^I.$$

---

[4]Notice, since the ideals $\{I \cdot B\}$ in B are *two-sided* (since $\phi(A)$ is contained in the center of B) that this collection of ideals makes B into a topological ring.

By Chapter 5, §2, Corollary 5.2.12, the t-torsion is bounded below in B iff the Note to Chapter 5, §2, Proposition 5.2.11 holds – i.e., iff

$$C(B^{(I)}) \subset B^I.$$

Therefore, the natural epimorphism:

$$C(B^{(I)}) \to \widehat{B^{(I)}}$$

is an isomorphism iff the t-torsion in B is bounded below. □

Example 5. Let A be a discrete valutions ring not a field, with topology given by powers of the maximal ideal (respectively: let $A = \mathbf{Z}[T]$, with the T-adic topology). Then A is a proper t-adic ring with parameter t, where t is any non-zero element of the maximal ideal of A (respectively: where $t = T$). Let B be the quotient A-agebra of the polynomial ring in denumerably many variables $A[T_1, T_2, \ldots]$, by the ideal generated by the elements $t \cdot T_1$, $t^2 \cdot T_2$, $\ldots$ (i.e., B is the A-algebra with generators T_i, $i \geq 1$, and relations: $t^i \cdot T_i = 0$, $i \geq 1$). Then the hypotheses of Corollary 6.4.7 hold, for the continuous ring homomorphism $\phi : A \to B$, but the five equivalent conditions, [1], ..., [5] all fail. (Since, by Example 4, these are equivalent to the t-torsion being bounded below in $\widehat{B}$. And this latter fails.)

Remark. In Example 5 above, A can be replaced by any t-adic ring such that the t-torsion is bounded below, and such that the topology of A is not the discrete topology.

Example 6. Let B be a commutative standard topological ring, such that there exists a sequence $a_1, a_2, \ldots$ of elements that converges to zero, but such that infinitely many of the a_i's are not in the closure of zero. (E.g., it suffices to take B to be any Hausdorff commutative topological ring that admits a denumerable neighborhood base at zero, and such that the topology of B is not the discrete topology.) Let A be the topological ring having the same underlying abstract ring as B, and with the discrete topology. Then the set-theoretic identity function: $\phi : A \to B$ obeys the hypotheses of Proposition 6.4.5 and of Corollary 6.4.6 (but not the hypotheses of Corollary 6.4.7). And the four equivalent conditions [1], [2], [3], [4] (of Proposition 6.4.5 and of Corollary 6.4.6) all fail.

Proof. The sequence $(a_1, a_2, \ldots)$ is an element of $\widehat{B^{(\omega)}}^B$ not in the image of $(\widehat{B}^B)^{(\omega)}$. Thus $C^A\left((\widehat{B}^B)^{(\omega)}\right) = (\widehat{B}^B)^{(\omega)} \neq \widehat{B^{(\omega)}}^B$, so condition [4] fails. □

Example 7. Let $\phi : A \to B$ be a homomorphism of Noetherian commutative rings. Let I be an ideal in A and let $J = I \cdot B$. Regard A, respectively B, as topological rings with the I-adic, respectively: the J-adic, topology. Then the hypotheses of Corollary 6.4.7 hold, and therefore the five equivalent conditions of Corollary 6.4.7 hold.

Proof. Entirely similar to the proof of Example 4 above. □

Remark. The most important applications today, to commutative algebra and to algebraic geometry, involve Noetherian commutative rings such that the topology is the I-adic for some ideal I. Therefore, Example 7 above essentially asserts that, in all "practical" cases, the five equivalent conditions [1], [2], [3], of Proposition 6.4.5, [4] of Corollary 6.4.6, and [5] of Corollary 6.4.7, hold. *However*, there is one point to be cautious about: Namely, given a continuous homomorphism $\phi : A \to B$ of such Noetherian topological rings, it certainly need not be true that the topology of B is the A-adic topology. And, when such is not the case, Example 7 does not apply. (And there are then easy counterexamples to the equivalent conditions of Proposition 6.4.5 holding — e.g, Example 5 above.)

Let us now return to the general situation of this section. That is, $\phi : A \to B$ is a continuous homomorphism of standard topological rings, neither of which is assumed to be commutative — (or, we can have the even more general hypotheses of Remark 1).

Definition 3. Let $\phi : A \to B$ be a continuous homomorphism of standard topological rings. Then, Lemma 6.4.1, we have the right exact functor $M \rightsquigarrow [B, M]$, "extension to B," from the abelian category $\mathscr{C}_A$ into the abelian category $\mathscr{C}_B$. The category $\mathscr{C}_A$ has enough projectives (Chapter 2, §4, Theorem 2.4.8, conclusion (3)). Given a C-complete A-module M, the ith *higher extension of M to B*, written $[B, M]_i$, $i \geq 0$, will denote the left derived functors $M \rightsquigarrow [B, M]_i$, $i \geq 0$, of the right exact functor $M \rightsquigarrow [B, M]$ from the abelian category $\mathscr{C}_A$ with enough projecctives into the abelian category $\mathscr{C}_B$.

An immediate consequence of Proposition 6.4.5 and Definition 3 is,

Corollary 6.4.8. *Let the hypotheses be as in Proposition 6.4.5. Suppose that the three equivalent conditions of Proposition 6.4.5 hold. Then there is an induced isomorphism of homological connected sequences of functors, from the category $\mathscr{C}_A$ into itself,*

$$(1): \ [B, M]_n \approx {}^{II}\mathrm{Tor}_n^{\mathscr{C}_A}(\widehat{B}^B, M),$$

for all C^A-complete A-modules M and for all integers n.

Proof. Both sides of the equation are systems of left derived functors, when considered as homological connected sequences of functors from $\mathscr{C}_A$ into $\mathscr{C}_A$. Therefore condition [1] of Proposition 6.4.5, which asserts that we have a natural equivalence of functors in dimension zero, completes the proof.   □

Theorem 6.4.9. *Let $\phi : A \to B$ be a continuous homomorphism of standard topological rings. Then for every abstract left A-module M, there are two induced first quandrant homological spectral sequences in the category of C^B-complete left B-modules,*

$$(1): \ E_{p,q}^2 = [B, C_q^A(M)]_p \Rightarrow K_n$$

$$(2): \ 'E_{p,q}^2 = C_p^B(\mathrm{Tor}_q^A(B, M)) \Rightarrow K_n.$$

These two spectral sequences converge to the same abutments K_n, $n \geq 0$, but with possibly different filtrations.

<u>Note</u>: The assignment, $M \rightsquigarrow K_n(M)$, $n \geq 0$, from the category of abstract left A-modules into the category $\mathscr{C}_B$, are the left derived functors of the right exact functor $M \rightsquigarrow C^B(B \underset{A}{\otimes} M)$, from Abs_A into $\mathscr{C}_B$. (For a universal mapping property characterization of this functor, see Exercise 1 below.)

Proof. Consider the diagram of abelian categories and right exact functors

$$(1): \qquad \begin{array}{ccc} Abs_A & \xrightarrow{\ C^A\ } & \mathscr{C}_A \\ {\scriptstyle M \rightsquigarrow B \underset{A}{\otimes} M} \downarrow & & \downarrow {\scriptstyle M \rightsquigarrow [B,M]} \\ Abs_B & \xrightarrow{\ C^B\ } & \mathscr{C}_B \end{array}$$

By the universal mapping property definition of $[B, C^A(M)]$ (Definition 2, p. 311), the universal mapping property definitions of C^A and C^B (Chapter 2, §5, Definition 1), and the (widely familiar) universal mapping property characterization of the functor $M \rightsquigarrow B \underset{A}{\otimes} M$ (which can be interpreted as a special case of Definition 2 — see Remark 8 above), for every C^B-complete left B-module N, we have canonical isomorphisms of abelian groups — [note that, by Definition 1, since N is in $\mathscr{C}_B$, N has a natural structure as a C^A-complete left A-module. Also, since N is in $\mathscr{C}_B$, respectively: $\mathscr{C}_A$, N is also, by definition (Chapter 2, §5, Definition 1) an abstract left B-, respectively: A-module] —

$$\mathrm{Hom}_{\mathscr{C}_B}([B, C^A(M)], N) \approx \mathrm{Hom}_{\mathscr{C}_A}(C^A(M), N) \approx \mathrm{Hom}_A(M, N)$$

$$\approx \mathrm{Hom}_B(B \underset{A}{\otimes} M, N) \approx \mathrm{Hom}_{\mathscr{C}_B}(C^B(B \underset{A}{\otimes} M), N).$$

Thus,

$$\mathrm{Hom}_{\mathscr{C}_B}([B, C^A(M)], N) \approx \mathrm{Hom}_{\mathscr{C}_B}(C^B(B \underset{A}{\otimes} M), N).$$

Moreover, these isomorphisms, for all N in $\mathscr{C}_B$, are a natural equivalence of covariant functors from $\mathscr{C}_B$ into the category of abelian groups. It follows that there is a unique isomorphism in the category $\mathscr{C}_B$,

$$(2): \ [B, C^A(M)] \approx C^B(B \underset{A}{\otimes} M),$$

that induces this natural equivalence of functors from $\mathscr{C}_B$ into the category of abelian groups.

Otherwise stated, the diagram (1) of right exact functors and abelian categories is commutative. The functor C^A maps projectives in Abs_A into projectives in $\mathscr{C}_A$ (Chapter 2, §4, Theorem 2.4.8, conclusion (2)), and the functor $M \rightsquigarrow B \underset{A}{\otimes} M$ maps projectives in Abs_A into projectives in Abs_B (in fact, this functor even maps free modules into free modules). Therefore, the hypotheses for the spectral sequence of the composite functor: $(M \rightsquigarrow [B, M]) \circ C^A$, respectively: $C^B \circ (M \rightsquigarrow B \underset{A}{\otimes} M)$, are satisfied. The left derived functors of the functors C^A, $M \rightsquigarrow [B, M]$, C^B and $M \rightsquigarrow B \underset{A}{\otimes} M$ (proceeding clockwise from the top in the diagram (1)), are, respectively, C_n^A, $n \geq 0$, $M \rightsquigarrow [B, M]_n$, $n \geq 0$, C_n^B, $n \geq 0$, and $M \rightsquigarrow \mathrm{Tor}_n^A(B, M)$, $n \geq 0$. Therefore we have the first quadrant homological spectral sequences (1) and (2) of the theorem, both of which converge to the same abutment (but for possibly different filtrations): the left derived functors of the composite functor, equation (2), $M \rightsquigarrow C^B(B \underset{A}{\otimes} M)$, from Abs_A into $\mathscr{C}_B$.   □

Exercise 1. Let $\phi : A \to B$ be a continuous homomorphism of standard topological rings. Let M be an abstract left A-module. Then prove that
(1) $C^B(B \underset{A}{\otimes} M)$, is a C^B-complete left B-module.
(2) The composite map

$$M \approx A \underset{A}{\otimes} M \overset{\phi \otimes M}{\underset{A}{\to}} B \underset{A}{\otimes} M \overset{\iota_{B \otimes M}^B}{\to} C^B(B \underset{A}{\otimes} M),$$

is a homomorphism, call it α_M, of abstract left A-modules, and that
(3) Given any C^B-complete left B-module N, and any homomorphism $j : M \to N$ of abstract left A-modules, prove that there exists a unique infinitely B-linear function $\tilde{j} : C^B(B \underset{A}{\otimes} M) \to N$ such that $\tilde{j} \circ \alpha_m = j$.

Exercise 1.1. Using the Note to Theorem 6.4.9, prove that, under the hypotheses of Theorem 6.4.9, the functors K_n, $n \geq 0$, from the category of abstract left A-modules into the category of C^B-complete left B-modules, can also be characterized as being: the left derived functors of the functor $M \rightsquigarrow \widehat{B \underset{A}{\otimes} M}^B$, from Abs_A into $\mathscr{C}_B$. Or, alternatively, as being: the left derived functors of the right-exact functor, $M \rightsquigarrow [B, C^A(M)]$. Or, yet again, as being: the left derived functors of the functor $M \rightsquigarrow [B, \widehat{M}^A]$.

Exercise 2. Using the same methods as in the proof of Theorem 6.4.3, and the universal mapping property that you have established in Exercise 1, prove that the functor $M \rightsquigarrow C^B(B \underset{A}{\otimes} M)$ from Abs_A into $\mathscr{C}_B$ preserves arbitrary direct limits of arbitrary covariant functors indexed by arbitrary set-theoretically legitimate categories. Show also that this functor admits as left adjoint, the "stripping functor" from $\mathscr{C}_B$ into Abs_A, which to every N in $\mathscr{C}_B$ associates the abstract left A-module, (by means of the abstract ring homomorphism ϕ) deduced from the underlying abstract left B-module of N.

Corollary 6.4.10. [Spectral Sequence of the Higher Extensions to B].
Let $\phi : A \to B$ be a continuous homomorphism of standard topological rings. Let M be an abstract left A-module, such that
(1) $C_n^A(M) = 0$ for $n \geq 1$.
Then there is induced a first quadrant homological spectral sequence in the category of C^B-complete left B-modules, such that
(2) $E_{p,q}^2 = C_p^B(\mathrm{Tor}_q^A(B, M)) \Rightarrow [B, C^A(M)]_n$.

Proof. In the first spectral sequence of Theorem 6.4.9, we have $E_{p,q}^2 = 0$ unless $q = 0$. Therefore the first spectral sequence degenerates, and the $E_{n,0}^2$-term is the abutment. That is,

$$K_n \approx [B, C^A(M)]_n, \ n \geq 0.$$

Substituting this last isomorphism into the second spectral sequence of Theorem 6.4.9 gives the corollary. □

Corollary 6.4.11. *Let* $\phi : A \to B$ *be a continuous homomorphism of standard topological rings. Suppose that the topological ring* A *is strongly C-O.K. Then for every* C^A*-complete left* A*-module* M*, we have that the first quadrant homological spectral sequence, in the category* $\mathscr{C}_B$*,*

$$E^2_{p,q} = C^B_p(\mathrm{Tor}^A_q(B, M)) \Rightarrow [B, M]_n.$$

Proof. Since A is strongly C-O.K., and since M is a C^A-complete left A-module,

$$(1): \ C^A_n(M) = \begin{cases} M, & n = 0 \\ 0, & n \geq 1. \end{cases}$$

Therefore the corollary follows from Corollary 6.4.10. □

Corollary 6.4.12. *Let* $\phi : A \to B$ *be a continuous homomorphism of standard topological rings. Suppose that* B *is flat as a right* A*-module. Let* M *be an abstract left* A*-module. Then there is induced a first quadrant homological spectral sequence, in the abelian category* $\mathscr{C}_B$*,*

$$E^2_{p,q} = [B, C^A_q(M)]_p \Rightarrow C^B_n(B \underset{A}{\otimes} M).$$

Note. The hypothesis "B is flat as a right A-module" can be replaced by the weaker hypothesis
(1) $\mathrm{Tor}^A_n(B, M) = 0$ for $n \geq 1$.

Proof. Substituting equation (1) into the second spectral sequence of Theorem 6.4.9, we see that

$$K_n \approx C^B_n(B \underset{A}{\otimes} M), \ n \geq 0.$$

Substituting this into the first spectral sequence of Theorem 6.4.9 gives the corollary. □

Corollary 6.4.13. *Let* $\phi : A \to B$ *be a continuous homomorphism of standard topological rings. Let* M *be an abstract left* A*-module such that*
(1) $C^A_n(M) = 0$ *for* $n \geq 1$*, and such that*
(2) $\mathrm{Tor}^A_n(B, M) = 0$ *for* $n \geq 1$*.*
Then there are induced canonical isomorphisms of C^B*-complete left* B*-modules*

$$[B, C^A(M)]_n \approx C^B_n(B \underset{A}{\otimes} M), \ n \geq 0.$$

Proof. Substituting equation (1) into the spectral sequence of Corollary 6.4.12 (or else, substituting equation (2) into the spectral sequence of Corollary 6.4.10), we see that the spectral sequence degenerates and that the edge homomorphism is an isomorphism in $\mathscr{C}_B$,

$$(n\text{th group of the abutment}) \xrightarrow{\approx} E^2_{n,0}.$$

$\square$

Corollary 6.4.14. *Let $\phi : A \to B$ be a continuous homomorphism of standard topological rings such that B is flat as a right A-module. Then the assignment*

$$M \rightsquigarrow C^B_n(B \underset{A}{\otimes} M), \ n \geq 0,$$

is the left derived functors of the functor: $M \rightsquigarrow \widehat{B \underset{A}{\otimes} M}^B$, from Abs_A into $\mathscr{C}_B$.

If also the topological ring A is strongly C-O.K., then, for C-complete left A-modules M, we have that

$$[B, M]_n \approx C^B_n(B \underset{A}{\otimes} M), \ n \geq 0,$$

as homological connected sequences of functors in M, from $\mathscr{C}_A$ into $\mathscr{C}_B$.

Proof. If B is right flat as an abstract right A-module, then the assignment: $M \rightsquigarrow C^B_n(B \underset{A}{\otimes} M)$ is an exact connected sequence of functors. And if M is a free abstract left A-module, then $B \underset{A}{\otimes} M$ is a free abstract left B-module, whence

$$C^B_n(B \underset{A}{\otimes} M) = \begin{cases} 0 & \text{for } n \geq 1, \\ \widehat{B \underset{A}{\otimes} M}^B, & \text{for } n = 0. \end{cases}$$

This proves the first paragraph.

Assume now that the topological ring A is strongly C-O.K., and let M be a C^A-complete left A-module. Then by Corollary 6.4.11 we have the spectral sequence of Corollary 6.4.11. Since by hypothesis B is right flat as an abstract right A-module, this spectral sequence degenerates, and the edge homomorphisms are isomorphisms,

$$[B, M]_n \xrightarrow{\approx} C^B_n(B \underset{A}{\otimes} M), \ n \geq 0.$$

$\square$

<u>Example 8.</u>   The special case of Theorem 6.4.9, respectively: of Corollary 6.4.12, in which $B = \widehat{A}$ (the completion of the ring A with its uniform topology), is as follows:

If A is any standard topological ring, and if we let $B = \widehat{A}$, and if we let $\phi : A \to B$ be the natural map, then clearly ϕ is continuous, so that we have the two spectral sequences of Theorem 6.4.9. However, since $B = \widehat{A}$, by Chapter 2, §5, Proposition 2.5.8, we have that $\mathscr{C}_B = \mathscr{C}_A$. It follows that, in this very special case, the functor: $M \rightsquigarrow [B, M]$ is the identity functor; and therefore also that the functors $[B, M]_i$, for $i \geq 1$, are identically zero. Therefore, for all abstract left A-modules M, the first spectral sequence of Theorem 6.4.9 degenerates, and the edge homomorphism in that spectral sequence: $C_n^A(M) \to K_n$ is an isomorphism in $\mathscr{C}_B = \mathscr{C}_A$, $n \geq 0$. Substituting into the second spectral sequence gives the spectral sequence

$$E_{p,q}^2 = C_p^{\widehat{A}}(\operatorname{Tor}_q^A(\widehat{A}, M)) \Rightarrow C_n^A(M),$$

in the category $\mathscr{C}_A$.

Similarly, using the identity

$$[\widehat{A}, M]_i = \left\{ \begin{array}{ll} M, & i = 0 \\ 0, & i \geq 1, \end{array} \right.$$

Corollary 6.4.12 becomes, in the special case in which $B = \widehat{A}$, that, if $\widehat{A}$ is flat as right A-module, then, for all abstract left A-modules M, we have that

$$C_n^A(M) = C_n^{\widehat{A}}(\widehat{A} \underset{A}{\otimes} M).$$

The next several theorems, corollaries, lemmas, propositions, exercises and remarks, are all connected with the question: If $\phi : A \to B$ is a continuous homomorphism of standard topological rings, and if M is a C^A-complete left A-module, such that either B is of finite presentation as an abstract right or left A-module; or else M is of finite presentation as an abstract left $\widehat{A}$-module — then to give conditions under which

$$[B, M]_i \approx \operatorname{Tor}_i^{\widehat{A}}(\widehat{B}^B, M), \ i \geq 0$$

— or, even, sometimes $\approx$ to $\operatorname{Tor}_i^A(B, M)$, $i \geq 0$.

One gets stronger results in the case in which M is left finitely presented as an abstract left $\widehat{A}$-module — e.g., Theorem 6.4.15 below. In the case in which B is finitely presented, (rather than M), it is difficult to obtain results unless also the topology of B (or of $\widehat{B}$) is the A-adic topology (or the $\widehat{A}$-adic topology, respectively). With these restrictions the answer to the question is generally "yes," in "reasonable" cases. (For example, see Corollary 6.4.17 below).

Theorem 6.4.15. *Let $\phi : A \to B$ be a continuous homomorphism of standard topological rings.*

(1) Let M be an abstract left $\widehat{A}$-module of finite presentation. Then the natural mapping

$$\widehat{B}^B \underset{\widehat{A}}{\otimes} M \to [B, M]$$

is an isomorphism of abstract lerft $\widehat{B}^B$-modules.

(2) More generally, let N be a non-negative integer, and let M be an abstract left $\widehat{A}$-module, such that M admits a chain of syzygies of length $\geq N+1$ as an abstract left $\widehat{A}$-module. (Chapter 5, §3, Definition 1.) Then the natural mapping

$$\mathrm{Tor}_i^{\widehat{A}}(\widehat{B}^B, M) \to [B, M]_i$$

is an isomorphism of abstract left $\widehat{B}^B$-modules, for all integers i, $N \geq i \geq 0$.

Proof. First, notice that, for every C^A-complete left A-module M, we have the $\widehat{A}$-bilinear function $(b, m) \to b \cdot \kappa_M(m)$, from $\widehat{B}^B \times M$ into $[B, M]$, from which we deduce a homomorphism of abstract left $\widehat{B}^B$-modules

$$\widehat{B}^B \underset{\widehat{A}}{\otimes} M \to [B, M].$$

Since $M \rightsquigarrow \mathrm{Tor}_i^{\widehat{A}}(\widehat{B}^B, M)$, $i \geq 0$, is a non-negative homological connected sequence of functors from $\mathscr{C}_A$ into $Abs_{\widehat{B}^B}$, and since $M \rightsquigarrow [B, M]_i$, $i \geq 0$, is the left derived functors of $M \rightsquigarrow [B, M]$, from $\mathscr{C}_A$ into $Abs_{\widehat{B}^B}$, we therefore deduce a natural map of homological connected sequences of functors from $\mathscr{C}_A$ into $Abs_{\widehat{B}^B}$,

$$\mathrm{Tor}_i^{\widehat{A}}(\widehat{B}^B, M) \to [B, M]_i,$$

for all integers $i \geq 0$ and all C^A-complete left A-modules M. Also, by Chapter 5, §3, Corollary 5.3.3, conclusion (3), we have that every finitely presented abstract left $\widehat{A}$-module M has a natural structure as a C^A-complete left A-module. Therefore, the maps described in (1) and (2) make sense, and are indeed homomorphisms of abstract left $\widehat{B}^B$-modules, for all integers $i \geq 0$. Next, let M be a free abstract left $\widehat{A}$-module of finite rank. If M is a free $\widehat{A}$-module of rank 1, then assertion (1) is true, by Example 1 above. Since the left and right sides of the map in assertion (1) are both additive functors of M, it follows that the map in (1) is an isomorphism for all free left $\widehat{A}$-modules of finite rank. And, in dimension $i \geq 1$, both connected sequences of functors in assertion (2) vanish at every free $\widehat{A}$-module M of finite rank. Therefore, the theorem is true for free abstract left $\widehat{A}$-modules M of finite rank.

The proof is by induction on N. If $N = 0$, then M is of finite presentation as an abstract left $\widehat{A}$-module. Therefore we have an exact sequence

$$(\widehat{A})^n \to (\widehat{A})^m \to M \to 0$$

of abstract left $\widehat{A}$-modules. This yields a commutative diagram, with exact rows

$$
\begin{array}{ccccccc}
[B, (\widehat{A})^n] & \longrightarrow & [B, (\widehat{A})^m] & \longrightarrow & [B, M] & \to & 0 \\
\alpha \uparrow & & \beta \uparrow & & \gamma \uparrow & & \\
\widehat{B}^B \underset{\widehat{A}}{\otimes} (\widehat{A})^n & \to & \widehat{B}^B \underset{\widehat{A}}{\otimes} (\widehat{A})^m & \to & \widehat{B}^B \underset{\widehat{A}}{\otimes} M & \to & 0
\end{array}
$$

Since the theorem is true for free abstract left $\widehat{A}$-modules of finte rank, α and β are isomorphisms. Therefore γ is an isomorphism. This proves the theorem in the case $N = 0$.

Next, assume that $N > 0$ and that the theorem is established for the integer $N - 1$. Then choose a short exact sequence

$$0 \to H \to (\widehat{A})^n \to M \to 0$$

in the category of abstract left $\widehat{A}$-modules, where H admits a chain of syzygies of length $N - 1$ as an abstract left $\widehat{A}$-module. Then we have a commutative diagram with exact rows:

$$\cdots \to [B, (\widehat{A})^n]_i \longrightarrow [B, M]_i \xrightarrow{d_i} [B, H]_{i-1} \longrightarrow [B, (\widehat{A})^n]_{i-1} \to \cdots$$
$$\alpha_i \uparrow \qquad\qquad \beta_i \uparrow \qquad\quad \gamma_{i-1} \uparrow \qquad\qquad\qquad \alpha_{i-1} \uparrow$$
$$\to \text{Tor}_i^{\widehat{A}}(\widehat{B}^B, (\widehat{A})^n) \to \text{Tor}_i^{\widehat{A}}(\widehat{B}^B, M) \xrightarrow{d_i} \text{Tor}_{i-1}^{\widehat{A}}(\widehat{B}^B, H) \to \text{Tor}_{i-1}^{\widehat{A}}(\widehat{B}^B, (\widehat{A})^n) \to$$

For $i \geq 1$, since $(\widehat{A})^n$ is a free abstract left $\widehat{A}$-module of finite rank, we have observed that both the domain and range of the map α_i is the zero group and by the inductive assumption, for $i \leq N$ the maps γ_{i-1} and α_{i-1} are isomorphisms. Therefore by the Five Lemma, for $1 \leq i \leq N$, the map β_i is an isomorphism. Also, since $N \geq 0$, in particular M is finitely presented, so by the case $N = 0$ covered above, β_0 is also an isomorphism. $\qquad\square$

Corollary 6.4.16. *Let $\phi : A \to B$ be a continuous homomorphism of standard topological rings. Suppose that A is left coherent. Suppose also that $\widehat{A}$ is flat as a right A-module (e.g., for this to be so, it suffices that A be complete). Let M' be any finitely presented abstract left A-module, and let $M = \widehat{A} \underset{A}{\otimes} M'$. Then there are induced canonical isomorphisms of abstract left $\widehat{B}^B$-modules,*

$$\text{Tor}_i^{\widehat{A}}(\widehat{B}^B, M) \overset{\cong}{\to} [B, M]_i,$$

for all integers $i \geq 0$.

Proof. Since A is left coherent and M' is finitely presented as a left A-module, therefore M' admits an infinite chain of syzygies P_* as an abstract left A-module. Since $\widehat{A}$ is flat as a right A-module, $\widehat{A} \underset{A}{\otimes} P_*$ is an infinite chain of syzygies for M as an abstract left $\widehat{A}$-module. The corollary follows from Theorem 6.4.15, conclusion (2). $\qquad\square$

Corollary 6.4.17. *Let A be a Noetherian commutative topological ring, such that the topology is given by denumerably many ideals, and such that the square of an open ideal is open. Let B be an arbitrary standard topological ring and let $\phi : A \to B$ be an arbitrary continuous ring homomorphism. Then for every finitely generated abstract A-module M', there are induced canonical isomorphisms of abstract left $\widehat{B}^B$-modules,*

$$\text{Tor}_i^{\widehat{A}}(\widehat{B}^B, M) \overset{\cong}{\to} [B, M]_i$$

$i \geq 0$, where $M = \widehat{A} \underset{A}{\otimes} M' \approx \widehat{(M')}^A$.

Proof. By Chapter 5, §4, Theorem 5.4.2, $\widehat{A}$ is flat as an A-module. And also by Chapter 5, §4, Theorem 5.4.2, $\widehat{A} \underset{A}{\otimes} M' \approx \widehat{(M')}^A$, canonically as abstract $\widehat{A}$-modules. Since the ring A is Noetherian commutative, the ring A is left coherent. Therefore the corollary follows from Corollary 6.4.16.   □

A related result is

Corollary 6.4.18. *Let A be a standard topological ring that is left coherent as an abstract ring. Then for every abstract left A-module M of finite presentation, and for every integer $n \geq 0$, the abstract left $\widehat{A}$-module $\mathrm{Tor}_n^A(\widehat{A}, M)$ has a natural structure as a C^A-complete left A-module.*

And if $\phi : A \to B$ is a continuous homomorphism from A into an arbitrary standard topological ring B, then we have a first quadrant homological spectral sequence in the category of abstract left $\widehat{B}^B$-modules,

$$E_{p,q}^2 = [B, \mathrm{Tor}_q^A(\widehat{A}, M)]_p \Rightarrow \mathrm{Tor}_n^A(\widehat{B}^B, M)$$

Proof. The assertion of the first paragraph follows from Chapter 5, §3, Corollary 5.3.4. Since A is left coherent, the category of finitely presented abstract left A-modules is an abelian category $\mathscr{A}$ with enough projectives. A collection of "enough projectives" in $\mathscr{A}$ is the class of free abstract left A-modules of finite rank. The functor, $M \rightsquigarrow \widehat{A} \underset{A}{\otimes} M$, maps free abstract left A-modules of finite rank into completions of free A-modules of finite rank — and these are projective as objects in the category $\mathscr{C}_A$ of C^A-complete left A-modules. (Chapter 2, §4, Theorem 2.4.8, conclusion 2.) Therefore the spectral sequence of the composite functor:

$$(1): \quad \mathscr{A} \xrightarrow{\;M \rightsquigarrow \widehat{A} \underset{A}{\otimes} M\;} \mathscr{C}_{\mathscr{A}} \xrightarrow{\;M \rightsquigarrow [B,M]\;} \mathscr{C}_B$$

applies. The composite of these functors is the functor: $M \rightsquigarrow [B, \widehat{A} \underset{A}{\otimes} M]$. For M of finite presentation as left A-module, $\widehat{B}^B \underset{A}{\otimes} M$ is of finite presentation as left $\widehat{B}^B$-module, and therefore, by Chapter 5, §3, Corollary 5.3.3, conclusion (3), has a natural structure as C-complete left $\widehat{B}^B$-module. Therefore, from the map in $\mathscr{C}_A$, $\widehat{A} \underset{A}{\otimes} M \to \widehat{B}^B \underset{A}{\otimes} M$ we deduce a map: $[B, \widehat{A} \underset{A}{\otimes} M] \to \widehat{B}^B \underset{A}{\otimes} M$. These maps, for all $M \in \mathscr{A}$, are a natural transformation of functors from the composite functor of (1) to the functor: $M \rightsquigarrow \widehat{B}^B \underset{A}{\otimes} M$; and this natural transformation is a natural equivalence of functors. (To prove this, note that the two functors claimed to be naturally equivalent are both right exact; and that by Example 2 that the natural transformation gives an isomorphism when M is a free abstract left A-module of finite rank.) The indicated spectral sequence follows.   □

Corollary 6.4.19. *Let A be a standard topological ring and let M be an abstract left A-module such that M admits an infinite chain of syzygies. Then for every integer $n \geq 0$, the abstract left $\widehat{A}$-module $\text{Tor}_n^A(\widehat{A}, M)$ has a natural structure as a C^A-complete left A-module.*

If $\phi : A \to B$ is a continuous homomorphism from A into any other standard topological ring, then there is induced a first quadrant homological spectral sequence in the category of abstarct left $\widehat{B}^B$-modules,

$$E_{p,q}^2 = [B, \text{Tor}_q^A(\widehat{A}, M)]_p \Rightarrow \text{Tor}_n^A(\widehat{B}^B, M).$$

Proof. Again, the first paragraph follows from Chapter 5, §3, Corollary 5.3.4.

Replacing B by $\widehat{B}$ if necessary, we can and do assume that the standard topological ring B is complete. Let F_* be an acyclic homological resolution of M as an abstract left A-module such that F_i is free of finite rank as an abstract left A-module.

Let $P_{*,*}$ be a projective resolution (in the sense of [CEHA], Chapter 17, §1.1, p. 363) of the chain complex: $\widehat{A} \underset{A}{\otimes} F_*$ in (the abelian category with enough projectives), the category $\mathscr{C}_A$ of C^A-complete left A-modules.

This means — see [CEHA], Chapter 17, §1.1, p. 363 — that $P_{*,*}$ is a first quadrant double complex in the category $\mathscr{C}_A$; that $P_{p,*}$ is a projective resolution of $\widehat{A} \underset{A}{\otimes} F_p$ in $\mathscr{C}_A$ for all $p \geq 0$; that

$(1) : H_I(P_{*,*})$ is a projective resolution of $H_I((\widehat{A} \underset{A}{\otimes} F_*))$ for all $p \geq 0$;

(In the above, following [CEHA], we use the notation "$H_I(P_{*,*})$, resp.: $H_{II}(P_{*,*})$" to mean "take homology on the first, resp.: second, index"); and that $\text{Im}(P_{p+1,*} \to P_{p,*})$ is a projective resolution of $\text{Im}(\widehat{A} \underset{A}{\otimes} F_{p+1} \to \widehat{A} \underset{A}{\otimes} F_p)$.

These conditions imply that, for every additive functor K from the abelian category $\mathscr{C}_A$ to any other abelian category, we have that $K(H_I(P_{*,*})) = H_I(K(P_{*,*}))$ — i.e., that K of the homology of the qth row of $P_{*,*}$ is equal to the qth row of the homology of $K(P_{*,*})$. This is in particular true for the functor $R \rightsquigarrow [B, R]$ from $\mathscr{C}_A$ to $\mathscr{C}_B$. Therefore $[B, H_I(P_{*,*})] = H_I([B, P_{*,*}])$.

Consider the first quadrant homological double complex $[B, P_{*,*}]$. By the above,

$(2): H_I([B, P_{*,*}]) = [B, H_I(P_{*,*})]$

By (1) above, $H_I(P_{*,*})$ is a projective resolution of $H_I((\widehat{A} \underset{A}{\otimes} F_*))$ in $\mathscr{C}_A$; therefore $H_{II}([B, H_I(P_{*,*})]) = [B, H_p((\widehat{A} \underset{A}{\otimes} F_*))]_q$. Combining with (2), we have that $H_{II}(H_I([B, P_{*,*}])) = [B, H_p((\widehat{A} \underset{A}{\otimes} F_*))]_q$. Since F_* is an acyclic free resolution of M in $\mathscr{A}bs_A$, we have that $H_i((\widehat{A} \underset{A}{\otimes} F_*)) = \text{Tor}_i^A(\widehat{A}, M)$. Therefore

$(3) : H_{II}(H_I([B, P_{i,j}])) = [B, \text{Tor}_i^A(\widehat{A}, M)]_j.$

The first spectral sequence of the double complex $[B, P_{*,*}]$ starts with $E_{p,q}^0 = [B, P_{p,q}]$,

$E^1_{p,q} = H_{II}([B, P_{p,*}]),$
$E^2_{p,q} = H_I(H_{II}([B, P_{*,*}])).$

Let $P'_{*,*}$ denote the *reverse* of the double complex $P_{*,*}$ — so $P'_{i,j} = P_{j,i}$; Then the *second* spectral sequence of the double complex $[B, P_{*,*}]$ is the *first* spectral sequence of the *reverse* double complex – that is, of the double complex $[B, P_{q,p}]$. This spectral sequence starts with

$`E^0_{p,q} = [B, P_{q,p}].$

$`E^1_{p,q} = H_{II}([B, P_{q,p}]),$ which by equation (2) above is $[B, H_{II}(P_{q,p})].$

And $`E^2_{p,q} = H_I(H_{II}([B, P_{q,p}])).$

The right side of this last equation is the same as the reverse of $H_{II}(H_I([B, P_{p,q}])).$ Hence by (3) above, the second spectral sequence starts with $'E^2_{p,q} = [B, \text{Tor}^A_q(\widehat{A}, M)]_p,$ and converges to the same abutment as the first spectral sequence of the double complex $[B, P_{*,*}].$

For every $p \geq 0$, $P_{p,*}$ is a projective resolution of $\widehat{A} \underset{A}{\otimes} F_p$. Therefore

(4): $H_{II}([B, P_{p,q}]) = [B, \widehat{A} \underset{A}{\otimes} F_p]_q.$

Since F_p is a free A-module of finite rank, $\widehat{A} \underset{A}{\otimes} F_p$ is a free $\widehat{A}$-module of finite rank, and therefore is a projective object in $\mathscr{C}_A$. Therefore

$$[B, \widehat{A} \underset{A}{\otimes} F_p]_q = \begin{cases} 0 & \text{for } q \geq 1, \\ [B, (\widehat{A} \underset{A}{\otimes} F_p)] \approx B \underset{A}{\otimes} F_p, & \text{for } q = 0. \end{cases}$$

(The isomorphism $[B, (\widehat{A} \underset{A}{\otimes} F_p)] \approx B \underset{A}{\otimes} F_p$ holds, e.g. using Example 2 above, since F_p is a free A-module of finite rank.) Combining with (4), we have that $H_I(H_{II}([B, P_{p,q}])) = 0$ for $q \geq 1$, and $H_I(H_{II}([B, P_{p,q}])) = H_p(B \underset{A}{\otimes} F_*)$ for $q = 0$. Since F_* is a free resolution of M is $\mathscr{A}bs_A$, this last group is $\text{Tor}^A_p(B, M)$. Therefore the first spectral sequence of the double complex degenerates, and the nth group of the abutment is $\text{Tor}^A_n(B, M)$. The second spectral sequence of the double complex has the same abutment, and therefore is of the form

$$`E^2_{p,q} = [B, \text{Tor}^A_q(\widehat{A}, M)]_p \Rightarrow \text{Tor}^A_n(B, M).$$

$\square$

Bibliography

[COC] Saul Lubkin, Cohomology of Completions, Mathematical Studies 42, North-Holland, 1980, 802 pp.

[EM] S. Eilenberg and J.C. Moore, "Limits and Spectral Sequences", Topology, Vol. 1, 1962, pp. 1-24.

[CEHA] H. Cartan and S. Eilenberg, Homological Algebra, Princeton University Press, 1956. 390 pp.

[PPWC] S. Lubkin, "A p-Adic Proof of the Weil Conjectures", Annals of Mathematics 87, Nos. 1-2, January - March (1968), pp. 105-255.

[FLI] S. Lubkin, "A Formula for lim inv 1", Proc. of NSF Regional Summer Conference on "Analytical Methods in Commutative Algebra," George Mason University, Fairfax, Virginia, September 1979, pp. 257-273.

[NB] N. Bourbaki, "Elements de Mathematique, Topologie Generale 1-4", 1971

Index

Printed in the United States
By Bookmasters